Principles of Semiconductor Processes and Device Technology

Principles of Semiconductor Processes and Device Technology

Zhigang Zang and Qingkai Qian

WILEY-VCH

Authors

Zhigang Zang
Chongqing University
Room B328
Information Technology
Research Building
Huxi Campus
Chongqing
China

Qingkai Qian
Chongqing University
Room B328
Information Technology
Research Building
Huxi Campus
Chongqing
China

Cover Design: Wiley
Cover Image: © da-kuk/Getty Images

Typesetting: Straive, Chennai, India
Printing and Binding: CPI Group (UK) Ltd, Croydon, CR0 4YY

Library of Congress Card No.: applied for

British Library Cataloguing-in-Publication Data A catalogue record for this book is available from the British Library.

Bibliographic information published by the Deutsche Nationalbibliothek
The Deutsche Nationalbibliothek lists this publication in the Deutsche Nationalbibliografie; detailed bibliographic data are available on the Internet at <http://dnb.d-nb.de>.

Print ISBN: 978-3-527-35410-8
ePDF ISBN: 978-3-527-84798-3
ePub ISBN: 978-3-527-84796-9
oBook ISBN: 978-3-527-84799-0

C9783527354108_150426

Contents

Preface

In many ways, modern technological progress is built upon the foundation of semiconductor science. From the ubiquitous smartphone to the powerful supercomputers driving scientific discovery, and from energy-harvesting solar panels to advanced medical imaging systems, the devices that define our modern world are enabled by our ability to understand and manipulate the properties of semiconducting materials. This book, *Semiconductor Process Principle and Device Technology*, is conceived as a comprehensive guide for students, researchers, and engineers seeking an in-depth and applied understanding of this dynamic field.

A comprehensive understanding of semiconductors necessitates atomic-level insight and involves the fundamental physics governing semiconductor behavior. Key concepts including crystal structures, energy band theory, doping, and carrier transport in semiconductors are introduced in this book, establishing the essential language and principles that underpin all subsequent discussions. This foundational knowledge is applied as we synthesize and characterize the high-quality optoelectronic materials that form the bedrock of semiconductor device fabrication.

In the context of device fabrication, lithographic processing serves as a cornerstone of modern semiconductor manufacturing. We provide a detailed overview of the cutting-edge extreme ultraviolet (EUV) technology that allows for the continued miniaturization of integrated circuits. Subsequently, we take a thorough investigation of the workhorse of the digital age: the transistor. A wide spectrum of transistor technologies is examined, ranging from fundamental bipolar and field-effect structures to advanced heterojunction and thin-film devices. The operational principles and performance characteristics of these technologies are analyzed.

Besides electronic devices, optoelectronic devices that interact with light represent an important category of semiconductor devices. The science and technology behind luminescent materials are illustrated, detailing the operation and manufacture of light-emitting diodes, including the prominent GaN-based systems and the rapidly emerging perovskite-based technologies. A specialized focus on superluminescent light-emitting diodes and their unique design for high-power, broadband applications is detailed. The inverse process, the conversion of light into electrical signals, i.e., photodetectors, is also examined, covering traditional and novel perovskite-based architectures for sensing across the electromagnetic spectrum.

Finally, we look to the research frontier by investigating the remarkable material graphene and its application in a new generation of devices. Its extraordinary properties, synthesis methods, and promising role in ultrahigh-speed transistors, sensitive sensors, and flexible nanoelectromechanical systems are discussed.

We hope that this integrated approach of this book—connecting core physical principles with advanced processing techniques and culminating in a broad survey of critical devices—will provide readers with not only knowledge but also the insight needed to contribute to the next wave of innovation in semiconductor and device technology.

September 25th, 2025
Chongqing, China

Zhigang Zang and Qingkai Qian
Huxi Campus, Chongqing University

1 Semiconductor Physics

1.1 Introduction

The definition of semiconductors is somewhat broad. For instance, semiconductors may be defined as materials possessing a bandgap (a fundamental concept to be explained later, signifying the minimum electronic excitation energy) ranging between 0 and 4 eV. Materials characterized by a zero bandgap are classified as metals or semimetals, whereas those featuring energy gaps exceeding 4 eV are termed insulators. Alternatively, semiconductors can be defined by their electrical conductivity, which lies between that of conductors and insulators, typically within the range of approximately 10^{-2}–10^{9} Ω cm. Unlike metals, the conductivity of semiconductors generally increases with elevated temperature. However, these definitions are not absolute. The conductivity of semiconductors, for example, can be drastically modified via doping—a technique involving the deliberate introduction of specific foreign atoms into the semiconductor lattice to modulate its conductive properties. Consequently, crystalline diamond, exhibiting a bandgap near 6 eV and typically considered an insulator, can be engineered into a functional semiconductor suitable for potential power electronic devices.

Common examples of semiconductors include silicon, germanium, and gallium arsenide, among others. Undoubtedly, silicon remains the most extensively studied and widely recognized semiconductor. It serves as the primary functional material in numerous electronic applications, including solar cells, analog circuits, and large-scale integrated circuits. The prominence of silicon stems from several key factors: its natural abundance, moderate energy bandgap, established doping techniques, and the exceptional stability of both silicon and its native oxide, silicon dioxide. Semiconductor properties critically depend on their chemical compositions and crystal structures, which fundamentally govern respective band structures, doping characteristics, and carrier transport mechanisms. As silicon-based electronic devices approach material performance limits while semiconductor markets demand enhanced device performance, research focus increasingly shifts toward alternative materials such as wide-bandgap semiconductors and oxide semiconductors. The fundamental physics of these semiconductors is comprehensively addressed within this chapter.

1.2 Properties of Semiconductors

The importance of semiconductors in electronics stems from their unique physical properties, determined by their atomic structure and governed by the physics of semiconductor materials. An intrinsic semiconductor exhibits very low conductivity. However, introducing a relatively small concentration of impurity atoms, known as doping, enables modulation of the material's conductivity over a large dynamic range. Crucially, depending on the type of dopant employed—such as phosphorus or boron in silicon—semiconductors are categorized as *n*-type or *p*-type, dictating whether electrical current conduction is mediated primarily by electrons or holes, respectively. Moreover, semiconductor conductivity can be significantly modulated by applied electric fields. Collectively, these properties render semiconductor materials exceptionally valuable for constructing electronic devices, enabling the realization of critical functionalities including unidirectional current flow, variable resistance, variable capacitance, photosensitivity, voltage-controlled conductivity, signal amplification, switching operations, and energy conversion.

1.2.1 Atomic Bonding of Semiconductors

The properties of semiconductors are predominantly governed by their chemical compositions and structures, which, in accordance with semiconductor physics, subsequently dictate their overall characteristics. Generally, atomic bonds can be classified into ionic, covalent, metallic, van der Waals, and hydrogen bonding, as illustrated in Figure 1.1, contingent upon electron transfers and the charge states of the constituent atoms. Certain atoms exhibit a strong proclivity to attain

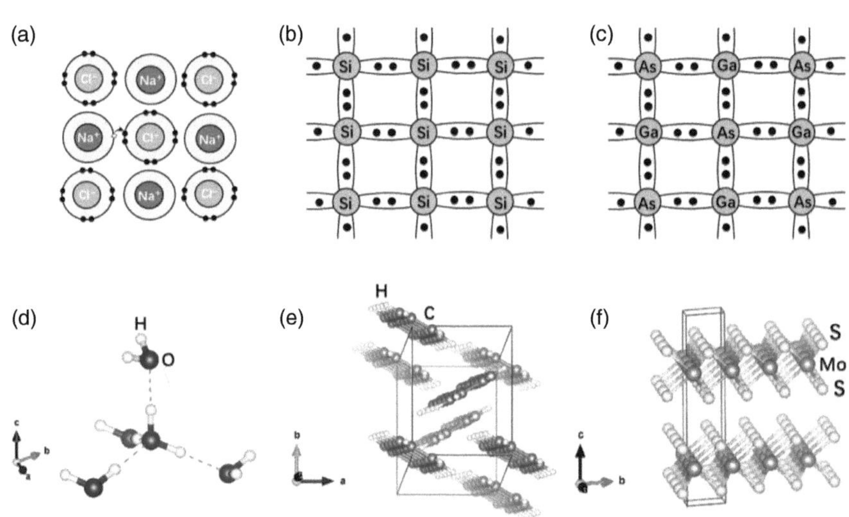

Figure 1.1 (a) Schematic of NaCl crystal with ionic bonding. (b) Schematic of Si crystal with covalent bonding. (c) Schematic of GaAs crystal with covalent bonding. (d) Hydrogen bonding between water molecules. (e) Pentacene and (f) MoS_2 with van der Waals bonding.

a completed outer electron shell by either gaining or losing a limited number of valence electrons, thereby forming charged ions. Ionic bonding—exemplified by the NaCl crystal structure shown in Figure 1.1a—arises from Coulombic attraction between positively and negatively charged ions. Ionic bonding also plays a significant role within semiconductors. For instance, in perovskite $CsPbBr_3$, ionic bonding occurs between Cs^+ ions and the surrounding $PbBr_3^-$ network, while it has been established that the bonding character between Pb and Br is metavalent, intermediate between covalent and metallic [1].

Covalent bonding is formed through the sharing of electrons between atoms, rather than through the gain or loss of electrons. The shared electrons create a bond that holds the atoms together. This bonding mechanism is essential for many important elemental and compound semiconductors (Figure 1.1b,c). Examples of elemental semiconductors involving covalent bonding include silicon (Si), germanium (Ge), and semiconducting single-walled carbon nanotubes (SWCNTs). In compound semiconductors, covalent bonding produces numerous critical materials. For instance, III–V semiconductors result from covalent bonding between Group III and Group V elements, such as gallium nitride (GaN), gallium phosphide (GaP), gallium arsenide (GaAs), and indium phosphide (InP). Similarly, II–VI semiconductors are formed by covalent bonding between Group II and Group VI elements, including zinc oxide (ZnO), zinc sulfide (ZnS), mercury telluride (HgTe), and cadmium telluride (CdTe). The bonding in these semiconductors typically exhibits a coordination number of 4, meaning each atom has four nearest neighbors, forming a tetrahedral structure.

When all atoms release their valence electrons, metallic bonding is formed, with the free electron fluid acting as the bonding agent. Although typically characteristic of metals, metallic bonding can also contribute to the lattice mechanical strength of semiconductors with a very high density of free carriers. Hydrogen bonding resembles ionic bonding and results from the attraction between a positively charged proton involved in a covalent bond and a nearby negatively charged atom (Figure 1.1d). This bonding type is highly localized due to the small size of the proton.

For atoms or molecules characterized by saturated covalent bonds, bonding can still arise via van der Waals interactions, including permanent dipole-dipole attractions and induced dipole interactions (Figure 1.1e,f). Many organic semiconductors, such as pentacene and polyacetylene, whose constituent building blocks are organic molecules, are primarily held together by van der Waals forces [2]. van der Waals forces are also responsible for the stacking of two-dimensional materials, such as transition metal dichalcogenides [3].

1.2.2 External Modulation Effects

Semiconductors possess a medium-sized bandgap, distinguishing them from metals and insulators. A precise understanding of the bandgap concept necessitates quantum physics, which describes electron motion within a crystal lattice. Broadly defined, the bandgap represents the minimum energy required for electronic excitation. Due to the presence of this bandgap, the intrinsic conductivity of a pure semiconductor is markedly low, as the excitation of free carriers necessitates overcoming the bandgap energy. This energy is typically substantial relative to thermal fluctuation energy at room temperature. Nevertheless, the conductivity of semiconductors can be profoundly modulated through the controlled introduction of foreign impurity atoms, known as dopants. Dopants significantly enhance the concentration of free charge carriers within the crystal lattice. Semiconductors accommodate two types of charge carriers, namely electrons and holes; doping processes associated with generating these carriers are designated as *n*-type and *p*-type, respectively. This versatility in controlling carrier type and concentration enables the fabrication of diverse semiconductor junctions. These junctions form the foundation for realizing electronic devices exhibiting specific functionalities, such as the rectifying properties characteristic of a PN junction.

Besides doping, the properties of semiconductors, particularly their electrical conductivity, can be modulated by various external forces, including temperature, applied voltage, magnetic field, light, mechanical stress, and chemical adsorption. These factors enhance the capability of semiconductors to serve as functional devices, as demonstrated in Figure 1.2. For example, numerous semiconductor properties can be significantly influenced by temperature. Elevated temperatures lead to lattice expansion, typically reducing the bandgap. Furthermore, high temperatures facilitate intrinsic carrier excitation and dopant activation, while diminishing the effect of Coulomb scattering. Consequently, semiconductors generally exhibit increased conductivity at heightened temperatures. However, excessively high temperatures induce more pronounced lattice oscillations, manifesting as heightened phonon excitation; this scatters carriers, reducing carrier mobility and thus

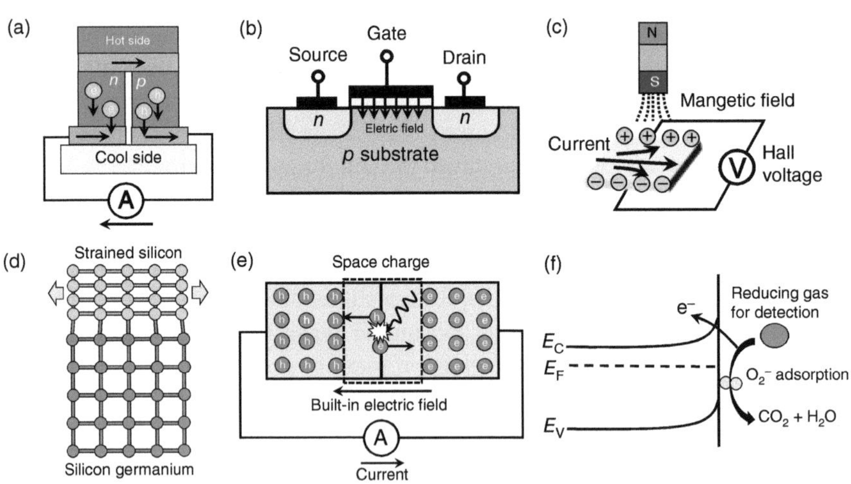

Figure 1.2 (a) Electricity generation via the thermoelectric effect (Seebeck effect) of semiconductors. (b) Metal-oxide-semiconductor field-effect transistor utilizing the electric field modulation effect of semiconductors. (c) Semiconductor characterization or magnetic field detection using the Hall effect. (d) Modulation of silicon by tensile strain introduced by the silicon–germanium substrate. (e) Photon detection based on the *p*–*i*–*n* diode. (f) Reducing gas sensing via the absorption on semiconductor and the subsequent modulation of Schottky barrier height.

decreasing conductivity. Other thermal effects include the thermoelectric phenomenon in semiconductors, arising from temperature gradients. Carriers migrate from the high-temperature region toward the low-temperature region, generating electric current, which can be exploited for electricity generation (Figure 1.2a), temperature measurement, or producing temperature differences via the reverse effect—the Peltier effect.

Owing to the relatively low carrier concentration within semiconductors, the carriers can be effectively modulated by an applied electric field. This field can be applied using various structures, such as a semiconductor junction or a metal-oxide-semiconductor capacitor. The electric field alters the carrier type and concentration, thereby influencing current conduction through mechanisms including drift, diffusion, or injection. Numerous devices exploit this modulation effect, such as the field-effect transistor (FET) and the bipolar junction transistor (BJT). The metal-oxide-semiconductor field-effect transistor (MOSFET), illustrated in Figure 1.2b, exhibits extremely high input impedance and can be manufactured with minimal dimensions. Consequently, it has become the foundation for modern digital integrated circuits. Conversely, the BJT operates as a current-controlled device, regulating a larger output current via a small input control current. Owing to its high operating speed, high transconductance, and low output resistance, the BJT remains a competitive device for amplification and switching applications, particularly in demanding analog circuits such as radio-frequency circuits for wireless systems.

The charge carrier transport within semiconductors is influenced by magnetic fields. When a magnetic field is applied perpendicular to the current direction in a semiconductor, a transverse voltage difference, known as the Hall voltage, develops across the electrical conductor (see Figure 1.2c). This phenomenon, termed the Hall effect, was first observed by Edwin Hall in 1879. The Hall voltage is fundamentally related to the carrier type and concentration, exhibiting opposite polarity for electron and hole conduction. Consequently, the Hall effect in semiconductors serves as a powerful characterization technique for determining carrier type, concentration, and mobility. Furthermore, this phenomenon is utilized in sensor applications for measuring magnetic fields or current, offering the advantages of non-contact measurement and high immunity to environmental contamination.

The properties of semiconductors can be modulated by strain. A typical example is the strain effect in silicon (Figure 1.2d), which finds important applications in the semiconductor industry to enhance further the performance of silicon-based devices. Strain influences silicon properties through the alteration of its interatomic distances. Silicon strain can be introduced through several techniques. Tensile strain can be created by depositing silicon over a silicon–germanium substrate, owing to the larger atomic size of germanium. This tensile strain increases silicon atom distances, resulting in higher electron mobility and thus lower energy dissipation in the fabricated transistor. Strain can similarly be created by doping the source and drain regions of a transistor with larger or smaller atoms. For instance, carbon doping in the source and drain regions of an n-type MOSFET induces tensile strain in the channel, enhancing electron mobility. Conversely, germanium doping induces compressive strain, thereby increasing hole mobility. Strain may also be introduced by depositing a strained dielectric, such as a stressed silicon nitride layer, onto the device channel.

Other factors, such as photon excitation or chemical adsorption, can also influence semiconductor conductivity, enabling the development of sensors like photodetectors and chemical gas sensors for gas detection. Photon excitation modifies semiconductor properties through distinct mechanisms, including the photoelectric effect and photothermal effects. For instance, a *p–n* junction acts as a photodetector via the photoelectric effect; following photon absorption, electron-hole pairs generated within the junction are separated by the electric field in the depletion region, inducing an increase in electrical current, as demonstrated in Figure 1.2e. Similarly, the chemical adsorption of molecules onto a semiconductor surface alters its conductivity. This phenomenon underpins chemical sensors, exemplified by the application shown in Figure 1.2f. To achieve enhanced sensitivity, semiconductor sensing materials require a substantial surface area exposed for gas adsorption. Consequently, diverse semiconductor materials, including metal oxides and polymers—particularly low-dimensional structures such as nanowires and nanotubes—have been investigated for this purpose. For example, tin oxide (SnO_2), a widely studied metal oxide, exhibits notable advantages for gas sensing due to its facile fabrication, low cost, high sensitivity, and stability.

1.3 Crystal Structure and Types of Semiconductors

1.3.1 Lattice and Crystal Systems

Although the chemical compositions of semiconductors primarily define key properties—for instance, silicon can be used to fabricate solar cells in amorphous, polycrystalline, or single-crystalline forms—their crystallinity significantly influences bandgap, optical absorption, carrier transport, carrier recombination, and related phenomena. The fundamental distinction among amorphous, polycrystalline, and single-crystalline semiconductors lies in the spatial scale of their long-range order. Single crystals exhibit extensive long-range order, enabling their properties to be modeled by considering infinite periodic structures wherein atomic arrangements remain invariant under finite translations along specific symmetry directions. In contrast, polycrystalline solids comprise numerous small single-crystalline domains exhibiting short-range order and localized periodicity. Amorphous solids lack long-range atomic order entirely; nevertheless, their atomic configurations deviate from true randomness and retain short-range similarities to crystalline counterparts, explaining their partial property resemblance to single crystals. The periodicity and long-range order in single crystals facilitate structural description via finite repeating unit cells and enable precise calculation and analysis of properties using quantum mechanical principles. Conversely, non-crystalline materials necessitate complex modeling of many-particle systems involving astronomically large numbers of atoms.

The description and understanding of semiconductor crystal structures form the foundation for understanding semiconductor properties. The most essential characteristic of semiconductor crystals is their periodicity, which imparts translational symmetry. In a bulk crystal, this translational symmetry extends along three orthogonal dimensions. Consequently, the entire crystal structure composes itself by the periodic replication of a finite unit cell arrangement in these three directions. While the dimensionality of repetition reduces in two-dimensional materials like graphene or one-dimensional systems such as carbon nanotubes, the descriptive formalism remains analogous; this discussion will therefore concentrate on the bulk crystal case. These three-dimensional repetitions generate a spatial framework termed the crystal lattice. The crystal lattice can be mathematically defined by equation,

$$\vec{\boldsymbol{R}} = n_1\vec{\boldsymbol{a}} + n_2\vec{\boldsymbol{b}} + n_3\vec{\boldsymbol{c}} \tag{1.1}$$

where $\vec{\boldsymbol{a}}$, $\vec{\boldsymbol{b}}$, and $\vec{\boldsymbol{c}}$ are the three primitive vectors defined by the repetition lengths and directions (Figure 1.3a), and n_1, n_2, and n_3 are the corresponding repetition numbers along those directions. The atomic structure is constructed by repeating a finite set of atoms over the crystal lattice (Figure 1.3b, Figure 1.3c, and Figure 1.3d). The smallest block of atoms replicated in this manner is termed the basis (Figure 1.3c), while the parallelepiped containing the atoms (Figure 1.3b) constituting the basis and defining the spatial periodicity is called the unit cell of the crystal.

For a given crystal structure, both the basis and the unit cell are non-unique. Using a two-dimensional crystal as an example illustrates the variety of possible unit cell definitions. The smallest possible unit cell, termed the primitive unit cell, itself lacks uniqueness. Nevertheless, conventional definitions of the primitive unit cell exist to better reflect inherent crystal symmetries. For certain crystals, symmetries remain obscured when viewed within a primitive cell. In these instances, a conventional unit cell—distinct from a primitive unit cell and employing a larger basis—is often adopted to facilitate symmetry recognition. Consider the body-centered cubic (bcc) lattice, routinely depicted as a cubic unit cell containing an additional lattice point at its center, as shown in Figure 1.4a. Its actual primitive unit cell, occupying half the volume, is defined by selecting the center point as the origin, with primitive vectors extending to three non-adjacent corner vertices. Analogously, the face-centered

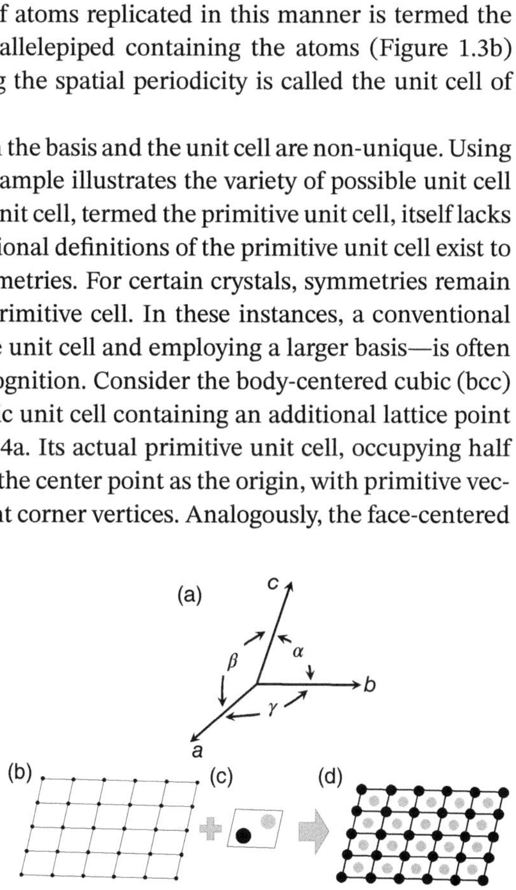

Figure 1.3 (a) Primitive lattice vectors and angles. (b) Bravais lattice grids, (c) the basis, and (d) the constructed atom structure.

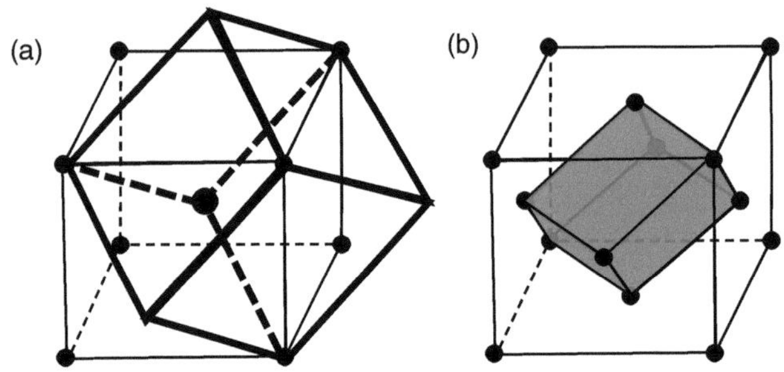

Figure 1.4 Cubic unit cell and the primitive unit cell of (a) the body-centered cubic lattice and (b) the face-centered cubic lattice.

cubic (fcc) lattice (Figure 1.4b) is typically represented as a cubic cell featuring additional lattice points at each face center. Here, the primitive vectors connect the origin (similarly defined) to three adjacent face centers. Symmetries for primitive unit cells in both bcc and fcc crystals derive from the rhombohedral lattice. Due to the ease with which cubic representations reveal higher symmetrical properties, they are preferred over rhombohedral descriptions.

Grids formed by translating the unit cell are termed Bravais lattices. These lattices can be classified based on their symmetry properties; lattices sharing the same symmetry belong to the same type. In three dimensions, only 14 distinct Bravais lattice types exist. These are further grouped into seven lattice systems, distinguished by the relationships among the lengths and angles of the conventional lattice vectors: triclinic, monoclinic, orthorhombic, tetragonal, rhombohedral (also known as trigonal), hexagonal, and cubic. Within the unit cells of these seven lattice systems, there may be more than one equivalent lattice point. The additional equivalent points can be located in different positions within the unit cell basis, leading to specific naming conventions: primitive (P) denotes lattices with points only at the unit cell corners; base-centered (A, B, or C, depending on the centered pair of parallel faces) has one additional point at the face center; body-centered (I) possesses one additional point at the body center; face-centered (F) features one additional point at the center of each face (Figure 1.5a–d). Accounting for these centered lattice points expands the seven basic lattice systems to the complete set of 14 three-dimensional Bravais lattice types (Table 1.1).

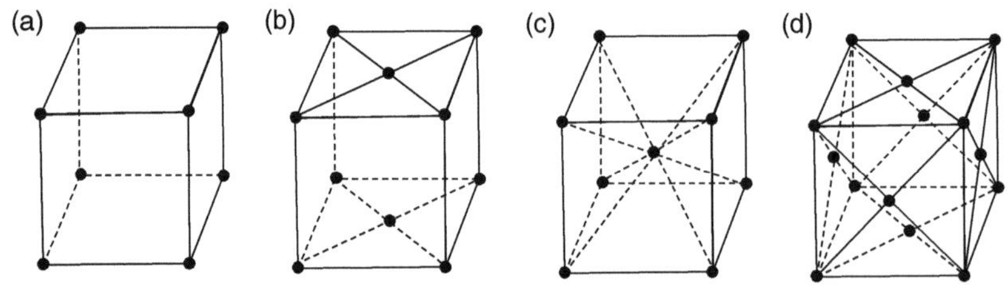

Figure 1.5 The four different Bravais lattices of the orthorhombic lattice system. (a) The primitive (P), (b) base-centered (S), (c) body-centered (I), (d) face-centered (F) Bravais lattices.

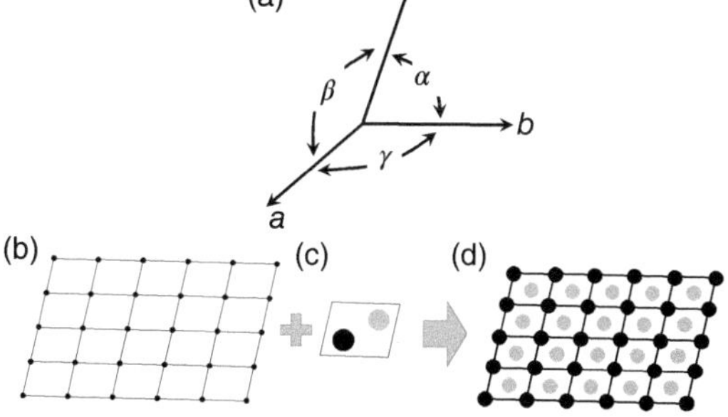

Table 1.1 Lattice systems, space groups, point groups, and crystal systems.

Lattice system	Lattice vector length relation	Lattice vector angle relation	Bravais lattice types	Space group types	Point group types	Crystal system
Triclinic			1	2	2	Triclinic
Monoclinic		$\alpha = \gamma = 90°$	2	13	3	Monoclinic
Orthorhombic		$\alpha = \beta = \gamma = 90°$	4	59	3	Orthorhombic
Tetragonal	$a = b$	$\alpha = \beta = \gamma = 90°$	2	68	7	Tetragonal
Rhombohedral	$a = b = c$	$\alpha = \beta = \gamma$	1	7	5	Trigonal
Hexagonal	$a = b$	$\alpha = \beta = 90°, \gamma = 120°$	1	18		
				27	7	Hexagonal
Cubic	$a = b = c$	$\alpha = \beta = \gamma = 90°$	3	36	5	Cubic
Total			14	230	32	7

The atomic structure of a crystal is constructed by repeating the basis throughout the Bravais lattice. The number of atoms within the basis may exceed one, thereby altering the lattice symmetry. The combination of basis atoms and lattice symmetry necessitates the classification of crystals into 230 space groups. These space groups encompass symmetry operations about points (e.g., rotation, reflection, and inversion) and incorporate translational symmetry, including operations combining translation with rotation. Specifically, the latter category comprises screw axes and glide planes: a screw axis represents a rotation combined with a translation, whereas a glide plane constitutes a reflection combined with a translation.

Symmetry operations centered on points—such as rotation, reflection, and inversion—constitute point groups. Only 32 distinct point groups exist among the 230 space groups. The 14 Bravais lattices represent a specific subset of space groups, exhibiting seven distinct point groups and categorized into seven crystal systems: triclinic, monoclinic, orthorhombic, tetragonal, trigonal, hexagonal, and cubic. Point groups are commonly designated using Schönflies notation, employing C, D, T, and O to denote cyclic, dihedral, tetrahedral, and octahedral symmetries, respectively, supplemented by subscripts indicating rotational symmetry order. For example, C_{3h} signifies a threefold cyclic rotational symmetry coupled with a mirror plane perpendicular to the rotational axis. Space groups are typically described using Hermann–Mauguin notation.

1.3.2 Reciprocal Lattice and Miller Indices

A direction connecting two distinct lattice points is uniquely specified by a corresponding parallel lattice vector,

$$\vec{\boldsymbol{R}} = n_1\vec{\boldsymbol{a}}_1 + n_2\vec{\boldsymbol{a}}_2 + n_3\vec{\boldsymbol{a}}_3 \tag{1.2}$$

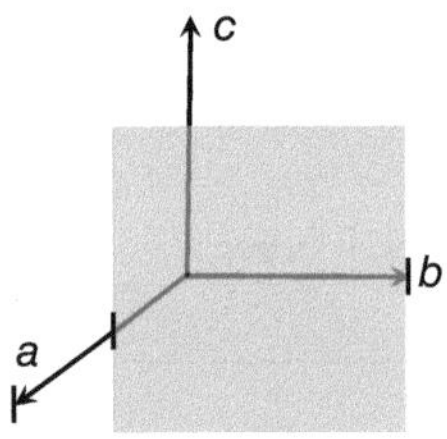

Figure 1.6 Identifying the (210) plane index by reading the interceptions of the plane on the lattice vectors.

which is generated by the lattice vectors $\vec{a}_1$, $\vec{a}_2$ and $\vec{a}_3$. For simplicity, the crystallographic direction can be denoted by the conventional notation $[n_1 n_2 n_3]$. For example, the [100] direction in a simple cubic lattice indicates the direction parallel to the positive x-axis. Due to lattice symmetry, the directions [100], [−100], [010], [0−10], [001], and [00−1] are crystallographically equivalent. The complete set of these equivalent directions corresponding to $[n_1 n_2 n_3]$ is conventionally indexed using angle brackets as $<n_1 n_2 n_3>$.

Certain lattices lie within a specific plane. Miller indices represent a symbolic notation for designating lattice planes. The lattice plane positioned closest to the origin intercepts the three lattice vectors at positions $\vec{a}_1/h$, $\vec{a}_2/k$ and $\vec{a}_3/l$. This plane is indexed as (hkl) using round brackets. For instance, the plane intercepting the a and b axes a $\vec{a}/2$ and $\vec{b}$, respectively, while being parallel to the c-axis (thus having an infinite intercept equivalent to $\vec{c}$), is designated as (210), as illustrated in Figure 1.6. The set of all crystallographically equivalent planes to (hkl) is denoted by{hkl} within curly brackets. The concept of Miller indices is more readily understood through the definition of a complementary set of vectors called the reciprocal lattice vectors, $\vec{b}_1$, $\vec{b}_2$ and $\vec{b}_3$. These reciprocal lattice vectors are defined by the relationship,

$$\vec{a}_i \cdot \vec{b}_j = 2\pi\delta_{ij} \tag{1.3}$$

where δ_{ij} is the Kronecker delta. δ_{ij} equals 1 only for $i = j$. Reciprocal lattice vector $\vec{b}_1$ can be directly constructed by

$$\vec{b}_1 = 2\pi \frac{\vec{a}_2 \times \vec{a}_3}{\vec{a}_1 \cdot \vec{a}_2 \times \vec{a}_3} \tag{1.4}$$

while $\vec{b}_2$ and $\vec{b}_3$ can be derived similarly. Reciprocal lattices serve as important tools for understanding the electronic band structure and the diffraction patterns of high-energy particles, such as X-rays, electrons, or neutrons. The reciprocal lattice of a simple cubic lattice is itself simple cubic. However, the reciprocal lattice of a body-centered cubic lattice is face-centered cubic, whereas the reciprocal lattice of a face-centered cubic lattice is body-centered cubic.

The orthogonality relationship between lattice and reciprocal lattice vectors renders the calculation of the inner product of a lattice direction vector and a reciprocal lattice direction vector particularly efficient. For example, for $\vec{R} = n_1\vec{a}_1 + n_2\vec{a}_2 + n_3\vec{a}_3$ and $\vec{G} = h\vec{b}_1 + k\vec{b}_2 + l\vec{b}_3$, their inner product will be

$$\vec{R} \cdot \vec{G} = 2\pi(n_1 h + n_2 k + n_3 l) \tag{1.5}$$

Utilizing this relationship, it follows that any vector $\vec{v}$ in the lattice plane (hkl), defined using the axial intercept method, satisfies $\vec{v} \cdot \vec{G} = 0$. Therefore, the lattice plane (hkl) is perpendicular to the reciprocal lattice vector $\vec{G} = h\vec{b}_1 + k\vec{b}_2 + l\vec{b}_3$, which can consequently be indexed as (hkl). The distance between adjacent (hkl) crystal planes can be calculated by

$$d = \frac{2\pi}{|h\vec{b}_1 + k\vec{b}_2 + l\vec{b}_3|} \tag{1.6}$$

For cubic crystal with lattice constant of a, this equation can be simplified as

$$d = \frac{a}{\sqrt{h^2 + k^2 + l^2}} \tag{1.7}$$

from which we can clearly see that crystal planes with higher Miller indices have smaller plane distance, usually making the corresponding crystal plane more difficult to cleavage due to stronger plane interaction.

1.3.3 Common Semiconductor Crystal Structures

The symmetry of the crystal structure significantly influences the properties of semiconductors. This symmetry is characterized by the space group. There exist 230 distinct space groups, categorized into seven lattice systems based on Bravais lattice symmetry or seven crystal systems based on point group symmetry. Lattice systems and crystal systems generally correspond. However, an exception occurs for crystals possessing a hexagonal Bravais lattice that crystallize within the trigonal system due to reduced symmetry resulting from inclusion of the basis. For instance, α-quartz exhibits this configuration (Figure 1.7 and Table 1.1).

Most significant semiconductors have cubic or hexagonal Bravais lattices and belong to the cubic, hexagonal, or trigonal crystal systems. For instance, the single crystals of elemental semiconductors, such as diamond, Si, and Ge, possess cubic Bravais lattices (Figure 1.8). Specifically, these materials exhibit face-centered cubic Bravais lattices. However, their basis consists of two atoms, meaning the primitive unit cell contains two atoms. Their atomic structures can be understood as the combination of two face-centered cubic lattices, displaced by one-quarter of the cubic body diagonal relative to each other. Structures of this type are referred to as diamond structures. Each atom is tetrahedrally coordinated by four nearest neighbors, resulting in a coordination number of 4, and these four atoms form a tetrahedron.

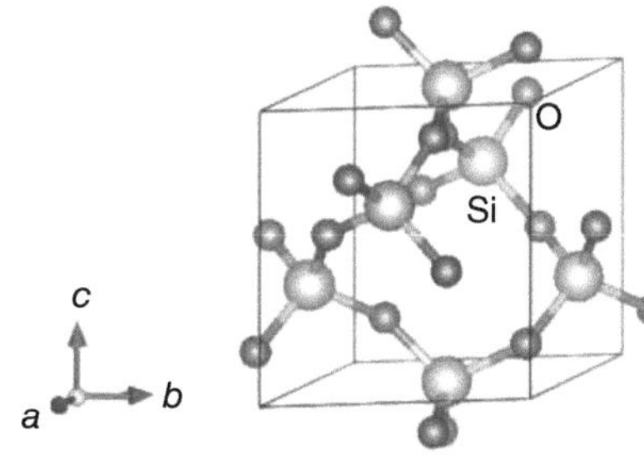

Figure 1.7 Atomic structure of α-quartz, which has hexagonal Bravais lattice but belongs to the trigonal crystal system.

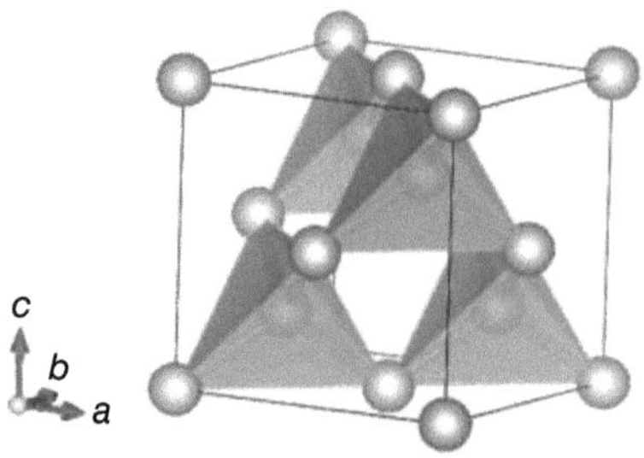

Figure 1.8 Diamond crystal structure of C, Si or Ge crystal with FCC Bravais lattice.

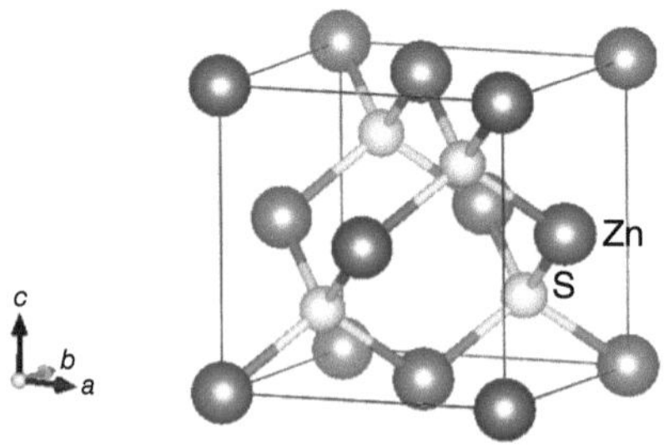

Figure 1.9 Zincblende crystal structure of ZnS.

In addition to elemental semiconductors from group IV, certain compound semiconductors exhibit crystal structures closely resembling the diamond structure. For example, binary AB compounds such as III–V or II–VI semiconductors can adopt the zincblende crystal structure. Similar to diamond, the zincblende structure possesses an fcc Bravais lattice and a basis of two atoms (Figure 1.9). However, in the zincblende basis, the two atoms are chemically distinct. Each central atom is tetrahedrally coordinated by four nearest-neighbor atoms of the opposite type. Semiconductors with zincblende structures include II–VI compounds (ZnS, ZnO, CdTe, HgTe) and III–V compounds (GaN, GaAs, InP). Furthermore, SiC, which is an IV-IV binary compound composed of elements from group IV, can also crystallize in the zincblende structure.

Besides the zincblende lattice, II–VI or III–V compounds can also adopt a hexagonal Bravais lattice. The simplest hexagonal structure is known as the wurtzite crystal structure (Figure 1.10). Both zincblende and wurtzite structures exhibit tetrahedral coordination with a coordination number of four. These structures represent different polytypes of the same compounds. The key distinction lies in the packing:

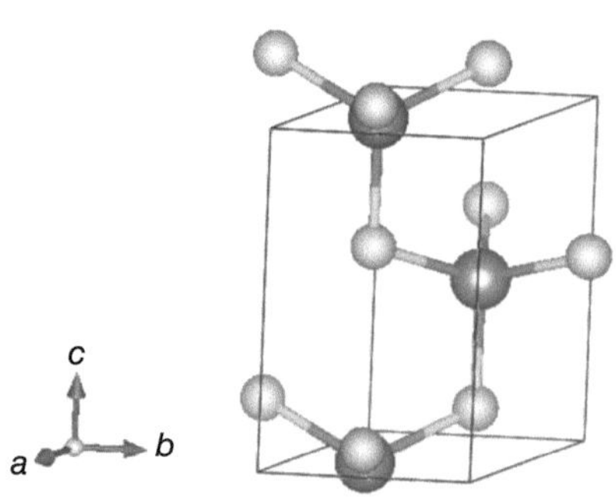

Figure 1.10 Wurtzite crystal structure of ZnS.

zincblende structures feature cubic close-packed (CCP) atoms in a face-centered cubic lattice, whereas wurtzite structures possess hexagonal close-packed (HCP) atoms in a hexagonal lattice. Both crystal types can be described using a hexagonal unit cell. Three different layer-by-layer stacking sequences exist for tetrahedrally coordinated structures. Zincblende crystals follow an ABCABCABC stacking pattern and possess a primitive unit cell with two atoms. Conversely, wurtzite crystals exhibit an ABABAB stacking pattern and contain four atoms per primitive unit cell. Zincblende is generally the low-temperature stable phase, while wurtzite becomes more stable at higher temperatures. Rapid cooling following heating can freeze-in and form the wurtzite structure at room temperature. Beyond zincblende and wurtzite, numerous other stacking sequences are possible for II–VI, III–V, or IV–IV compounds, resulting in additional polymorphs. For instance, SiC exhibits several polytypes such as 2H, 3C, 4H, and 6H [4]. Understanding these fundamental structures facilitates the construction and understanding of more complex semiconductor crystals. $Al_xGa_{1-x}N$ serves as a specific example; its structure is based on the wurtzite lattice of GaN. Al and Ga, being homologous group III metals, are randomly distributed in $Al_xGa_{1-x}N$, with Ga sites occupied by a mixture of Al and Ga atoms consistent with their respective concentrations [5].

Another important crystal structure for semiconductors is the perovskite structure. Perovskite has a chemical formula of ABX_3, in which A and B are two cations (positively charged ions), and X is a negatively charged ion (i.e., an anion). A and B are typically of significantly different sizes, with A ions larger than B ions. A representative example is $SrTiO_3$. ABX_3 adopts a cubic structure where the B cation is surrounded by an octahedron of X anions in sixfold coordination, and the A cation is surrounded by 12 X anions in cuboctahedral coordination. Perovskite structures occur in an extensive range of materials. All-inorganic cesium lead halide perovskites ($CsPbX_3$, where X = I, Br, or Cl) exhibit crystal structures analogous to $SrTiO_3$. Considering $CsPbBr_3$ as an example (Figure 1.11), it adopts the simple

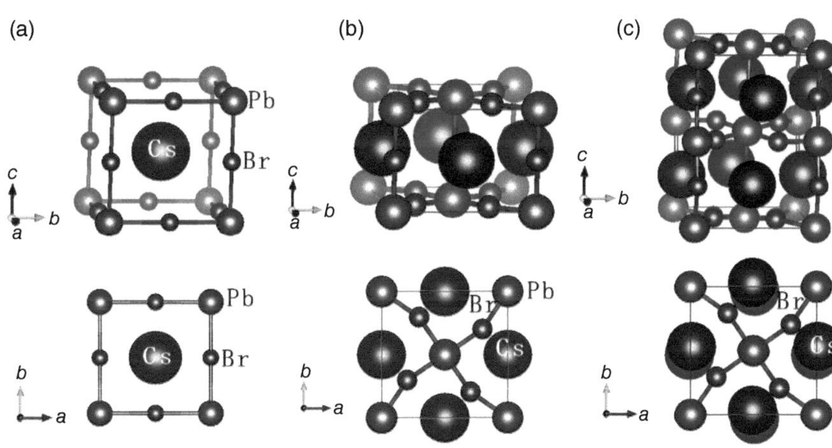

Figure 1.11 Side and top views of $CsPbBr_3$ in (a) cubic phase, (b) Pnma phase and (c) P4/mbm phase.

cubic Bravais lattice with $Pm\bar{3}m$ space group symmetry solely at temperatures exceeding 130 °C. $CsPbBr_3$ in the $Pm\bar{3}m$ phase contains 5 atoms per unit cell. Between 88 and 130 °C, the stable phase of $CsPbBr_3$ is tetragonal, belonging to the P4/mbm space group with 10 atoms in the unit cell; this phase can be derived from a $\sqrt{2} \times \sqrt{2} \times 1$ supercell of the cubic structure with distortion. At temperatures below 88 °C, the tetragonal phase transforms to the orthorhombic phase with Pnma space group symmetry [6]. The orthorhombic phase can be constructed based on a $\sqrt{2} \times \sqrt{2} \times 2$ supercell of cubic $CsPbBr_3$ with distortion. The cubic $Pm\bar{3}m$ phase possesses the highest symmetry, characterized by 48 symmetry operations; the tetragonal P4/mbm phase exhibits the second highest symmetry with 16 symmetry operations; and the orthorhombic Pnma phase has the lowest symmetry with 8 symmetry operations. $CsPbBr_3$ with the orthorhombic Pnma phase, stable at room temperature, achieves high carrier mobility and excellent optoelectronic properties, rendering it suitable for applications in various optoelectronic devices, such as light-emitting diodes, solar cells, photodetectors, lasers, among others.

1.4 Carrier Transport

Semiconductors contain two types of charge carriers: electrons and holes. An accurate description of electrons and holes requires the framework of electronic band structure, which will be presented in subsequent sections. Broadly, if an intrinsic semiconductor possesses additional electrons, these electrons occupy the conduction band and transport similarly to free electrons, albeit with a distinct effective mass. Conversely, if an intrinsic semiconductor loses electrons, holes exist within the valence band. These holes transport as positively charged particles with properties analogous to electrons but with a positive charge and a different effective mass. Electrons and holes coexist within the semiconductor simultaneously. However, they can recombine upon encountering each other, leading to the annihilation of both carriers. This recombination releases energy in the form of light emission, lattice excitations, or excitations of other particles. Conversely, an electron-hole pair can be generated through sufficient energy input, for example, via light absorption, thermal activation, or impact ionization. Due to the variability in carrier types and underlying physical mechanisms within semiconductors compared to metallic conductors, carrier transport exhibits more complex behaviors, including drift, diffusion, recombination, generation, thermionic emission, and tunneling.

1.4.1 Carrier Drift and Diffusion

An electric field exerts a force on both electrons and holes, as they are charged carriers, thereby accelerating them and inducing drift transport. In the absence of scattering mechanisms, these carriers would accelerate continuously, achieving arbitrarily high velocities. However, scattering events arising from disturbances such as lattice vibrations (phonons) and crystal defects decelerate the carriers. This subsequent deceleration dissipates the acquired kinetic energy of the carriers, converting it into heat. Consequently, an average drift velocity is attained, which is described by the relationship given below,

$$v_e = \mu_e E \tag{1.8}$$

$$v_h = \mu_h E \tag{1.9}$$

where v_e and v_h are the average velocities of electrons and holes, respectively, E is the electric field strength, and μ_e and μ_h are coefficients termed carrier mobility. The velocity has units of cm/s, while the electric field strength has units of V/cm; therefore, the unit of carrier mobility is $\mathrm{cm^2/(V \cdot s)}$. Mobility characterizes the rate at which a carrier moves when subjected to an electric field. It is noted that electrons and holes possess opposite charges; consequently, they experience opposing forces and transport in opposite directions within the semiconductor. However, despite their opposing transport directions, the opposite charges of electrons and holes therefore result in their electric currents having the same direction. If the carrier densities for electrons and holes are n_e and n_h, respectively, the total electric current density caused by carrier drift is

$$J_{\text{drift}} = en_e v_e + en_h v_h = en_e \mu_e E + en_h \mu_h E \tag{1.10}$$

The drift current can be rewritten as

$$J_{\text{drift}} = \sigma_e E + \sigma_h E = \sigma E \tag{1.11}$$

$$\sigma_e = en_e \mu_e \tag{1.12}$$

$$\sigma_h = en_h \mu_h \tag{1.13}$$

where σ_e and σ_h are the electrical conductivity of electrons and holes, respectively, and e is the fundamental electron charge. The electrical conductivity is related to resistivity by $\sigma = 1/\rho$. Carrier scattering imposes a limit on their drift velocity, which gives rise to Joule heating. The volumetric Joule heating power density generated by an electric field is given by

$$P = J_{\text{drift}} E = \sigma E^2 \tag{1.14}$$

The carrier mobility is fundamentally governed by the lifetime and mean-free-path of carriers during transport and scattering events. According to the Drude model, the carrier's momentum is accelerated by the applied electric field and dissipated through momentum relaxation during scattering events. Assuming a carrier scattering lifetime of τ, the equation of motion accelerated by the electric field is expressed as:

$$m\frac{\mathrm{d}v}{\mathrm{d}t} = eE - m\frac{v}{\tau} \tag{1.15}$$

where m is the carrier mass and e is the carrier charge, depending on the electron or hole carrier type. At steady state, $\mathrm{d}v/\mathrm{d}t = 0$, we can get

$$v = \frac{e\tau}{m} E \tag{1.16}$$

In associate with previous equation, the carrier mobility can be written as

$$\mu = \frac{e\tau}{m} \tag{1.17}$$

and the electric conductivity will be

$$\sigma = \frac{e^2\tau}{m}n \tag{1.18}$$

Multiple scattering mechanisms exist, exhibiting distinct dependences on temperature and thereby influencing the scattering lifetime. For acoustic phonon scattering via the deformation potential, mobility varies proportionally to $T^{-3/2}$. In contrast, for acoustic phonon scattering via the piezoelectric mechanism, mobility varies proportionally to $T^{-1/2}$. Ionized impurity scattering results in a temperature dependence of mobility scaling with $T^{3/2}$. At low electric fields, typically below 10^3 V/cm in silicon and gallium arsenide, the drift velocity increases linearly in accordance with Ohm's law. However, at higher fields, the carrier velocity exhibits a sublinear increase with field strength and eventually saturates at approximately 10^7 cm/s, a phenomenon termed velocity saturation. Certain semiconductors, such as GaAs, InP, and InGaAs, exhibit more complex behavior characterized by a velocity maximum preceding saturation as illustrated for fields between 3×10^3 V/cm and 2×10^5 V/cm. This maximum constitutes velocity overshoot, leading to a decrease in current with increasing electric field strength, known as negative differential resistance.

Besides the drift motions induced by the electric field, carrier transport also arises due to the gradient of carrier density, known as diffusion current. The diffusion current is proportional to the magnitude of the carrier density gradient, expressed by the following equation,

$$J_{\text{diff}} = eD_e\frac{\mathrm{d}n_e}{\mathrm{d}x} - eD_h\frac{\mathrm{d}n_h}{\mathrm{d}x} \tag{1.19}$$

which has considered the diffusion of both electrons and holes. D_n and D_p represent the diffusion coefficients of electrons and holes, respectively. The differing signs in the equation arise from the opposite charge polarity between electrons and holes. From a microscopic perspective, the diffusion coefficient can be determined using

$$D = \frac{v_{\text{rms}}^2}{3}\tau \tag{1.20}$$

where v_{rms} is the root-mean-square velocity of either electrons or holes, and τ is the same scattering lifetime as mentioned in the mobility. Remember that $mv_{\text{rms}}^2/2 = 3kT/2$, we can get the Einstein relationship between the diffusion coefficient and the carrier mobility,

$$D = \mu\frac{kT}{e} \tag{1.21}$$

The Einstein relation characterizes the diffusion of charge carriers, specifically electrons and holes governed by Boltzmann statistics. This relation can also be derived from the equilibrium condition occurring in a semiconductor possessing a built-in electric field, E. Under such conditions, the net current attributable to any particular carrier type, electrons or holes, must vanish to maintain equilibrium.

Using electron transport as an example, the combined contributions of the drift current and the diffusion current imply that the following equation must hold

$$\mu_e n_e E = -D_e \frac{dn_e}{dx} \tag{1.22}$$

While at equilibrium the carrier concentration, because of the electric field induced energy difference, according to the Boltzmann distribution, the spatial electron density distribution should follow,

$$n_e(x) = n_e(x_0)e^{-\frac{eE(x-x_0)}{kT}} \tag{1.23}$$

$$\frac{dn_e}{dx} = -n_e \frac{eE}{kT} \tag{1.24}$$

where x_0 is a reference coordinate with electron density of $n_e(x_0)$. Similar equations hold for holes as well. Combining these equations, we can get the Einstein relation.

The total current inside a semiconductor will be the combination of drift current and diffusion current, as indicated below,

$$J_{tot} = J_{drift} + J_{diff} = en_e\mu_e E + eD_e \frac{dn_e}{dx} + en_h\mu_h E - eD_h \frac{dn_h}{dx} \tag{1.25}$$

which can be separated into electron and hole contributions respectively as below (Figure 1.12),

$$J_e = en_e\mu_e E + eD_e \frac{dn_e}{dx} \tag{1.26}$$

$$J_h = en_h\mu_h E - eD_h \frac{dn_h}{dx} \tag{1.27}$$

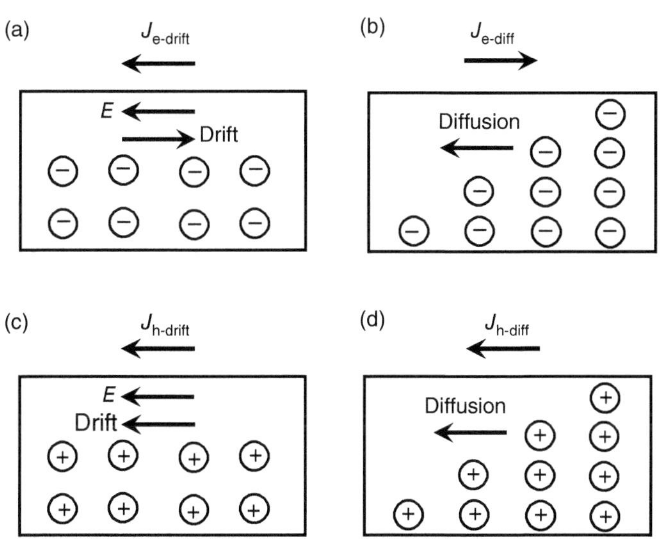

Figure 1.12 (a) Drift current caused by electric field and (b) diffusion current caused by concentration gradient for electrons. (c) Drift current caused by electric field and (d) diffusion current caused by concentration gradient for holes.

1.4.2 Quasi-Fermi Levels and Electrochemical Potentials

After excitations such as optical illumination or electron/hole injection, electron and hole densities deviate from their thermodynamic equilibrium values. Nonetheless, quasi-Fermi levels can still describe these densities; for instance, the Boltzmann distribution applies to nondegenerate cases. Assuming the electrostatic potential induced by the electric field is $\varphi(x)$, the electric field strength E is given by $E = -\mathrm{d}\varphi/\mathrm{d}x$. Considering the quasi-Fermi levels E_{Fe} and E_{Fh} for electrons and holes, respectively, the electron and hole densities can be expressed according to the Boltzmann distribution as

$$n_e(x) = n_e(x_0)e^{\frac{E_{\mathrm{Fe}}(x)+e\varphi(x)-E_{\mathrm{Fe}}(x_0)-e\varphi(x_0)}{kT}} \tag{1.28}$$

$$n_h(x) = n_h(x_0)e^{\frac{-E_{\mathrm{Fh}}(x)-e\varphi(x)+E_{\mathrm{Fh}}(x_0)+e\varphi(x_0)}{kT}} \tag{1.29}$$

where x_0 is a reference coordinate with an electron density of $n_e(x_0)$ and a hole density of $n_h(x_0)$. Since the electrostatic potential $-e\varphi(x)$ influences both the conduction and valence band positions by $-e\varphi(x)$, the relative position between the quasi-Fermi level and the conduction or valence band edge is given by $E_{\mathrm{Fe}}(x) + e\varphi(x)$ or $E_{\mathrm{Fh}}(x) + e\varphi(x)$. For electrons, a higher value of $E_{\mathrm{Fe}}(x) + e\varphi(x)$ indicates that the quasi-Fermi level is closer to the conduction band minimum, resulting in a higher electron density. In contrast, for holes, a higher value of $E_{\mathrm{Fe}}(x) + e\varphi(x)$ indicates that the quasi-Fermi level is more distant from the valence band maximum, resulting in a lower hole density. More specifically, we can get

$$\frac{\mathrm{d}n_e}{\mathrm{d}x} = n_e\left(\frac{\mathrm{d}E_{\mathrm{Fe}}}{\mathrm{d}x} + e\frac{\mathrm{d}\varphi}{\mathrm{d}x}\right) \tag{1.30}$$

$$\frac{\mathrm{d}n_h}{\mathrm{d}x} = n_h\left(-\frac{\mathrm{d}E_{\mathrm{Fh}}}{\mathrm{d}x} - e\frac{\mathrm{d}\varphi}{\mathrm{d}x}\right) \tag{1.31}$$

Inserting these two equations in the electron and hole current equations respectively, we can get the following equations for the total current after considering $E = -\mathrm{d}\varphi/\mathrm{d}x$ and Einstein relation $D = \mu kT/e$,

$$J_e = \frac{eD_e n_e}{kT}\frac{\mathrm{d}E_{\mathrm{Fe}}}{\mathrm{d}x} = n_e\mu_e\frac{\mathrm{d}E_{\mathrm{Fe}}}{\mathrm{d}x} \tag{1.32}$$

$$J_h = \frac{eD_h n_h}{kT}\frac{\mathrm{d}E_{\mathrm{Fh}}}{\mathrm{d}x} = n_h\mu_h\frac{\mathrm{d}E_{\mathrm{Fh}}}{\mathrm{d}x} \tag{1.33}$$

Similar to the electrostatic potential, the electrochemical potentials for electrons and holes are defined by

$$E_{\mathrm{Fe}} = e\varphi_{\mathrm{Fe}} \tag{1.34}$$

$$E_{\mathrm{Fh}} = e\varphi_{\mathrm{Fh}} \tag{1.35}$$

Consequently, the total electron and hole current densities become,

$$J_e = en_e\mu_e\frac{\mathrm{d}\varphi_{\mathrm{Fe}}}{\mathrm{d}x} = \sigma_e\frac{\mathrm{d}\varphi_{\mathrm{Fe}}}{\mathrm{d}x} \tag{1.36}$$

$$J_h = en_h\mu_h\frac{\mathrm{d}\varphi_{\mathrm{Fh}}}{\mathrm{d}x} = \sigma_h\frac{\mathrm{d}\varphi_{\mathrm{Fh}}}{\mathrm{d}x} \tag{1.37}$$

This formalism is analogous to the description of conduction current in a conductor or drift current in a semiconductor. The crucial modifications involve replacing the electrostatic potential with the quasi-Fermi levels for electrons and holes, respectively. From the preceding equations, it is evident that the gradient of the electrochemical potential governs carrier flow. In semiconductors under thermal equilibrium conditions, no net current exists, and the quasi-Fermi levels for electrons and holes are equal. Therefore, $J_e + J_h = 0$ and $\varphi_{\mathrm{Fe}} = \varphi_{\mathrm{Fh}}$, implying that the electrochemical potential remains constant throughout the semiconductor material.

1.4.3 Carrier Generation and Recombination

At the thermal-equilibrium condition, the electron concentration p and hole concentration n satisfy the following equation,

$$pn = n_i^2 \tag{1.38}$$

where n_i is the intrinsic electron and hole concentration. During carrier transport, electron and hole concentrations may deviate from thermal equilibrium conditions. Electron-hole recombination dominates when $pn > n_i^2$, whereas thermal generation dominates when $pn < n_i^2$. During electron-hole recombination, energy is conserved through photon emission, excitation of another free carrier (Auger recombination), or phonon emission. Conversely, the energy required for the thermal generation of electron-hole pairs originates from the lattice thermal energy. Recombination and generation mechanisms are classified as either band-to-band transitions or trap-mediated transitions (Figure 1.13). Band-to-band transitions primarily occur in direct-bandgap semiconductors, such as GaAs and other III–V compounds. For indirect-bandgap semiconductors like Si and Ge, however, trap-mediated transitions predominate.

For the band-to-band transitions, the net recombination or generation rate can be calculated by

$$U = R\left(pn - n_i^2\right) \tag{1.39}$$

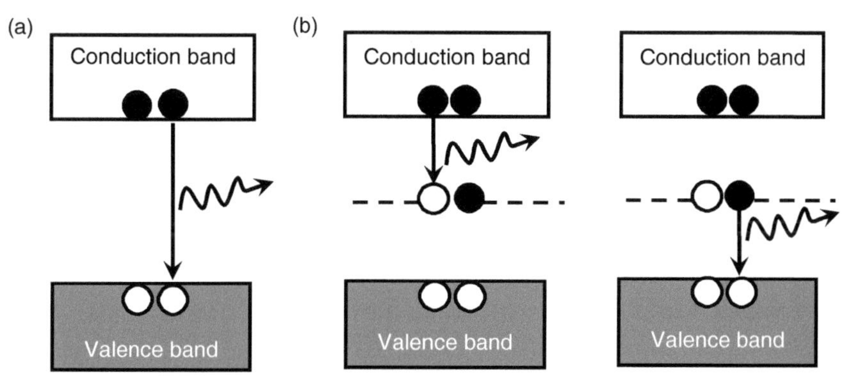

Figure 1.13 Schematics of (a) band-to-band recombination process and (b) Shockley-Read-Hall recombination process.

where R is the recombination coefficient, net recombination of electron-hole pairs occurs when $pn > n_i^2$; otherwise, net generation occurs. Under low-level injection conditions—for instance, in an n-type semiconductor with $n \approx N_D$ and injected hole density $p = \Delta p + p_0$, where the equilibrium concentrations satisfy $np_0 = n_i^2$—the net recombination rate arising from band-to-band transitions is

$$U \approx \frac{\Delta p}{\tau_p} \tag{1.40}$$

$$\tau_p = \frac{1}{RN_D} \tag{1.41}$$

Similarly, for a low-level injection of n-type semiconductor, the net recombination rate due to band-to-band transitions is

$$U \approx \frac{\Delta n}{\tau_n} \tag{1.42}$$

$$\tau_n = \frac{1}{RN_A} \tag{1.43}$$

where $p \approx N_A$ is the hole and acceptor doping density.

For recombination through trap levels, the capture of electrons and holes by the trap levels occurs sequentially. The capture probabilities for electrons and holes can be characterized by the capture cross sections, carrier thermal velocities, and trap density, respectively. The electron capture rate is determined

$$\frac{n}{\tau_{n,\text{cap}}} = \sigma_n v_{\text{th}} n(N_{t,\text{tot}} - N_{t,\text{occ}}) \tag{1.44}$$

where $N_{t,\text{tot}}$ and $N_{t,\text{occ}}$ are the total and occupied trap densities, respectively, and τ_n is the electron lifetime due to capture by unoccupied traps, characterized by the density $N_{t,\text{tot}} - N_{t,\text{occ}}$. Here, σ_n and v_{th} are the electron capture cross-section and average thermal velocity, respectively. Similarly, the hole capture probability is proportional to,

$$\frac{p}{\tau_{p,\text{cap}}} = \sigma_p v_{\text{th}} p N_{t,\text{occ}} \tag{1.45}$$

where $N_{t,\text{occ}}$ is the occupied trap density. Besides electron and hole capture processes, charges can also undergo thermal excitation to the conduction and valence bands, respectively. The corresponding generation rates can be derived from the detailed balance principle.

$$\frac{N_{t,\text{occ}}}{\tau_{n,\text{gen}}} = \sigma_n v_{\text{th}} n_i N_{t,\text{occ}} e^{\frac{E_t - E_i}{kT}} \tag{1.46}$$

$$\frac{N_{t,\text{tot}} - N_{t,\text{occ}}}{\tau_{p,\text{gen}}} = \sigma_p v_{\text{th}} n_i (N_{t,\text{tot}} - N_{t,\text{occ}}) e^{\frac{E_i - E_t}{kT}} \tag{1.47}$$

Considering all these factors, the net recombination rate can be described using Shockley-Read-Hall statistics

$$U = \frac{\sigma_n \sigma_p v_{\text{th}} N_t \left(pn - n_i^2\right)}{\sigma_n \left(n + n_i e^{\frac{E_t - E_i}{kT}}\right) + \sigma_p \left(p + n_i e^{\frac{E_i - E_t}{kT}}\right)} \tag{1.48}$$

Recombination due to surface trap states can be described by analogous/similar equations. However, the surface trap density N_t has units of cm^{-2}; consequently, the net recombination rate U has unit of $cm^{-2} \cdot s^{-1}$ rather than $cm^{-3} \cdot s^{-1}$. The equation above demonstrates that a net recombination or generation rate exists only if $pn \neq n_i^2$, and this rate is proportional to $pn - n_i^2$. Furthermore, the trap-assisted recombination rate is maximized when the trap level E_t lies close to the mid-bandgap, i.e., $E_t \approx E_i$. Thus, the net Shockley-Read-Hall recombination rate is given by:

$$U = \frac{\sigma_n \sigma_p v_{th} N_t \left(pn - n_i^2\right)}{\sigma_n (n + n_i) + \sigma_p (p + n_i)} \tag{1.49}$$

Under low-level injection conditions, such as in an n-type semiconductor with doping concentration $n \approx N_D$ and an injected hole density $p = \Delta p + p_0$ satisfying $np_0 = n_i^2$, the net Shockley-Read-Hall recombination rate is expressed as

$$U \approx \frac{\Delta p}{\tau_p} \tag{1.50}$$

$$\tau_p = \frac{1}{\sigma_p v_{th} N_t} \tag{1.51}$$

Similarly, for low-level injection in an n-type semiconductor, the net Shockley-Read-Hall recombination rate is

$$U \approx \frac{\Delta n}{\tau_n} \tag{1.52}$$

$$\tau_n = \frac{1}{\sigma_n v_{th} N_t} \tag{1.53}$$

For high-level injections $\Delta n = \Delta p > n_i$, the recombination lifetimes due to band-to-band recombination and trap-assisted recombination, respectively, are derived as

$$\tau_n = \tau_p = \frac{1}{R\Delta n} \tag{1.54}$$

$$\tau_n = \tau_p = \frac{\sigma_n + \sigma_p}{\sigma_n \sigma_p v_{th} N_t} \tag{1.55}$$

For trap-assisted recombination. These equations indicate that the band-to-band recombination lifetime decreases with injection level, while the trap-assisted recombination lifetime increases by approximately a factor of two when $\sigma_n = \sigma_p$. The increase in trap-assisted recombination lifetime may be explained by the half-occupation of trap states under high-level injection. Impurities or lattice defects typically introduce deep-level defect states within the bandgap, which shorten the minority-carrier lifetime. For instance, gold impurities in silicon introduce trap levels that can reduce the lifetime from several microseconds to several nanoseconds. This lifetime reduction can be advantageous for high-speed devices.

Besides band-to-band recombination and Shockley-Read-Hall recombination, Auger recombination represents another crucial recombination process. This nonradiative mechanism is significant in semiconductors, occurring when an electron and hole recombine by transferring their energy to a third charge carrier—either another electron or hole—instead of emitting a photon. The net Auger recombination rate is described by the following equation

$$U = (C_n n + C_p p)\left(pn - n_i^2\right) \tag{1.56}$$

where C_n is the Auger coefficient for electron-dominated processes (cm^6/s), and C_p is the Auger coefficient for hole-dominated processes (cm^6/s). Auger recombination becomes increasingly prominent at high carrier densities as it competes with radiative recombination. This competition significantly reduces the efficiency of optoelectronic devices, including light-emitting diodes (LEDs), solar cells, and lasers.

1.5 Doping Process

Intrinsic semiconductors exhibit very low carrier densities, resulting in negligible conductivity. The reason semiconductors are immensely valuable in electronic devices is that their electronic properties can be precisely modified by introducing a relatively small concentration of defects, such as impurity dopants, typically less than one part per million. For example, silicon can be doped to form an n-type or p-type semiconductor by adding P atoms or B atoms, respectively, enabling conduction through electron or hole transport. More complex structures can be fabricated by spatially varying the doping profile within semiconductors, leading to significant applications and realizing functionalities unattainable with conventional devices. Typical examples include the diode and the bipolar junction transistor. A diode achieves unidirectional conduction; its fundamental structure is the PN junction, which conducts current under forward bias and blocks current under reverse bias. The BJT, based on PNP or NPN junctions, functions as a current amplifier. These unique properties of doped semiconductor devices underpin information processing in crucial circuits.

1.5.1 Point Defects and Energy Levels

Understanding defects and achieving controllable doping are of fundamental importance for utilizing semiconductors in electronic device applications. As previously emphasized, the required density of targeted defects is typically very low. Moreover, not every defect type contributes beneficially to device performance. Consequently, the creation and study of specific defects rely on the ability to first synthesize a semiconductor crystal of exceptionally high purity exhibiting negligible intrinsic defect density. Upon achieving this pristine base material, controlled introduction and subsequent investigation of the desired defect types and quantities become feasible. Based on spatial extent, defects are categorized as point defects, defect complexes, or line defects. Point defects involve only a few isolated atoms localized within the host crystal lattice. Defect complexes consist of several aggregated point defects. Line defects, primarily dislocations, involve one-dimensional rows of atoms. The functional point defects critical for the doping process will be the focus of this section.

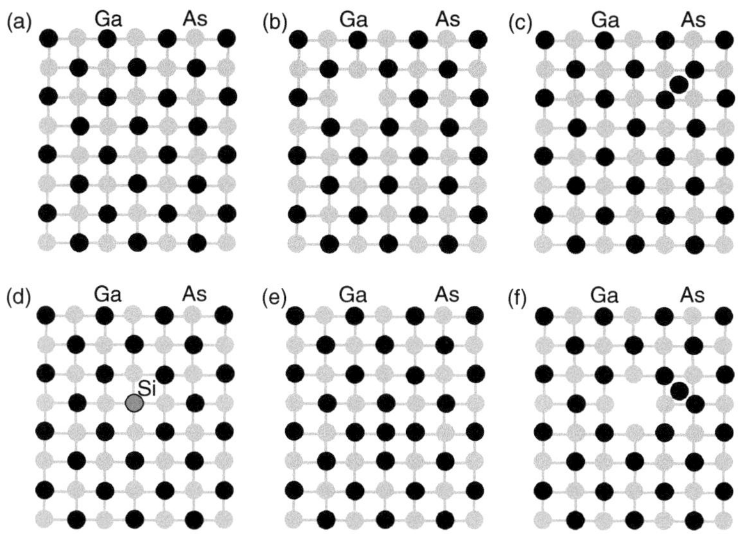

Figure 1.14 Schematic representation of various defects in GaAs in a 2D plane. (a) Pristine GaAs, (b) Ga vacancy defect (V_{Ga}), (c) Ga interstitial defect (I_{Ga}), (d) Ga substitution defect by a Si atom (Si_{Ga}), (e) Ga antisite defect (Ga_{As}), and (f) a Frenkel complex defect with Ga atom displaced from the lattice site but occupying a nearby interstitial site ($V_{Ga}-I_{Ga}$).

Point defects are further categorized into vacancy, interstitial, substitutional, antisite, and Frenkel pair. We exemplify these defects using GaAs (Figure 1.14). Figure 1.14a is the pristine GaAs. A vacancy in GaAs occurs due to a missing host atom, such as a missing Ga atom denoted V_{Ga} (Figure 1.14b). An interstitial defect arises when a host atom occupies an interstitial site, as exemplified by a Ga atom in an interstitial position, denoted I_{Ga} (Figure 1.14c). Substitutional defects result from the replacement of a host atom by an impurity atom, as when a Ga atom is substituted by a Si atom, denoted Si_{Ga} (Figure 1.14d). An antisite defect represents a specific substitutional type where one host atom replaces another host atom; for example, a Ga atom occupies an As site, denoted GaAs (Figure 1.14e). A Frenkel pair is a complex defect comprising a host atom displaced from its lattice site into a nearby interstitial position, illustrated by V_{Ga}–I_{Ga}, which involves a displaced Ga atom (Figure 1.14f). Defects involving impurity atoms are termed extrinsic defects. In contrast, defects such as vacancies, interstitials, and antisites, which lack impurity atoms, are referred to as intrinsic or native defects.

Electrically active defects, which can generate free carriers following thermal excitation, are categorized as donors or acceptors. Donors provide free electrons, whereas acceptors generate holes. Illustrating with silicon, substitutional defects involving Group V atoms (e.g., P, As, Sb) or interstitial defects involving monovalent atoms (e.g., Li, Na) act as donors (Figure 1.15a,b). Conversely, substitutional defects involving Group III atoms (e.g., B, Al, Ga, In) serve as acceptors (Figure 1.15c,d). Group V atoms possess one more valence electron than Group IV atoms, explaining their donor behavior, while Group III atoms lack one electron, accounting for their acceptor properties. Similarly, substitutional Group VI atoms (e.g., S, Se, Te)

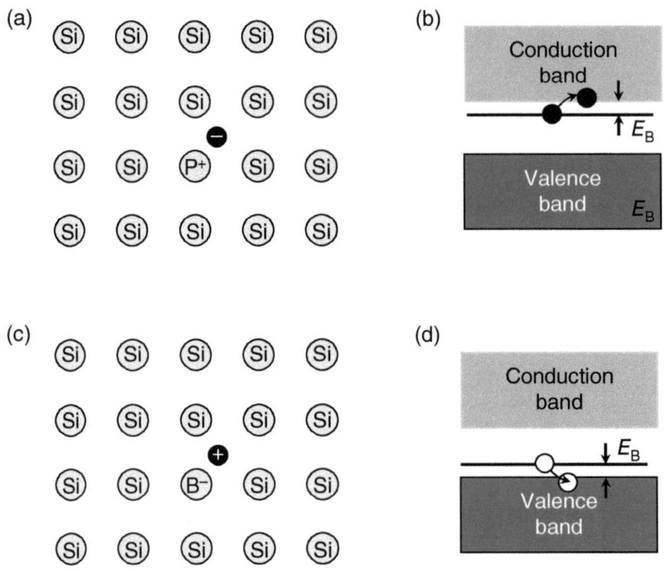

Figure 1.15 (a) Schematic representation of substitutional P doping of Si. (b) The donor level and free electron excitation in the conduction band of substitutional P defect. (c) Schematic representation of substitutional B doping of Si. (d) The acceptor level and free hole excitation in the valence band of substitutional B defect.

function as double donors, and substitutional Group II atoms (e.g., Be) function as double acceptors in silicon, releasing two free electrons or holes, respectively, upon excitation. Substitution by a Group IV foreign atom, such as C_{Si}, creates an isovalent center characterized by the same number of valence electrons as the host silicon lattice. Generally, isovalent centers remain electrically inactive; however, under some conditions, they can function as donors or acceptors.

The doping mechanism can be further elucidated through the effective mass approximation. For phosphorus (P) substitutional defects in silicon (Si), the P nucleus possesses one extra positive charge compared with the Si nucleus. However, the P atom also contributes an additional valence electron, rendering the defect charge-neutral. A Coulomb attraction exists between this extra positive charge and the additional electron. This interaction can be modeled using the effective mass approximation, wherein the extra electron is treated as a quasi-free electron in the conduction band, characterized by an effective mass (m^*) determined by the conduction band dispersion discussed later. This electron becomes bound to the positively charged P ion, analogous to the electron in a hydrogen atom. Crucially, the presence of the Si host crystal introduces a dielectric screening effect, distinct from the vacuum Coulomb potential in hydrogen. Incorporating this dielectric screening and the effective mass approximation, the binding energy for the electron bound to the positively charged donor is expressed as

$$E_B(n) = \frac{R_\infty}{n^2} \frac{m^*}{m_e} \frac{\varepsilon_0^2}{\varepsilon_s^2} \tag{1.57}$$

where $R_\infty = 13.6$ eV is the Rydberg constant, n is the principal quantum number for the bound states, m_e is the free electron mass, m^* is the conduction band effective mass, ε_0 is the vacuum permittivity, and ε_s is the static permittivity of the semiconductor. For the ground state ($n = 1$), the binding energy is maximal, and the corresponding Bohr radius of the electron probability density is

$$a^* = a_B \frac{m_e}{m^*} \frac{\varepsilon_s}{\varepsilon_0} \tag{1.58}$$

where $a_B = 0.529$ Å is the hydrogen Bohr radius. Given the anisotropy of silicon effective masses, a density-of-states average effective mass ($m^* = 0.36m_e$) and a static permittivity of $\varepsilon_s = 11.7\varepsilon_0$ yield a calculated binding energy of 36 meV. This value is reasonably close to the experimentally determined donor ionization energy of 45 meV. The calculated Bohr radius is 17.2 Å, significantly larger than the silicon lattice constant (5.4 Å), validating the application of the effective mass approximation. At room temperature, the thermal energy $kT = 25$ meV is sufficient to ionize this weakly bound electron into the conduction band, rendering P in Si an efficient donor. A parallel analysis applies to boron (B) doping in Si; a key distinction is that the B acceptor is negatively charged and binds a positively charged hole, analogous to an antihydrogen atom model.

Some defects, such as phosphorus-doped silicon (P_{Si}) or boron-doped silicon (B_{Si}), can be interpreted using the effective mass approximation as binding free electrons or free holes from the conduction band and valence band, respectively. These defects are commonly termed shallow impurities or shallow defects. However, certain defects possess properties that cannot be correctly explained by this effective mass approximation, and these are referred to as deep centers. Substitutional impurities resembling the host atoms, like P_{Si} in silicon or germanium-doped gallium arsenide (Ge_{Ga}) in GaAs, typically behave as shallow donors or acceptors. In contrast, if the substituting atoms differ significantly from the host atoms and induce a strongly localized potential, such as a substantial strain field, the resulting impurity introduces a deep-level center.

1.5.2 Doping Techniques

Doping can be implemented either during or after growth processes. For example, during bulk crystal growth or the epitaxial deposition of thin films, impurity dopants may be introduced concurrently with the source materials. However, such in situ doping inherently results in bulk or large-area doping distributions, which offer insufficient design flexibility for advanced device fabrication requirements. To fully unlock the potential of semiconductors, post-growth controllable doping techniques have been developed. Diffusion and ion implantation constitute the most extensively utilized and studied doping methodologies within the semiconductor industry, which will be detailed in the remainder of this section.

1.5.2.1 Thermal Diffusion

Diffusion is a doping technique that utilizes a concentration gradient to introduce dopant atoms from a high-concentration source into the semiconductor material.

Although the thermal motion of dopant atoms is inherently random, their net diffusivity depends exponentially on temperature and varies with specific dopant and semiconductor types. A concentration gradient induces net dopant transport from regions of high concentration to regions of low concentration. Conventional diffusion typically produces deep doping profiles. However, using doped oxides (e.g., borosilicate glass, BSG, for boron) or nitrides as diffusion sources combined with rapid thermal annealing (RTA) enables the realization of relatively shallow doping profiles. Selective doping, such as that required for the source and drain regions of MOSFETs, is essential for semiconductor device fabrication. Selective doping within a defined window region can be achieved via diffusion by patterning a mask layer (e.g., thick silicon dioxide or silicon nitride) to form the diffusion window.

Original sources containing dopants at high concentrations may exist in gaseous, liquid, or solid phases, as illustrated in Figure 1.16. For boron (B) doping of silicon, gaseous trimethylboron $(CH_3)_3B$ or diborane B_2H_6, liquid boron tribromide BBr_3, or a solid boron nitride (BN) disc coated with B_2O_3 may serve as the source. However, the actual source material facilitating boron diffusion is a layer of B_2O_3 deposited on the silicon surface. This intermediary B_2O_3 layer is introduced by oxidizing the gaseous, liquid, or solid precursors. The B_2O_3 subsequently reacts with elemental silicon, generating boron atoms according to the following reaction:

$$2B_2O_3 + 3Si \rightarrow 3SiO_2 + 4B \tag{1.59}$$

Similarly, gaseous phosphine PH_3, liquid phosphorus oxychloride $POCl_3$, or solid phosphorus pentoxide P_2O_5 functions as sources for phosphorus (P) doping of silicon. Consequently, the effective diffusion source is P_2O_5 deposited on the silicon surface. In all cases, the reaction between the precursor source material and silicon results in the simultaneous growth of an oxide layer.

The diffusion of dopants relies on defects for assistance (Figure 1.17). Diffusion mechanisms can be categorized into vacancy-assisted and interstitial-assisted diffusion. In vacancy-assisted diffusion, the impurity migrates via transition into

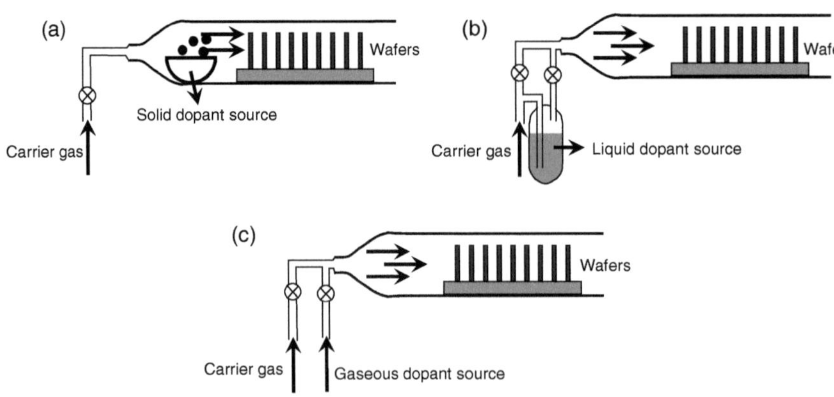

Figure 1.16 Schematic representations of diffusion systems with (a) solid, (b) liquid, and (c) gaseous dopants.

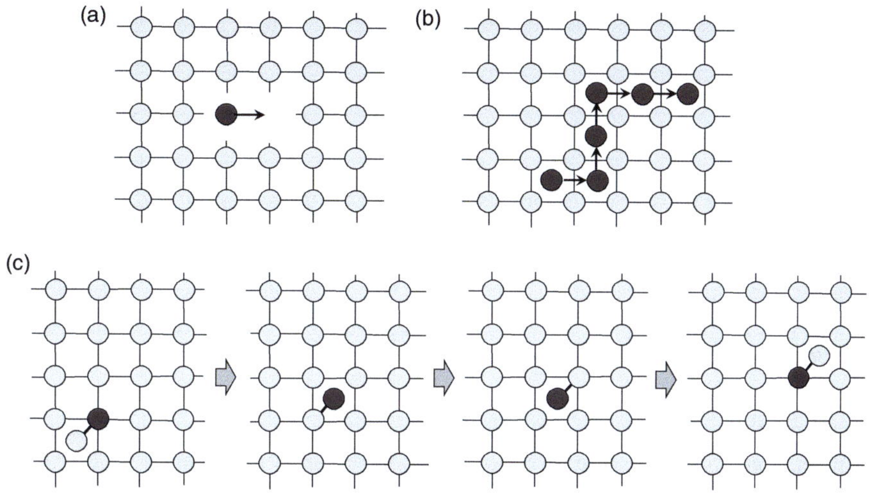

Figure 1.17 Diffusions of impurity atoms by (a) impurity jumping into a vacancy site, (b) the interstitial sites, and (c) impurity-defect pair (interstitialcy).

a vacant lattice site. Conversely, interstitial-assisted diffusion entails an impurity migrating from one interstitial position to another. Small-sized elements—such as hydrogen, helium, argon, lithium, sodium, and potassium (Group I and II)—exhibit rapid diffusion kinetics in silicon crystals. This phenomenon occurs consequently to interstitial-assisted diffusion. The interstitial mechanism also prevails for diffusion of boron, phosphorus, aluminum, or gallium in silicon. Conversely, the diffusion of arsenic and antimony is dominated by vacancy-assisted diffusion.

Dopant atoms are introduced and redistribute within the semiconductor wafer inside a furnace at elevated temperatures. The resultant doping profiles can be controlled by the source concentration, diffusion temperature, and diffusion time. Increasing the temperature enhances the diffusivity of the dopant species. A higher source concentration, higher diffusion temperature, or longer diffusion time generally increases the junction depth. To quantitatively model the diffusion process, Fick's laws of diffusion are applied. According to the diffusion flux equation derived from this theory, which remains valid under dilute impurity concentration conditions ($<10^{18}/\text{cm}^3$), the flux J, defined as the net rate of particle flow through a unit area perpendicular to the diffusion direction, is given by

$$J = -D\frac{\partial n}{\partial x} \tag{1.60}$$

where D is the diffusion coefficient (diffusivity) in units of cm^2/s, and $\partial n/\partial x$ is the impurity concentration gradient. The negative sign indicates that diffusion occurs down the concentration gradient, from regions of high concentration to regions of low concentration. The magnitude of the diffusivity governs the speed of the diffusion process and depends on the specific impurity type and the host crystal structure.

The diffusivity exhibits a strong temperature dependence described by the following Arrhenius relationship:

$$D = D_0 e^{-\frac{E_A}{kT}} \tag{1.61}$$

where D_0 is an impurity-specific constant and E_A is a measure of the diffusion energy barrier. For B and P diffusion in silicon, the D_0 values are 0.76 and 3.85 cm^2/s, respectively, and the E_A values are 3.46 and 3.66 eV, respectively. The diffusion will change the local impurity concentration in the semiconductor by the following continuity equation.

$$\frac{\partial n}{\partial t} = -\frac{\partial J}{\partial x} = D\frac{\partial^2 n}{\partial x^2} \tag{1.62}$$

This equation, obtained by substituting Fick's first law into the continuity condition, states that the rate of concentration change is determined by the spatial gradient of the flux. Subject to appropriate initial and boundary conditions, this partial differential equation can be solved to determine the evolving dopant distribution profile.

Under specific conditions, analytical solutions exist for the diffusion profile. For diffusion driven by a dopant source maintaining a constant surface concentration (as depicted in Figure 1.18a), the time-evolving concentration profile is given by

$$n(x, t) = n_0 \mathrm{erfc}\left(\frac{x}{2\sqrt{Dt}}\right) \tag{1.63}$$

where n_0 represents the constant dopant concentration at the surface, and erfc(x) is the complementary error function. In practice, dopant introduction during furnace diffusion initially forms a predeposition layer near the semiconductor surface. The surface dopant concentration is typically established by either the dopant partial pressure in the gaseous source or the solid solubility limit, thus validating the assumption of constant interface concentration used in the solution. In Eq. (1.63), the diffusion length is defined as $L = \sqrt{Dt}$. As time proceeds, L increases, signifying greater penetration depth of the impurities into the semiconductor. Integrating the

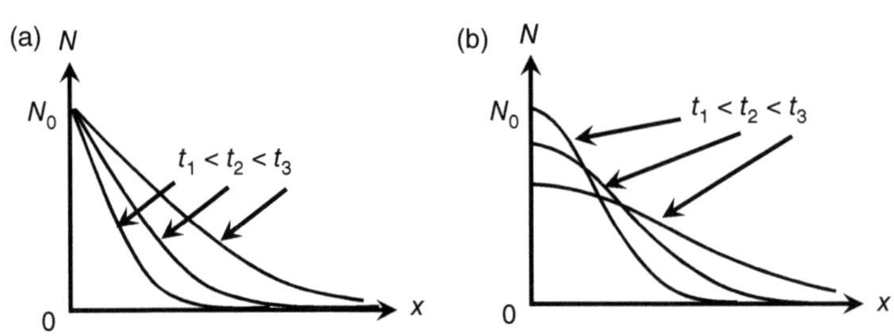

Figure 1.18 Impurity profiles with increasing time for (a) constant source diffusion and (b) constant dose diffusion.

impurity concentration yields the total dopant amount per unit area, Q, within the semiconductor after a specific diffusion time.

$$Q(t) = \frac{2n_0\sqrt{Dt}}{\sqrt{\pi}} \tag{1.64}$$

Analytical solutions for diffusion can also be derived when the dopant source is located at the semiconductor surface with a fixed amount of impurity concentration per unit area (Figure 1.18b). These surface dopants are subsequently driven deeper into the semiconductor during the diffusion process. In accordance with Fick's diffusion laws, the diffused dopant concentration conforms to a Gaussian distribution function.

$$n(x,t) = \frac{Q}{\sqrt{\pi Dt}} e^{-\frac{x^2}{4Dt}} \tag{1.65}$$

1.5.2.2 Ion Implantation

Besides diffusion, ion implantation serves as another fundamental process for doping semiconductors. In contrast to diffusion, implantation is a low-temperature technique that introduces dopants by accelerating dopant ions to high energies and bombarding the semiconductor. The penetration depth of the ions depends on both the dopant species and the incident ion energy. Ion implantation currently plays a central role in semiconductor circuit fabrication.

An implanter is utilized for ion implantation, which consists of an ion source, an accelerating column, a mass analyzer, a scanning system, and a wafer chamber (Figure 1.19). The final implantation performance is influenced by the ion type, kinetic energy, and dose. To generate energetic ions, source materials in gaseous, liquid, or solid phases are first vaporized or sputtered and introduced as gases into the source chamber. These gases are subsequently ionized within a plasma system generated via a hot filament or microwave discharge. Desired positive ions within the plasma are then extracted and accelerated to high energies (typically ranging from 10 to 100 keV) by the potential difference applied between the source chamber and the extraction electrode. Singly charged ions, such as B^+, BF_2^+, P^+,

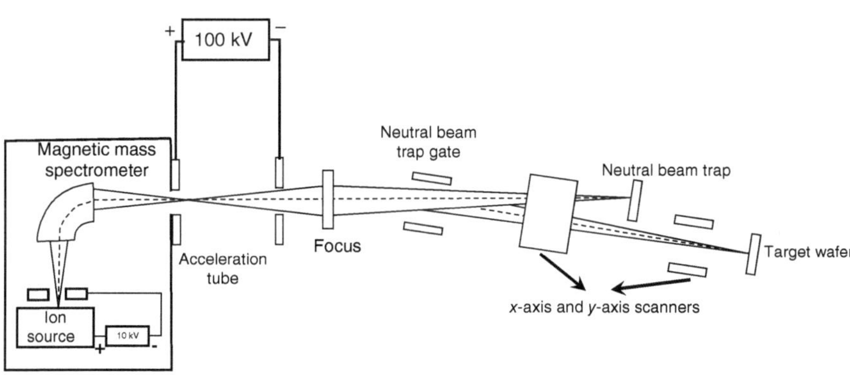

Figure 1.19 Schematic structure of an implantation system.

and Sb^+, constitute the most abundant species in the plasma, while doubly and triply charged ions are significantly less prevalent. For a given applied acceleration voltage, multiply charged ions attain higher kinetic energies than their singly charged counterparts. This facilitates achieving deeper doping profiles, albeit at the cost of a reduced implantation flux. For instance, an ion energy of 200 keV can be attained by accelerating doubly charged ions using a 100 keV accelerator.

The desired ions are selected by the mass analyzer. A magnetic field is applied perpendicular to the direction of ion transport, thereby deflecting the ions according to their mass-to-charge ratio. The deflection radius R is calculated by

$$R = \sqrt{\frac{2mV}{qB^2}} \tag{1.66}$$

where V is the accelerating voltage, B is the magnetic field strength, and m/q is the mass-to-charge ratio of the ions.

The deflected ions are selected by a slot aperture, which usually has the desired ion species, and can be further focused and accelerated to higher energies before being directed to the target substrate for implantation. To avoid Coulombic repulsion forces during transport, trapped electrons are often injected into the ion beam. Additionally, a low-energy electron shower is applied to the target wafer to maintain charge-neutrality during implantation, minimizing beam disturbance. Since the ion beam possesses a relatively narrow diameter, homogeneous implantation across the semiconductor wafer requires uniform sweeping of the beam over the wafer surface within the target chamber. Common sweeping techniques include the use of electrostatic deflection plates, mechanical motion of the wafer, or a combination of both scanning methods.

Ion implantation offers several advantages. Firstly, it enables doping profiles inaccessible via diffusion. Diffusion, driven by concentration gradients, inherently produces higher dopant concentrations near the semiconductor surface. In contrast, implantation provides superior profile control, allowing the realization of designed impurity distributions unattainable through simple diffusion. It can introduce dopants deep within the substrate, forming buried layers where the highest concentration resides internally or creating retrograde profiles through implants at single or multiple energies. Following ion implantation, the dopants are typically electrically inactive, necessitating high-temperature annealing for activation. However, the annealing duration for implantation is substantially shorter, resulting in a significantly lower thermal budget. RTA is commonly employed, capable of achieving temperature ramping rates of ~150 °C/s to ~1000 °C, compared with furnace processing rates of ~1 °C/s. Consequently, ion implantation is the predominant technique for doping gallium arsenide (GaAs), as the compound semiconductor decomposes substantially at the elevated temperatures required for diffusion. Furthermore, the lower temperatures involved permit the use of temperature-sensitive materials, such as photoresist, as implantation masks for selective doping within window regions.

Ion implantation does exhibit certain disadvantages. Primarily, high-energy ions induce crystalline radiation damage. These defects enhance the diffusivity

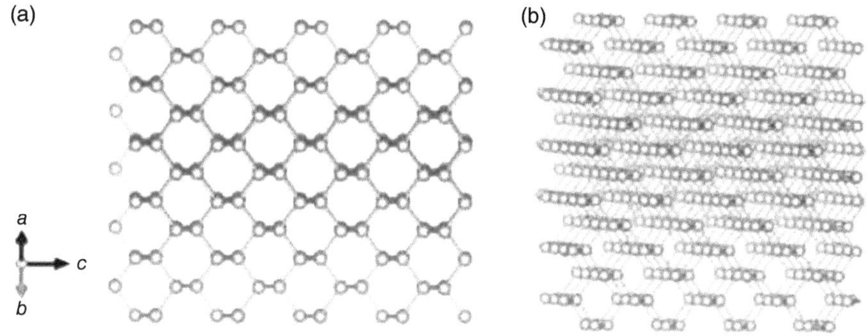

Figure 1.20 (a) Silicon crystal structure viewed from <110> direction with strong channeling effect during implantation. (b) Silicon structure tilted 10° from <110> direction.

of specific species, rendering the fabrication of ultra-shallow doping profiles challenging, even at comparatively low annealing temperatures. Additionally, residual radiation defects inevitably remain within the crystal lattice, increasing junction leakage currents in fabricated devices. Furthermore, energetic ions cause sputtering of metal surfaces, potentially contaminating the wafer. Unlike implantation in amorphous materials, high-energy ions penetrating crystalline semiconductors along specific crystallographic directions undergo the channeling effect, leading to deviations from the intended doping concentration. Controlling the incident ion beam angle or forming an amorphous surface layer mitigates this effect. For <110> wafers, perpendicular incidence induces significant channeling with substantial ion penetration depths; however, tilting the wafer 10° off-normal yields an approximately random ion distribution (Figure 1.20). The substantial kinetic energy carried by implanted ions thermally dissipates upon striking mask materials, necessitating restricted implantation fluxes to prevent overheating thermally sensitive components such as photoresists, whose operational temperatures typically must not exceed ~100 °C. Finally, the inherently low ion flux compromises throughput compared with furnace diffusion techniques.

1.6 Energy Band Theory

1.6.1 Quantum Mechanical Equation and Bloch Wavefunction

The investigation of electron behaviors within semiconductors constitutes the most fundamental approach to understanding their electronic properties. However, semiconductor crystals consist of atoms arranged periodically along specific dimensions, resulting in electron densities typically on the order of 10^{23} electrons/cm^3. Solving problems involving such a vast number of interacting electrons and nuclear ions proves formidable. Consequently, approximations become necessary to simplify this complexity.

Atomic ions are much heavier than electrons. For example, the proton in a hydrogen atom is approximately 1836 times heavier than an electron. Since the kinetic energy E of a particle depends on momentum p according to $E = p^2/2m$ and the velocity v is given by $v = p/m$, both kinetic energy and velocity are inversely proportional to the particle mass m. Consequently, for identical momentum, both the kinetic energy and velocity of an electron are significantly larger than those of the heavier proton.

Owing to the considerably smaller energies and slower response times of ions, electrons respond instantaneously to ionic displacements, whereas ions cannot track the rapid electron motion and instead experience a time-averaged electron potential. For investigations of electronic properties, ions may therefore be regarded as stationary. This approximation, which is fundamental in quantum mechanics, is termed the Born–Oppenheimer or adiabatic approximation.

Accurate understanding of the electronic properties of a semiconductor necessitates the application of quantum mechanics to solve the kinetics and dynamics of electrons within the crystal lattice. Utilizing the Born–Oppenheimer approximation, the electronic Hamiltonian can be expressed as

$$\hat{H}_e = \sum_i \frac{\hat{p}_i^2}{2m_e} + \frac{1}{2}\sum_{i,i'} \frac{e^2}{4\pi\varepsilon_0 |\hat{r}_i - \hat{r}_{i'}|} - \sum_{i,j} \frac{Z_j e^2}{4\pi\varepsilon_0 |\hat{r}_i - R_{j0}|} \tag{1.67}$$

where $\hat{p}_i$ and $\hat{r}_i$ represent the momentum and position vectors of electron i, ε_0 denotes the vacuum permittivity, and Z_j corresponds to the atomic number of the positively charged ion at equilibrium position R_{j0}. In this equation, the first term signifies the kinetic energy of the electrons, the second term describes the Coulomb interaction energy between electrons, and the third term accounts for the Coulomb interaction energy between electrons and ions. The ion–ion interactions can be treated as essentially stationary due to the Born–Oppenheimer approximation (Figure 1.21a). The state of the electrons is characterized by

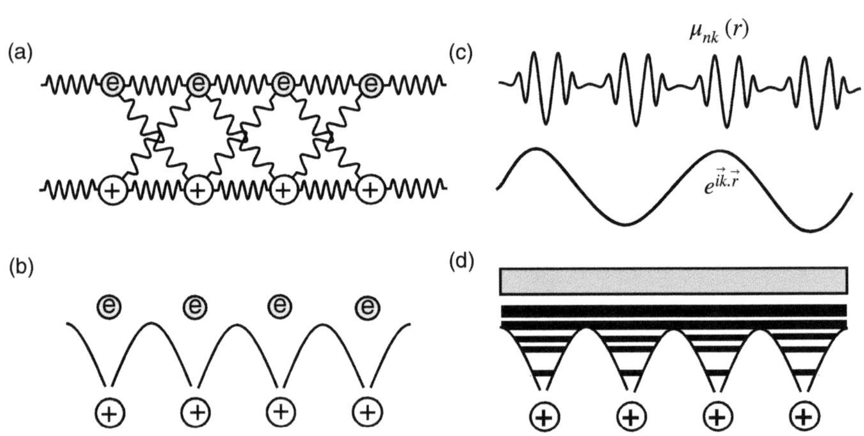

Figure 1.21 (a) Electron–ion, electron–electron, and ion–ion interactions in a crystal. (b) Mean-field potential experienced by electrons as independent particles with fixed ions in the Born–Oppenheimer approximation. (c) The periodic part and plane wave part of the Bloch wavefunction. (d) Filling of the energy levels from low energy to high energy according to Pauli principle.

the many-particle wavefunction $\Phi(r_1, r_2, \dots, r_N)$ for an N-electron system. This wavefunction is determined by solving the eigenvalue problem derived from the Schrödinger equation.

$$\hat{H}_e \Phi(r_1, r_2, \dots, r_N) = E\Phi(r_1, r_2, \dots, r_N) \tag{1.68}$$

where E represents the total energy of the electron system.

Solving this many-body problem in quantum mechanics remains intractable. Consequently, another approximation, known as the mean-field approximation, simplifies this many-body problem to that of a single particle. Within this approximation, the complex electron–electron interactions in the Hamiltonian are represented by the interactions of independent electrons with a mean field generated by the electron ensemble. Therefore, all electrons become independent under the influence of the same average potential $V(r)$, which combines both the electron-derived potential (e.g., the Hartree potential) and the ionic potential. The Schrödinger equation describing the motion of each electron is thus expressed as

$$\hat{H}\Phi_n(r) = \left(\frac{\hat{p}^2}{2m_e} + V(r)\right)\Phi_n(r) = E_n\Phi_n(r) \tag{1.69}$$

where $\Phi_n(r)$ denotes the wavefunction for the n_{th} electron and E_n represents its corresponding energy eigenvalue. The wavefunctions and eigenvalues are solved independently using this mean-field approach.

The one-electron potential $V(r)$ is determined by the atomic species, atomic positions, and the electron density. The atomic nucleus generates a Coulombic potential, which can lead to complex effects such as rapid oscillations in the wavefunction near the nucleus, particularly for atoms with high atomic numbers and wavefunctions characterized by large quantum numbers. To circumvent these difficulties, a smoothly varying pseudopotential is typically employed as an approximation. This substitution enables the description of the wavefunction using fewer Fourier components, thereby lowering the cutoff energy required for plane wave basis sets and facilitating numerical convergence with computationally feasible resources. The pseudopotential yields a pseudo-wavefunction that accurately reproduces the true wavefunction outside the core region. After defining the potential of the atomic nuclei, the electron density must also be determined to account for the mean-field electron–electron interaction. This allows the final one-electron potential $V(r)$ to be obtained for solving the one-electron wavefunctions (Figure 1.21b). Because the electron density is itself determined by the occupied one-electron wavefunctions, this calculation must be performed self-consistently, typically through iterative cycles updating the wavefunctions and the electron density.

After getting the one-electron potential $V(r)$, solving the Schrödinger equation for a crystal with so many electrons is still a complicated task. Fortunately, if the crystal is large enough, it can be treated as a system with an infinite size and a translation symmetry, which can greatly simplify the problem. The translation symmetry means that the system is unchanged after translation along the Bravais lattice. We can introduce the translation operator T_R, which transform a function $f(x)$ by

$$\hat{T}_R f(x) = f(x + R) \tag{1.70}$$

where $R = n_1\vec{a}_1 + n_2\vec{a}_2 + n_3\vec{a}_3$ is a Bravais lattice vector. The translational symmetry of the crystal system implies that the operators $\hat{T}_R$ and $\hat{H}$ commute, denoted by the operator equation.

$$\hat{H}\hat{T}_R\Phi_n(r) = \hat{T}_R\hat{H}\Phi_n(r) = E_n\hat{T}_R\Phi_n(r) \tag{1.71}$$

Therefore, we recognize that $\hat{T}_R\Phi_n(r)$ is also an eigenfunction of the Hamiltonian operator $\hat{H}$ with the identical eigenvalue E_n. Since the translation operator $\hat{T}_R$ and $\hat{H}$ commute, eigenfunctions $\Phi_n(r)$ can always be constructed to be simultaneous eigenfunctions of both $\hat{T}_R$ and $\hat{H}$. Consequently,

$$\hat{T}_R\Phi_n(r) = \Phi_n(r+R) = e^{i\vec{k}\cdot\vec{R}}\Phi_n(r) \tag{1.72}$$

where the wave vector $\vec{k}$ arises from the phase difference induced by the lattice translation vector $\vec{k}$. Consequently, in $\Phi_{nk}(r)$, n and k serve as quantum indices characterizing the eigenfunctions $\Phi_{nk}(r)$ and eigenvalues E_{nk}. The eigenfunction $\Phi_{nk}(r)$ can be expressed in the form

$$\Phi_{nk}(r) = e^{i\vec{k}\cdot\vec{r}}\mu_{nk}(r) \tag{1.73}$$

where $\mu_{nk}(r)$ is a periodic lattice function satisfying $\mu_{nk}(r) = \mu_{nk}(r + R)$. This expression is known as a Bloch function (Figure 1.21c), indicating that the electronic wavefunction within a crystal, while similar to a plane wave characteristic of free electrons, is spatially modulated by a periodic function. We note that electrons are fermions governed by the Pauli exclusion principle; therefore, each distinct wavefunction can accommodate a maximum of two electrons possessing opposite spin orientations. In the ground state of the electron ensemble, electronic states are populated sequentially starting from the lowest available energy levels (Figure 1.21d).

1.6.2 Band Structures and Carrier Transport

For a reciprocal lattice vector $\vec{G} = m_1\vec{b}_1 + m_2\vec{b}_2 + m_3\vec{b}_3$, we know that $\vec{G}\cdot\vec{R} = 2\pi(m_1n_1 + m_2n_2 + m_3n_3)$, which results in no net phase shift. Consequently, both the wave vectors $\vec{k}$ and $\vec{k}+\vec{G}$ yield identical phase shifts upon lattice translation and can serve as quantum numbers to index the wavefunction. Conventionally, only the equivalent wave vector $\vec{k}$ with the smallest magnitude is employed. The region encompassing all such possible $\vec{k}$ vectors is termed the first Brillouin zone. A plot of the eigenvalue E_{nk} versus $\vec{k}$ within the first Brillouin zone gives the electronic band structure, which serves as a fundamental tool for characterizing and understanding the electronic properties of crystals. The first Brillouin zone corresponds to the Wigner–Seitz cell of the reciprocal lattice. It is defined as the set of points in reciprocal space closer to the origin than to any other reciprocal lattice point. This region, containing the origin point, is constructed by bisecting all reciprocal lattice vectors (Figure 1.22). Examples of first Brillouin zones for typical lattices are depicted in Figure 1.23. The reciprocal lattices inherit the symmetries of their corresponding real-space lattices. Consequently, electronic band diagrams are

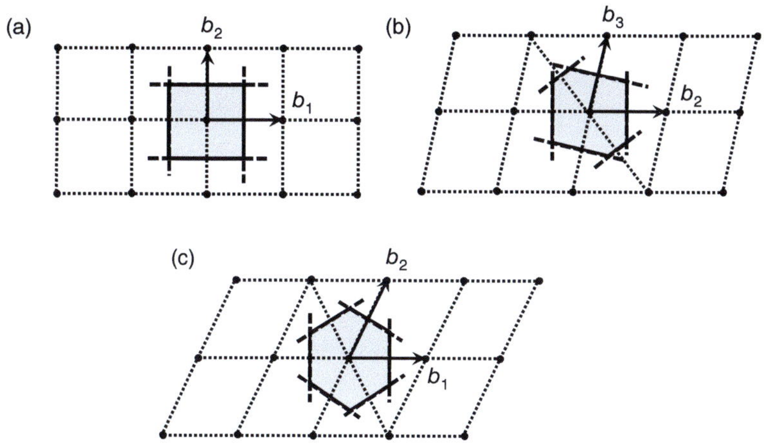

Figure 1.22 Construction of first Brillouin zones of (a) simple cubic lattice, (b) monoclinic lattice, and (c) hexagonal lattice by bisecting the reciprocal lattice vectors.

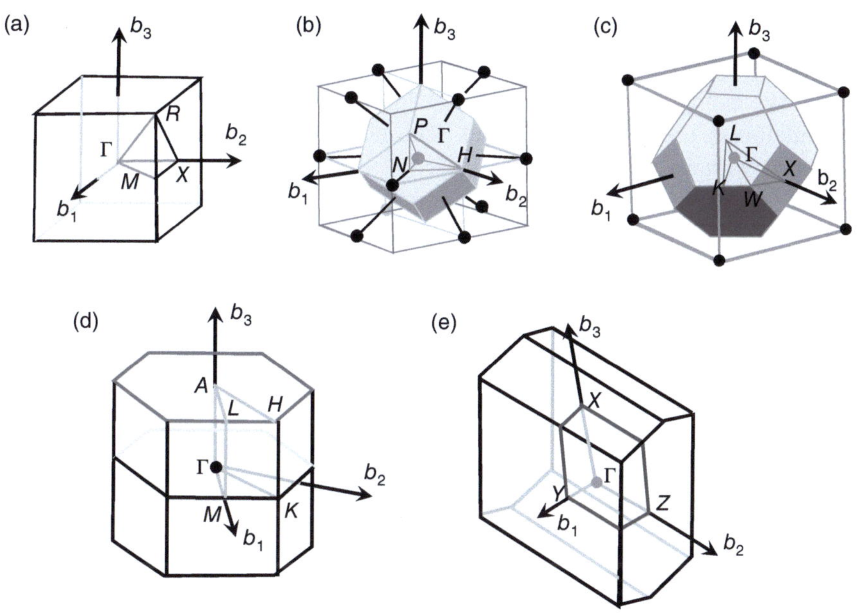

Figure 1.23 First Brillouin zones and high-symmetry point notations of (a) simple cubic lattice, (b) body-centered cubic lattice, (c) face-centered cubic lattice, (d) hexagonal lattice, and (e) monoclinic lattice.

typically plotted along $\vec{k}$-space paths connecting high-symmetry points within the Brillouin zone. Conventionally, high-symmetry points and lines inside the Brillouin zone are denoted by Greek letters, while points on the zone surfaces are denoted by Roman letters. Γ invariably designates the zone origin. For cubic lattices, X commonly represents the intersection point of the Brillouin zone surface with any of the main axes.

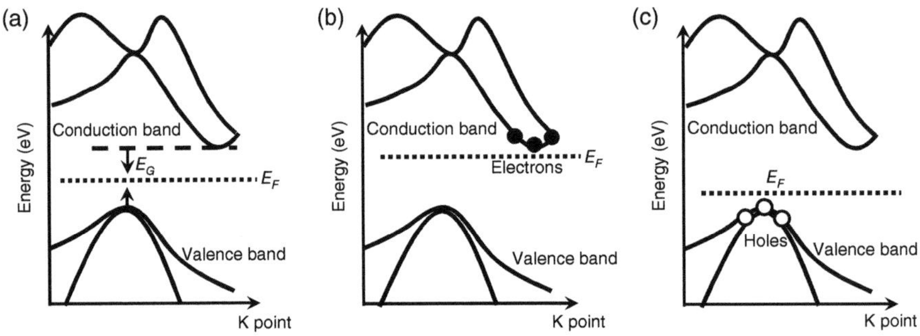

Figure 1.24 Schematic band diagrams and carrier types in (a) the intrinsic, (b) *n*-type, and (c) *n*-type semiconductors.

Electrons occupy the electronic band structure from low to high energies. Each energy band accommodates up to two electrons in the primitive cell, accounting for the spin degeneracy of electrons. Depending on the band filling, crystals are classified as metals, semiconductors, or insulators. Metals possess partially filled bands, with the Fermi level traversing one or more bands. In contrast, semiconductors and insulators have fully occupied valence bands and unoccupied conduction bands, separated by a bandgap, as shown in Figure 1.24a. Semiconductors are distinguished from insulators by their smaller bandgap. Semimetals, such as graphene, constitute an exception with an absence of a bandgap. The presence or absence of a bandgap profoundly influences the electron transport properties of a crystal. Electrons in completely filled bands cannot sustain net current flow, explaining the low conductivity of insulators and intrinsic semiconductors at cryogenic temperatures.

However, after doping with impurities, electrostatic modulation, or more directly, optical or thermal excitation, the formerly completely occupied or unoccupied bands can become partially occupied, enabling electrical conduction. The previously unoccupied bands are termed conduction bands, while the previously occupied bands are termed valence bands. Carrier transport in the conduction or valence bands can be understood by treating the carriers analogously to classical particles, referred to as electrons in the conduction bands or holes in the valence bands, respectively (Figure 1.24b,c). This analogy arises because electrons and holes are predominantly located at the conduction band minimum or valence band maximum, which exhibit nearly parabolic energy dispersion akin to classical particles, albeit with distinct effective masses. Specifically, the effective mass near the conduction band minimum can be derived from the electronic band structure using the relation:

$$m^* = \frac{\hbar^2}{\frac{\partial^2 E}{\partial k^2}} \tag{1.74}$$

where an isotropic band dispersion along all wave vector directions is assumed for simplicity. In contrast, anisotropic band dispersions necessitate a description using anisotropic effective masses.

Electrons with wave vector $\vec{k}$ in the conduction band can be treated as a wave packet with momentum $\vec{p} = \hbar\vec{k}$, where $\hbar$ is the Planck constant. If an external force $\vec{F}$, such as that produced by an electric field $\vec{E}$ or a magnetic field $\vec{B}$, acts on the electron, the wave vector evolves according to the following equation:

$$\hbar \frac{\mathrm{d}\vec{k}}{\mathrm{d}t} = \vec{F} = q\vec{E} + q\vec{v} \times \vec{B} \tag{1.75}$$

where q is the elementary charge and $\vec{v}$ is the group velocity. For a band dispersion $E(\vec{k})$, the group velocity can be calculated by

$$\vec{v} = \frac{1}{\hbar} \nabla_k E(\vec{k}) \tag{1.76}$$

As an example, for one-dimensional band dispersion along the x axis, the group velocity is given by $v_x = \partial E/\hbar\ \partial k_x$. For an electron with an effective mass m^*, the band dispersion near the conduction band minimum can be approximated as $E = \hbar^2 k^2/2m^*$, yielding a group velocity $v = \hbar k/m^*$, analogous to a classical particle. Holes in the valence band can similarly be described as classical particles. For this interpretation to hold, the charge, momentum, and energy assigned to a hole must exhibit signs opposite to those of the corresponding electron states within the band structure. For instance, a hole possesses a charge of the same magnitude as an electron, albeit opposite in sign (positive).

1.7 PN Junctions

A PN junction is formed by bringing p-type and n-type semiconductors into contact (Figure 1.25a). Studying the current conduction characteristics of PN junctions is

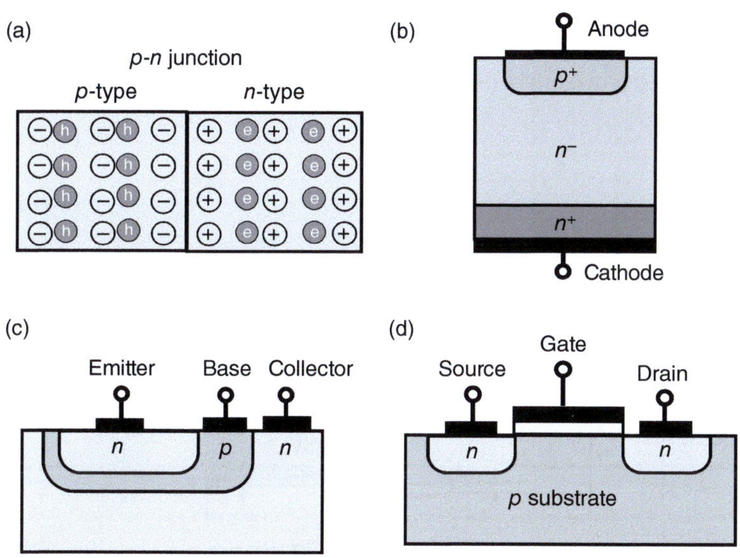

Figure 1.25 (a) Schematic of a PN junction formed by bringing n-type and n-type semiconductors into contact. Cross-sectional structures of (b) a p^+–n^-–n^+ junction diode, (c) NPN bipolar junction transistor, and (d) an n-type metal–oxide–semiconductor field-effect transistor. The PN junctions can be found in these devices.

crucial for implementing and understanding semiconductor devices. An essential and fundamental device based on the PN junction is the PN diode (Figure 1.25b). This two-terminal device exhibits nonlinear current–voltage characteristics, conducting with low resistance in one direction, while presenting high resistance in the reverse direction. Consequently, diodes are widely used as rectifiers to convert alternating current (AC) to direct current (DC). Additionally, specific diodes are engineered to emit light efficiently for use as illumination sources. Furthermore, PN junctions constitute the fundamental building blocks of BJTs and MOSFETs (Figure 1.25c,d).

1.7.1 Depletion Region and Built-in Potential

To understand the properties of the PN junction, it is necessary to first consider the depletion region formed due to charge transfer. The abrupt junction serves as an illustrative model for this phenomenon. An abrupt junction consists of semiconductors undergoing an abrupt transition from acceptor doping to donor doping (Figure 1.26). Owing to the concentration gradients of electrons and holes, electrons diffuse from the *n*-type region into the *p*-type region, while holes diffuse from the *p*-type region into the *n*-type region. Since electrons are negatively charged and holes are positively charged, this diffusion process establishes space-charge regions. The resulting space charge generates an electric field, which induces a drift current of carriers in the opposite direction to the diffusion currents (Figure 1.26a). Consequently, the carrier dynamics within the PN junction ultimately reach thermal equilibrium.

At thermal equilibrium without an applied voltage, no net current flows. From the perspective of carrier transport, the electron and hole currents are directly related to the gradients of their respective quasi-Fermi levels, as expressed by

$$J_e = n_e \mu_e \frac{dE_{\mathrm{Fe}}}{dx} \tag{1.77}$$

$$J_h = n_h \mu_h \frac{dE_{\mathrm{Fh}}}{dx} \tag{1.78}$$

from which we know at equilibrium the quasi-Fermi level will be equal and constant in the whole sample, i.e., $E_{\mathrm{F}} = E_{\mathrm{Fe}} = E_{\mathrm{Fh}}$. The alignment of Fermi level is achieved by

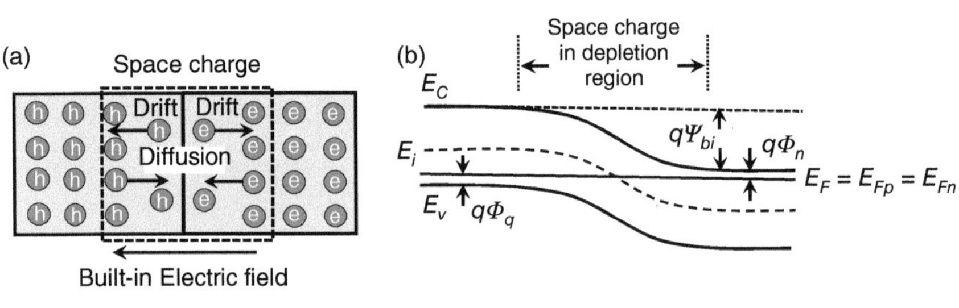

Figure 1.26 (a) The static equilibrium formed due to the balance of diffusion and drift currents in a PN junction. (b) Energy band diagram of a PN junction at equilibrium with aligned quasi-Fermi levels.

the built-in potential in the depletion region. According to the notation in the band diagram of Figure 1.26b, the built-in potential strength will simply be

From these equations, it follows that at equilibrium, the quasi-Fermi levels are equal and constant throughout the entire sample, denoted by $E_F = E_{Fe} = E_{Fh}$. This alignment of the Fermi level is facilitated by the built-in potential within the depletion region. Referring to the notation in the band diagram of Figure 1.26b, the magnitude of the built-in potential is given by

$$\psi_{bi} = E_G - \phi_n - \phi_p \tag{1.79}$$

For nondegenerate semiconductors, this equation can be further expressed as

$$\psi_{bi} = \frac{kT}{q}\ln\left(\frac{n_{n0}}{n_i}\right) + \frac{kT}{q}\ln\left(\frac{p_{p0}}{n_i}\right) \tag{1.80}$$

where n_{n0} and p_{p0} are the electron concentration in the n-region and the hole concentration in the p-region, respectively, and n_i is the electron or hole concentration of the intrinsic semiconductor. Since the relationships $n_{n0}p_{n0} = n_i^2$ and $p_{p0}n_{p0} = n_i^2$ hold for carriers in the n- and p-regions, the expression can be equivalently represented as

$$\psi_{bi} = \frac{kT}{q}\ln\left(\frac{n_{n0}}{n_{p0}}\right) = \frac{kT}{q}\ln\left(\frac{p_{p0}}{p_{n0}}\right) \tag{1.81}$$

where n_{p0} and p_{n0} are the minority-carrier concentrations in the n-type and p-type semiconductor, respectively. The aforementioned derivations rely on the Boltzmann statistics for nondegenerate semiconductors. For degenerate semiconductors, Fermi–Dirac statistics must be applied.

The built-in potential arises from the space charge within the depletion region. To facilitate analysis, it is commonly assumed that the space charge adopts a box profile in the depletion region, wherein all donors and acceptors are fully depleted (Figure 1.27a). No space charge exists outside this region. Given the donor and acceptor concentrations N_D and N_A, respectively, the depletion width W_D and W_A in the n-type and p-type semiconductors satisfy $N_DW_D = N_AW_A$ due to charge conservation. The electric potential is governed by the Poisson equations:

$$\frac{d^2\varphi}{dx^2} = \frac{qN_A}{\varepsilon_S} \quad \text{for } -W_A < x < 0 \tag{1.82}$$

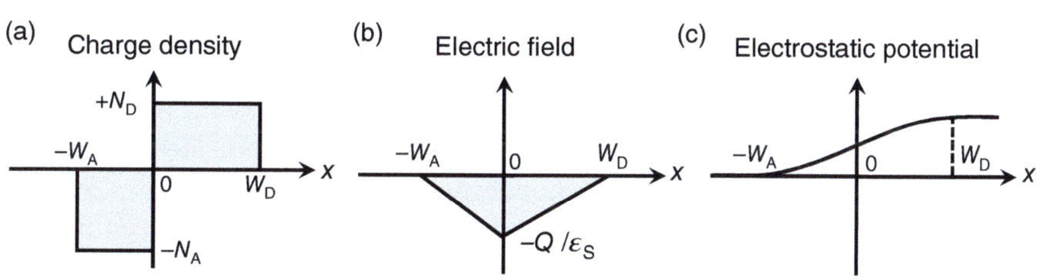

Figure 1.27 (a) Space-charge distribution in the depletion region of PN junction according to the box-profile approximation. (b) Electric field distribution generated by the space charge. (c) Electrostatic potential obtained by integration of the electric field.

$$\frac{d^2\varphi}{dx^2} = -\frac{qN_D}{\varepsilon_S} \quad \text{for } 0 < x < W_D \tag{1.83}$$

Integrating these equations yields the electric field distribution (Figure 1.27b), which exhibits its maximum magnitude at $x = 0$.

$$E_{max} = -\frac{d\varphi}{dx}(x = 0) = -\frac{qN_A W_A}{\varepsilon_S} = -\frac{qN_D W_D}{\varepsilon_S} \tag{1.84}$$

Further integration of the electric field (Figure 1.27c) provides the total potential difference across the PN junction as

$$\psi_{bi} = \frac{qN_A {W_A}^2}{2\varepsilon_S} + \frac{qN_D {W_D}^2}{2\varepsilon_S} \tag{1.85}$$

From this expression, combined with $N_D W_D = N_A W_A$, the depletion widths are derived as

$$W_D = \sqrt{\frac{2\varepsilon_S \psi_{bi} N_A}{qN_D(N_D + N_A)}} \tag{1.86}$$

$$W_A = \sqrt{\frac{2\varepsilon_S \psi_{bi} N_D}{qN_A(N_D + N_A)}} \tag{1.87}$$

A special case is the one-sided abrupt junction, featuring a heavily doped *p*-type region or *n*-type region; the depletion width equations then simplify for p^+-n and n^+-p junctions, respectively.

$$W_D = \sqrt{\frac{2\varepsilon_S \psi_{bi}}{qN_D}} \tag{1.88}$$

$$W_A = \sqrt{\frac{2\varepsilon_S \psi_{bi}}{qN_A}} \tag{1.89}$$

1.7.2 Current-Voltage Characteristics

When a voltage is applied across the terminals of a PN junction, the static equilibrium condition is disrupted, yielding carrier transport and net current flow. The current flow behavior in a PN junction under ideal conditions is described by the Shockley equation. The derivation of the Shockley equation relies upon several key assumptions. First, the depletion region is assumed to remain abrupt even under applied bias, with the semiconductor remaining neutral outside this region. Second, the semiconductor is treated as nondegenerate so that Boltzmann statistics are applicable. Third, the magnitude of minority-carrier injection induced by the applied voltage is significantly smaller than the majority-carrier concentration on either side. Finally, owing to the relatively short depletion width compared to the diffusion length, carrier recombination and generation within the depletion region are negligible.

Even though the carrier statistics depart from static equilibrium conditions, quasi-Fermi levels remain valid for describing the nonequilibrium state. Analogous to thermal equilibrium conditions, the electron and hole concentrations corresponding to quasi-Fermi levels E_{Fe} and E_{Fh} are expressed by

$$n = n_i e^{\frac{E_{\mathrm{Fe}} - E_i}{kT}} \tag{1.90}$$

$$p = n_i e^{\frac{E_i - E_{\mathrm{Fh}}}{kT}} \tag{1.91}$$

The quasi-Fermi levels across a PN junction under forward and reverse bias can be qualitatively represented (Figure 1.28a,b). At forward bias, $pn = n_i^2 e^{(E_{\mathrm{Fe}} - E_{\mathrm{Fh}})/kT}$ exceeds n_i^2, indicating that injected minority carriers undergo recombination. In contrast, under reverse bias, pn falls below n_i^2, signifying that electrons and holes undergo generation within the space-charge region.

Under the low-injection assumption, within the charge-neutral region outside the space-charge region, there is no drift current; thus, only diffusion current and carrier recombination require consideration. For the minority-carrier holes in the n-type region, at steady state, the hole density n_p satisfies the following equation:

$$-\frac{\mathrm{d}}{\mathrm{d}x}\left(-D_n \frac{\mathrm{d}n_p}{\mathrm{d}x}\right) = \frac{n_p - n_{p0}}{\tau_n} \tag{1.92}$$

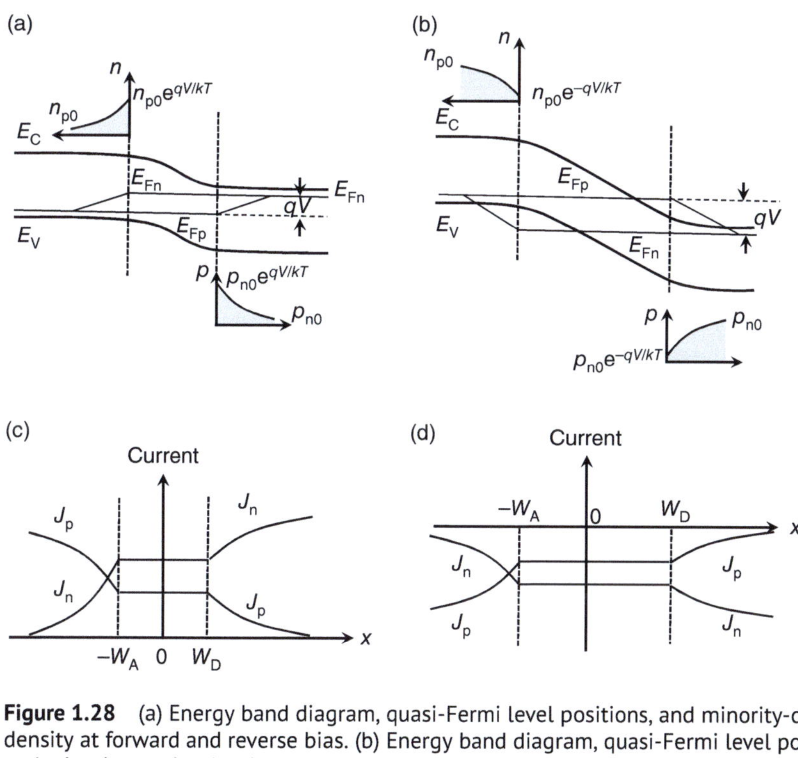

Figure 1.28 (a) Energy band diagram, quasi-Fermi level positions, and minority-carrier density at forward and reverse bias. (b) Energy band diagram, quasi-Fermi level positions, and minority-carrier density at reverse bias. Current contributions at (c) forward and (d) reverse biases.

where D_n is the hole diffusivity and τ_n is the hole recombination lifetime. Depending on the recombination mechanism—whether band-to-band or Shockley–Read–Hall recombination—the hole recombination lifetime is inversely proportional to the acceptor concentration or the trap density, respectively. This equation expresses the balance between diffusion current and recombination current. The boundary conditions are at the space-charge region edge ($x = 0$) $n_p = n_{p0}e^{qV/kT}$ and far from the space-charge region ($x = -\infty$) $n_p = n_{p0}$. The solution to the equation is

$$n_p = n_{p0}\left(e^{\frac{qV}{kT}} - 1\right)e^{\frac{x}{\sqrt{D_n\tau_n}}} + n_{p0} \tag{1.93}$$

where we define $L_n = \sqrt{D_n\tau_n}$ as the hole diffusion length in the n-type semiconductor. At $x = 0$, the hole diffusion current density is

$$J_n = qD_n\frac{dn_p}{dx} = \frac{qD_n n_{p0}}{L_n}\left(e^{\frac{qV}{kT}} - 1\right) \tag{1.94}$$

Similarly, for minority-carrier electrons in the n-type semiconductor region, the electron diffusion current density is

$$J_p = qD_p\frac{dp_n}{dx} = \frac{qD_p p_{n0}}{L_p}\left(e^{\frac{qV}{kT}} - 1\right) \tag{1.95}$$

From the aforementioned equation, we know that the diffusion current at forward bias is mainly caused by the minority-carrier injection (Figure 1.28c). The intensity of this diffusion current diminishes as carriers diffuse deeper into the semiconductor bulk. Assuming that carrier recombination and generation within the depletion region are negligible, the total current density simplifies to

$$J = J_n + J_p = J_0\left(e^{\frac{qV}{kT}} - 1\right) \tag{1.96}$$

where

$$J_0 = \frac{qD_n n_{p0}}{L_n} + \frac{qD_p p_{n0}}{L_p} \tag{1.97}$$

This expression represents the Shockley ideal diode equation. Although derived for forward bias, the same governing equation applies under reverse-bias conditions. In reverse bias, carrier generation within the neutral region follows an equation analogous to that describing recombination in forward bias, shown in Figure 1.28d. Given the acceptor N_A and donor N_D concentrations, J_0 can be reformulated as

$$J_0 = \frac{qD_n n_i^2}{L_n N_A} + \frac{qD_p n_i^2}{L_p N_D} \tag{1.98}$$

Temperature significantly influences PN junction diode current through several parameters. First, the terms $D_n/L_n = \sqrt{D_n/\tau_n}$ and $D_p/L_p = \sqrt{D_p/\tau_p}$ exhibit a temperature dependence proportional to $T^{\gamma/2}$, where γ is a material-specific exponent. Furthermore, the intrinsic carrier density n_i relates to temperature via $n_i \propto T^{3/2}e^{-E_G/2kT}$. This exponential term overwhelmingly dominates the weaker power-law dependence. Consequently, for a constant forward bias voltage V, the current depends exponentially on temperature through $e^{\frac{qV-E_G}{kT}}$. Under reverse bias,

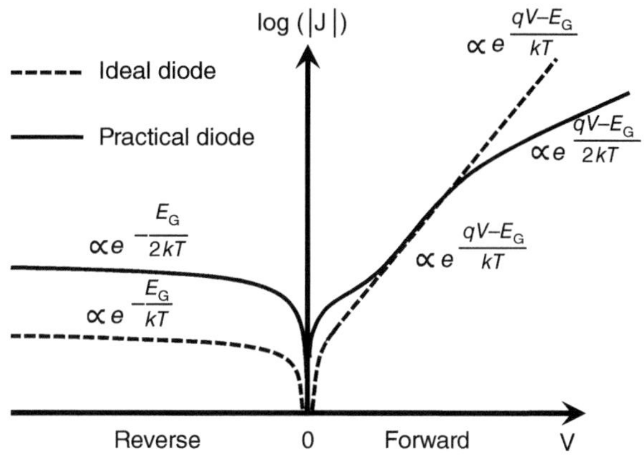

Figure 1.29 Current–voltage characteristics of the ideal diode and the practical diode.

however, the saturation current density scales with $e^{\frac{-E_G}{kT}}$. This analysis indicates that semiconductor PN junctions exhibiting smaller bandgaps support both larger forward and larger reverse saturation currents. For instance, Ge PN junctions with a bandgap of 0.66 eV exhibit higher currents than Si PN junctions, which possess a bandgap of 1.12 eV.

The Shockley equation can quite well predict the current–voltage of real PN junction qualitatively. However, because of the approximations that have been used during the derivation of Shockley equation, deviations from the ideal Shockley equation can be observed (Figure 1.29). The generation or recombination processes in the depletion region can lead to departures from the ideal Shockley equation even at small-injection conditions. At reverse bias, from the quasi-Fermi level positions, both the electron and hole densities in the depletion region is far smaller than the intrinsic carrier density, i.e., $p \ll n_i$ and $n \ll n_i$. The generation will dominate because $pn \ll n_i^2$. If the generation is mainly dominated by band-to-band transition, the generation rate will be proportional to $Rn_i^{\,2} \propto e^{-\frac{E_G}{kT}}$. However, usually the trap-assisted generation is faster and more important in semiconductors, and the generation rate will be roughly proportional to $\sigma v_{\text{th}} N_t n_i / 2 \propto e^{-\frac{E_G}{2kT}}$. Compared with the diffusion saturation current at reverse bias that is proportional to $e^{-\frac{E_G}{kT}}$, the trap-assisted generation can be more important in small-bandgap semiconductors, even at room temperature. For this reason, at room temperature, the reverse saturation current of Ge diode deviates more from the Shockley equation than Si diode. At forward bias, if carrier injection is comparable to the doping density or even larger, the trap-assisted recombination rate will be proportional to $\sigma v_{\text{th}} N_t n_i e^{\frac{qV}{2kT}}/2$, which depends differently on V than the $e^{\frac{qV}{kT}}$ term in the ideal Shockley equation. Considering this, an empirical form is usually used to describe the diode current at forward bias.

$$J_F \propto e^{\frac{qV}{\eta kT}} \tag{1.99}$$

where the ideality factor η equals 1 when diffusion currents dominate and 2 when recombination currents dominate. Note that under high-injection conditions, the current also scales approximately with $e^{\frac{qV}{2kT}}$. However, such high currents typically induce significant resistive voltage drops due to series resistance, which must be considered.

1.7.3 Breakdown Mechanisms in PN Junctions

Under reverse bias, the PN junction current exhibits saturation. However, at sufficiently high bias voltages, a strong electric field develops within the depletion region. This field ultimately leads to PN junction breakdown, accompanied by a sudden current increase (Figure 1.30). Several mechanisms can cause breakdown. Thermally induced breakdown occurs when power dissipation within the depletion region elevates its temperature (Figure 1.30a). A substantial reverse bias generates significant power. Limited heat dissipation causes this power dissipation to increase the junction temperature. Higher temperature, in turn, increases the reverse saturation current, subsequently generating further heat. When the reverse-bias voltage is sufficiently large, this positive feedback loop becomes uncontrollable, resulting in breakdown. This thermally driven process is generally irreversible, typically destroying the device due to the resultant high current. Protective measures, such as incorporating current-limiting resistors, are therefore necessary. Thermally induced breakdown is more probable in small-bandgap semiconductors operating at relatively high ambient temperatures. At lower temperatures, alternative breakdown mechanisms may dominate.

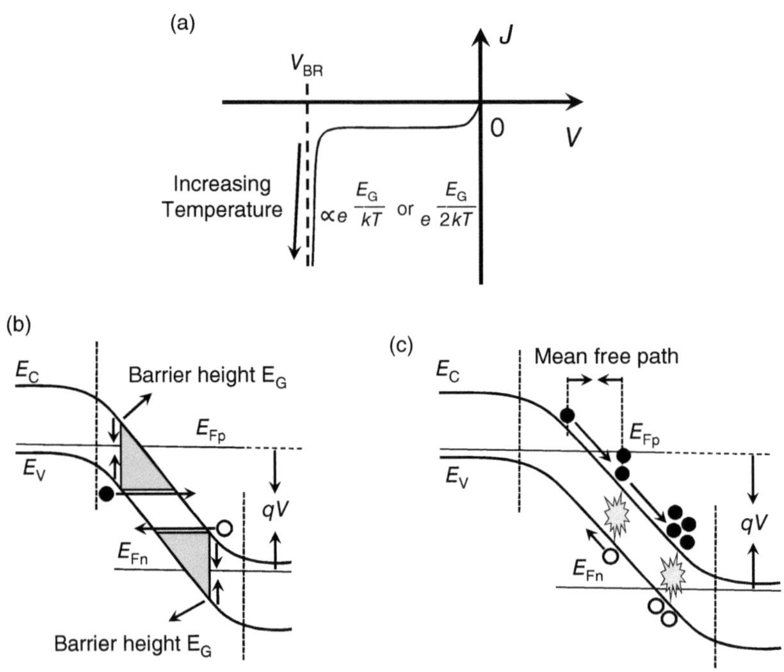

Figure 1.30 (a) Reverse current–voltage characteristics of thermal breakdown. (b) Energy band diagram of tunneling breakdown. (c) Energy band diagram of avalanche breakdown.

Besides thermal excitation, PN junction breakdown can also be induced by tunneling or avalanche effects (Figure 1.30b,c). This section presents both mechanisms and provides a systematic comparison. Tunneling results from a sufficiently strong electric field narrowing the tunneling barrier. As illustrated in Figure 1.30, electrons and holes are confined by the bandgap barrier within the conduction band and valence band, respectively. Under sufficiently large reverse-bias voltages, however, the barrier thickness decreases, as indicated by the triangular markers. Achieving tunneling necessitates an intense electric field, e.g., 10^6 V/cm for silicon. Such high field strengths require substantial doping concentrations to minimize the space-charge width. In contrast to band-to-band tunneling, breakdown can also result from impact ionization by high-energy carriers, the most prevalent mechanism observed experimentally in devices such as PN diodes, BJTs, and MOSFETs. When a free electron or hole gains sufficient kinetic energy ($>E_G$) during its free mean path before scattering within the depletion region, it can ionize the lattice and generate an electron–hole pair. These newly generated carriers may themselves be accelerated by the field to create further ionizations. Avalanche breakdown typically requires a critical field strength. Achieving this field alone is insufficient, as carrier acceleration and the resulting multiplicative impact ionization require a finite distance within the depletion region.

The tunneling breakdown and avalanche breakdown mechanisms differ in several aspects. Although both require a high electric field, tunneling breakdown typically occurs in heavily doped PN junctions and exhibits a relatively low breakdown voltage ($<4E_G/q$). In contrast, avalanche breakdown necessitates the maintenance of a high electric field over a distance and possesses a relatively high breakdown voltage ($>6E_G/q$). Another distinction lies in their temperature coefficients. Tunneling breakdown demonstrates a negative temperature coefficient for semiconductors such as Si and GaAs, meaning the breakdown voltage decreases with rising temperature. This phenomenon occurs because elevated temperatures typically reduce the bandgap, thereby lowering the barrier height and facilitating tunneling. Conversely, avalanche breakdown exhibits a positive temperature coefficient, resulting in an increased breakdown voltage at higher temperatures. This characteristic is attributed to increased scattering of energetic carriers, such as by phonons, at elevated temperatures. Such scattering diminishes the average kinetic energy of electrons and holes, consequently requiring a stronger accelerating electric field to reach the impact ionization energy threshold.

References

1 Wuttig, M., Schön, C.F., Schumacher, M. et al. (2022). *Adv. Funct. Mater.* 32: 2110166.

2 Cornil, J., Calbert, J.P., and Brédas, J.L. (2001). *J. Am. Chem. Soc.* 123: 1250–1251.

3 Radisavljevic, B., Radenovic, A., Brivio, J. et al. (2011). *Nat. Nanotechnol.* 6: 147–150.

4 Park, C.H., Cheong, B.H., Lee, K.H., and Chang, K.J. (1994). *Phys. Rev. B* 49: 4485–4493.

5 Ito, K., Hiramatsu, K., Amano, H., and Akasaki, I. (1990). *J. Cryst. Growth* 104: 533–538.

6 Stoumpos, C.C., Malliakas, C.D., Peters, J.A. et al. (2013). *Cryst. Growth Des.* 13: 2722–2727.

2

Synthesis and Characterization of Optoelectronic Materials

2.1 Introduction

The synthesis of semiconductors, metals, insulators, etc., whether in the form of bulk crystals, crystalline or polycrystalline materials, or amorphous thin films, is fundamentally important for their applications in optoelectronic devices. For instance, thin-film deposition is now an indispensable process within the semiconductor industry for fabricating optical and electronic devices. Numerous methodologies have been established for material synthesis, varying significantly according to the governing thermodynamics and kinetics. Nutrient materials in liquid phase (melt or solution), vapor phase, or solid phase can be employed to grow materials. Depending on whether the chemical composition changes between the nutrient materials and the synthesized product, deposition processes are broadly classified as physical or chemical methods. For example, sputtering, also termed cathode sputtering, is a typical physical vapor deposition (PVD) process where atoms are ejected from a solid target via high-energy ion bombardment and subsequently deposited onto a substrate through vapor transport. In contrast, chemical vapor deposition (CVD) is a chemical process in which one or more volatile precursors react and/or decompose on the substrate surface to yield the desired deposit. Various physical and chemical deposition methods, particularly epitaxial growth techniques, are introduced in this chapter.

The physical, chemical, electronic, and optical properties of synthesized materials must be rigorously characterized to ensure they meet the requirements for optoelectronic device applications. Although a plethora of material characterization techniques exist, with many continuously evolving and new methods emerging, this chapter, due to space constraints, concentrates on techniques for characterizing electronic and optical properties. For electronic properties, four-point probe measurement, Hall effect measurement, and C–V and I–V measurements will be discussed. For optical properties, photoluminescence (PL) spectroscopy, ultraviolet–visible (UV–vis) absorption spectroscopy, Fourier transform infrared (FTIR) spectroscopy, and X-ray photoelectron spectroscopy (XPS) will be elaborated upon. Numerous other significant characterization techniques not covered here, including scanning electron microscopy (SEM), electron energy loss spectroscopy (EELS), ellipsometry, Raman spectroscopy, etc., are detailed in monographs dedicated to material characterization.

2.2 Synthesis of Semiconductors

2.2.1 Chemical Method

2.2.1.1 Solid/Liquid Phase Chemical Synthesis

Various chemical methods are utilized for semiconductor growth to meet the demands of optoelectronic applications. During chemical growth, the composition of source materials can be adjusted. These source materials may exist in solid, liquid, or gas phases. Reactions occur more readily for source materials in liquid and gas phases. However, it remains feasible to directly employ solid-phase materials in chemical reactions for semiconductor synthesis. For example, mixing MAI and PbI_2 powders followed by a ball-milling process yields $MAPbI_3$ powders, as shown in Figure 2.1. Despite the advantages of low-temperature and solvent-free processing, this method typically produces only powders with unsatisfactory optical performance due to high defect density. Consequently, semiconductors are more frequently synthesized using solution or gas-phase sources, which constitute the focus of this subsection.

Crystals can be grown from solution. Solvents commonly employed in semiconductor synthesis encompass water, various organic liquids, melts of specific chemical compounds, and their mixtures. A fundamental guideline for solvent selection is the chemical similarity between the solvent and the solute destined for semiconductor growth. For instance, crystals of nonpolar organic compounds readily grow from nonpolar organic solvents. This chemical similarity also governs solute solubility within the solvent. Crystal growth must occur below the boiling point of the solvent. For solutions based on water or organic liquids, growth temperatures are relatively low, typically not exceeding 200 °C. In contrast, solutions utilizing chemical compound melts require significantly higher temperatures. Solution growth methods offer advantages of simpler, lower-cost apparatus and avoid material decomposition at high temperatures or the development of thermal stresses upon cooling to room temperature.

Achieving precursor precipitation requires creating a state of supersaturation within the solution. Several strategies facilitate this. One strategy involves altering

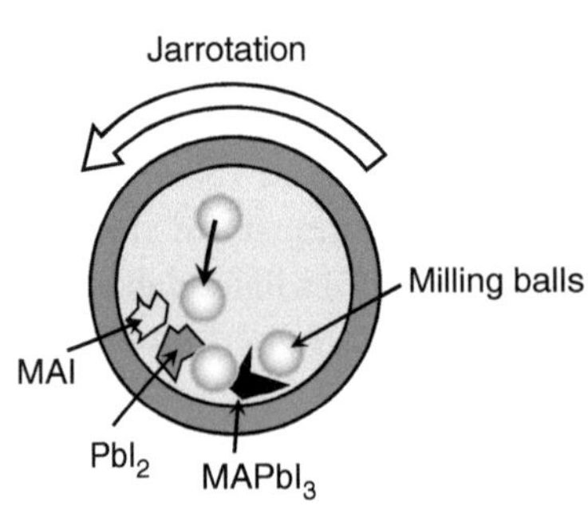

Figure 2.1 Schematic of ball-milling method to synthesize $MAPbI_3$ using chemical reaction between solid MAI and PbI_2 reactants.

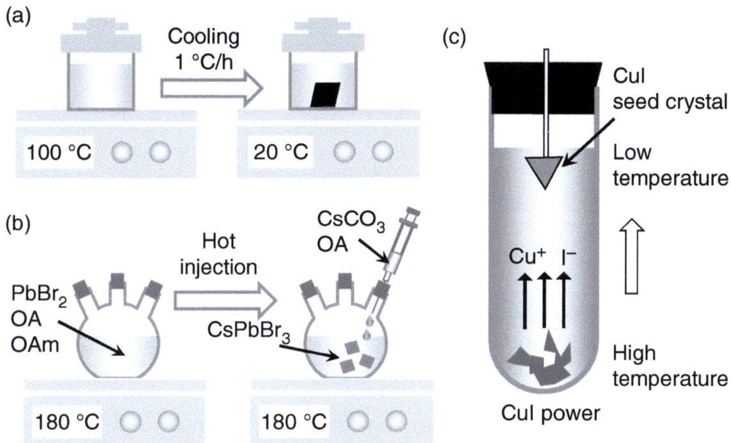

Figure 2.2 Semiconductor synthesis using solution methods with supersaturation condition realized by temperature control. (a) Schematic of the temperature-decreasing method. (b) Schematic of hot-injection method. (c) Schematic of temperature-gradient method.

the solution temperature, leveraging temperature-dependent solubility. Crucially, decreasing the temperature will saturate the solution for substances exhibiting lower solubility at reduced temperatures. Given a nucleation seed—either intentionally introduced or arising from impurities—supersaturated solute precipitates onto the seed, enabling single-crystal growth. An alternative strategy entails adding an anti-solvent. This agent either reduces solute solubility or diminishes solvent volume, thereby inducing supersaturation. Similar supersaturation effects arise from the controlled, gradual evaporation of the solvent.

The temperature-decreasing method is employed for materials whose solubility varies significantly within a specific temperature range, thereby maintaining the supersaturation necessary for growth (Figure 2.2a). Most substances possess positive temperature coefficients of solubility, making them suitable for this method. Temperature reduction occurs slowly, at a controlled rate of several degrees per hour, regulated by specialized programming. If the saturation solubility is c_s, the precipitation mass Δm resulting from a temperature decrease ΔT is expressed as

$$\Delta m = \mathrm{V}\frac{\mathrm{d}c_s}{\mathrm{d}T}\Delta T \tag{2.1}$$

where V represents the solution volume. Seeds either nucleate spontaneously or are placed in solution prior to growth. Solution containers are typically sealed to prevent uncontrolled evaporation and enhance controllability. This method is widely used commercially for crystal growth. Its primary drawbacks are impurity adsorption and incorporation during the crystal growth.

The hot-injection method, a less controllable variant, is frequently used for synthesizing nanocrystals, quantum dots, and nanowires. Here, a "cold" precursor solution is rapidly injected into a high-temperature reaction bath. This triggers rapid nanocrystal nucleation and growth. For instance, synthesizing $CsPbBr_3$

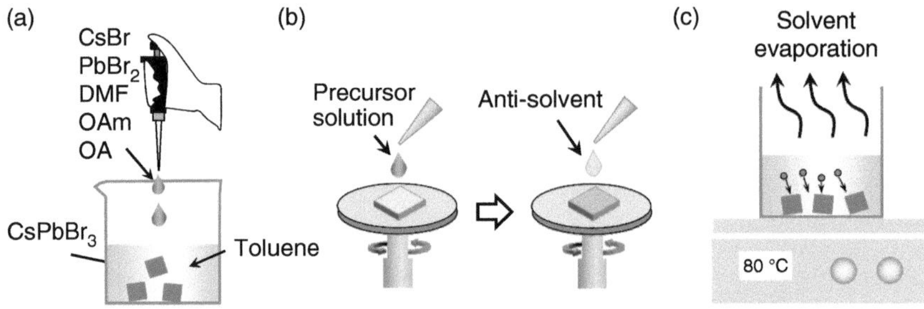

Figure 2.3 (a) Schematic of anti-solvent method to synthesize $CsPbBr_3$ nanocrystals. (b) Spin-coating with anti-solvent to synthesize perovskite thin film. (c) Schematic of solvent evaporation method to reach the supersaturation condition for semiconductor synthesis.

nanocrystals involves first mixing $PbBr_2$ in 1-octadecene (Figure 2.2b). Trace amounts of oleylamine and oleic acid aid $PbBr_2$ dissolution. This mixture is heated to 180 °C. Simultaneously, a Cs–oleate mixture is prepared by dissolving Cs_2CO_3 in oleic acid and 1-octadecene. Injecting this cooler (around 120 °C) Cs–oleate solution into the hot $PbBr_2$ solution initiates a chemical reaction, forming $CsPbBr_3$ nanocrystals. The reaction is quenched within seconds using an ice-water bath, after which the nanocrystals are centrifuged and stored.

The temperature gradient technique also creates supersaturation for semiconductor growth (Figure 2.2c). This approach utilizes two solution zones maintained at different temperatures. The high-temperature zone dissolves the solute to saturation. This saturated solution is then transported to a lower-temperature zone, where supersaturation occurs. Crystal growth proceeds on a seed introduced into this cooler zone. Solution transport between zones can be driven by thermal convection or forced pumping. When the high-temperature zone is situated below the low-temperature zone, thermal convection facilitates transport and growth. Forced convection can be employed using separate tanks at different temperatures; solution is pumped from the high-temperature tank to the low-temperature tank to deliver solute for crystallization.

The anti-solvent method adds a chemically different solvent to the solution, reducing solubility and causing precipitation and crystallization. For example, $CsPbBr_3$ nanocrystals can be synthesized this way. $PbBr_2$ and CsBr are dissolved in DMF (Dimethylformamide) (Figure 2.3a). Small quantities of oleylamine and oleic acid are added as stabilizers to the precursor. Rapid injection of toluene into this solution functions as the anti-solvent. Under vigorous stirring, $CsPbBr_3$ nanocrystals form. The same principle applies to spin-coating perovskite thin films for solar cells (Figure 2.3b). Here, a precursor solution of PbI_2 and CH_3NH_3I (MAI) in dimethyl sulfoxide (DMSO) is prepared. This dispersion is spin-coated onto a substrate, and toluene, acting as an anti-solvent, is added just before the spin-coating concludes, promoting $MAPbI_3$ nucleation. Subsequent high-temperature annealing completes the perovskite layer growth.

The solvent evaporation method establishes supersaturation for precipitation and growth (Figure 2.3c), particularly suited for materials whose solubility remains relatively constant over the working temperature range. Supersaturation arises from controlled solvent evaporation at a constant temperature. Thus, growth rate is governed by the evaporation rate. This growth rate is proportional to both the mass of solvent evaporated and the supersaturation concentration of the solute. Evaporation occurs primarily near the solution surface, which holds higher solute concentrations, potentially inducing spurious nucleation. Nontoxic solvents, such as water vapor, may safely evaporate into the atmosphere if solvent recovery is economically impractical. For other solvents, leakage must be prevented for environmental and economic reasons. An important consideration is the increase in impurity concentration within the solution as the solvent diminishes, leading to elevated impurity levels in the crystal. Using large initial solvent volumes can mitigate this effect.

2.2.1.2 Vapor Phase Chemical Synthesis

Besides solid/solution-based methods, chemicals in the gaseous phase can also serve as sources for material growth. CVD employs chemical vapor reactions to fabricate diverse materials. Various epitaxial thin films or bulk crystals can be synthesized via CVD, a technique with significant applications in the modern semiconductor industry. The initial precursor sources may exist as gases or liquids but are vaporized, forming volatile chemical species subsequently transported by a carrier gas into the reaction chamber. These volatile gas-phase precursors decompose and undergo reactions, forming the desired material deposited onto the target substrate or seed crystal. Unwanted reaction byproducts are removed from the system via the exhaust gas stream. For instance, reaction between Cd vapor and H_2S gas yields considerably large CdS crystals.

CVD reactors operate across a wide pressure range, from atmospheric pressure to ultrahigh vacuum (UHV). Low-pressure CVD (LPCVD) reduces unwanted chemical reactions and enhances growth uniformity. While atmospheric-pressure CVD (APCVD) offers a high chemical reaction rate and, consequently, a high growth rate, it exhibits poorer uniformity and film quality. Thus, APCVD is more suitable for depositing thick films. Radio-frequency (RF) plasma can assist precursor decomposition in CVD (Figure 2.4a), a process termed plasma-enhanced CVD (PECVD). PECVD significantly reduces the required growth temperature because the plasma supplies energy beyond thermal activation. For example, PECVD enables the deposition of amorphous Si_3N_4 at relatively low temperatures using SiH_4 and NH_3 precursor gases. CVD facilitates material growth on diverse substrates. When the growth occurs epitaxially on a single-crystal substrate, the process is commonly referred to as vapor-phase epitaxy (VPE).

The CVD technique can produce numerous materials, including metals, oxides, nitrides, oxynitrides, silicon carbide, diamond, carbon nanotubes, fibers, and various semiconductor compounds. Carbon nanotubes, for instance, are fabricated via catalytic decomposition of hydrocarbon precursors (e.g., methane or ethanol) using CVD. Nanoclusters of iron often serve as catalysts. Factors like catalyst size, composition, gas flow parameters, and other growth conditions determine whether single-walled or multi-walled carbon nanotubes form, resulting in horizontally aligned arrays, vertically aligned bundles, or random networks.

Using metal–organic chemical vapors for CVD growth defines the method as metal–organic CVD (MOCVD). MOCVD is extensively employed for growing

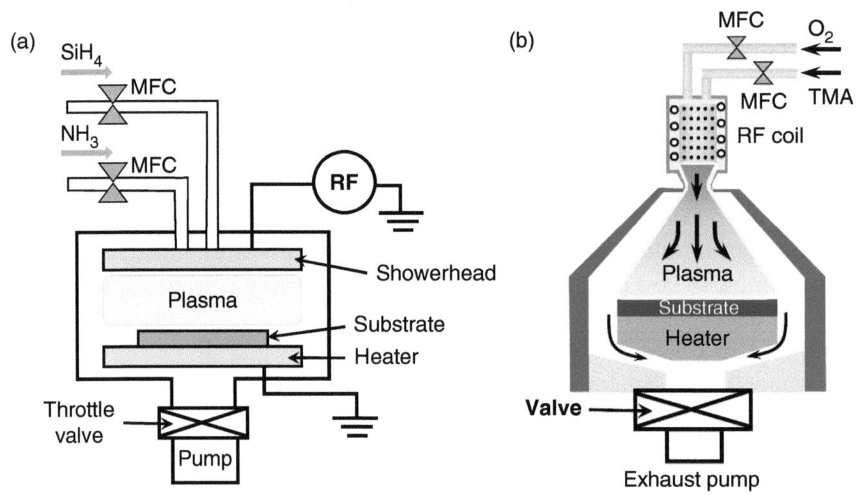

Figure 2.4 (a) Schematic of PECVD to grow amorphous Si_3N_4. (b) Schematic of PEALD to grow amorphous Al_2O_3.

Group III–V semiconductor compounds, prized for its excellent growth uniformity, precise dopant control, and sharp heterointerface definition. For example, high-quality AlN grown by MOCVD exhibits superior properties compared to those obtained via alternative methods. Trimethylaluminum (TMA) and ammonia (NH_3) serve as the aluminum and nitrogen precursors, respectively, with sapphire commonly used as the substrate. Hydrogen acts as the carrier gas, delivering TMA and NH_3 to the substrate surface for reaction and AlN deposition. Beyond AlN, GaN is also industrially produced via MOCVD, finding critical applications in light-emitting diodes (LEDs), laser diodes, power electronics, and other semiconductor devices. For GaN growth, trimethylgallium (TMGa) and NH_3 provide gallium and nitrogen. However, the significant lattice mismatch (around 16%) between the c-plane of GaN and c-plane sapphire substrates complicates direct growth of low-defect-density GaN. Consequently, a thin AlN buffer layer is typically deposited first to mitigate lattice mismatch strain before initiating GaN growth. Due to unintentional oxygen and/or silicon contamination, the grown GaN commonly exhibits *n*-type conductivity. To achieve high-quality GaN with low background doping, controlled *n*-type doping is implemented using silicon impurities, while *p*-type doping utilizes magnesium impurities followed by low-energy electron-beam irradiation treatment to activate acceptors.

Atomic layer deposition (ALD) is another vapor-phase deposition technique. Unlike CVD, ALD employs a self-limiting growth mechanism, enabling exceptional control over film thickness and composition. Rather than continuously injecting precursors simultaneously, the vapors are introduced alternately into the reaction chamber via discrete pulses. Each pulse deposits at most a self-limiting monolayer onto the substrate surface. An inert gas purge phase follows each precursor pulse, removing excess reactants and reaction byproducts until surface adsorption and reaction reach equilibrium, thus preventing gas-phase interactions.

Applying RF plasma to aid precursor decomposition during the process is possible (Figure 2.4b), termed plasma-enhanced ALD (PEALD). Similar to PECVD, PEALD allows reduced growth temperatures by providing activation energy beyond thermal means. For instance, PEALD can deposit Al_2O_3 using TMA as the Al precursor and O_2 plasma as the oxygen source. The final film thickness is directly proportional to the number of growth cycles. ALD's self-limiting mechanism enables the deposition of films with near-perfect stoichiometry, uniformity, and conformality. A broad spectrum of materials can be deposited via ALD, encompassing oxides, nitrides, carbides, chalcogenides, metals, and more. These attributes make ALD indispensable in the electronics industry for depositing insulating high-k dielectric films (e.g., for complementary metal–oxide–semiconductor [CMOS] transistors) and barrier/seed layers for interconnects (e.g., in dynamic random-access memory [DRAM] devices). A primary limitation of ALD is its inherently slow deposition rate.

Successful ALD hinges on precise optimization of parameters like growth temperature, precursor pulse duration, and purge time. Ensuring that precursor adsorption occurs in a self-limiting monolayer fashion and that reaction byproducts are completely evacuated before introducing the next precursor is critical. Failure to meet these conditions can cause gas-phase reactions or multilayer adsorption, effectively transitioning the process into CVD-like deposition. Ideal ALD occurs within a specific temperature window where the growth rate per cycle exhibits minimal temperature dependence. Operating outside this window compromises the self-limiting layer-by-layer growth. At excessively low temperatures, precursors or reaction products may condense physically, increasing growth rates or causing nonuniformity. Conversely, inadequate thermal energy may impede surface reactions, reducing the deposition rate. Excessively high temperatures risk inducing precursor thermal decomposition, leading to uncontrolled CVD-like growth and higher rates. Elevated temperatures can also cause desorption of adsorbed precursors before they react with the subsequent co-reactant, decreasing the growth rate.

As an example of ALD, high-k dielectrics such as Al_2O_3 or HfO_2 can be deposited, which are usually in amorphous form. The deposition temperature is typically at 300 °C. Lower-temperature growth is also possible to make the ALD compatible with other semiconductor processes. For example, to grow Al_2O_3, TMA and water can be used as the precursors to provide Al and O sources. During the growth, a pulse of TMA source carried by H_2 is injected into the chamber, which adsorbs on the substrate through van der Waals interaction or chemical reaction with the hydroxyl radical on the substrate, forming a single Al atomic layer. After purging the excess TMA and methyl wastes, a pulse stream of water is introduced, which reacts with the Al monolayer. The excess water and wastes are purged again. Repeating this cycle, thin film of Al_2O_3 is grown in a layer-by-layer manner.

2.2.2 Physical Method

The physical method refers to a material growth approach that does not alter the chemical composition of source materials. Source materials for physical deposition may exist in liquid or gas phases prior to solid deposition. Methods involving the

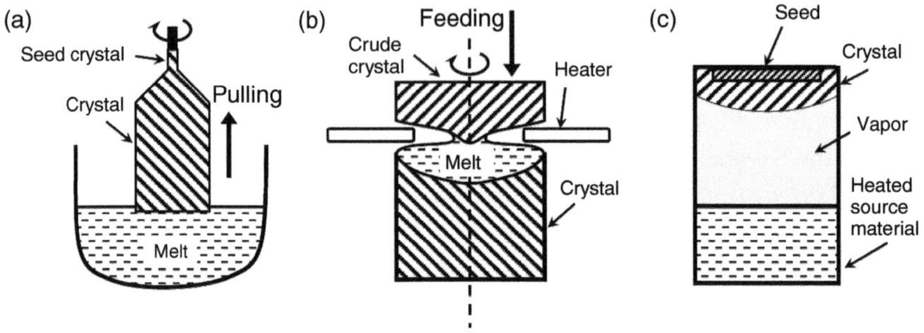

Figure 2.5 Physical methods to grow single crystals. (a) Schematic setup of CZ pulling. (b) Schematic setup of FZ method. (c) Physical vapor transport method to grow single crystal.

liquid-to-solid phase transition primarily comprise Czochralski (CZ) growth and floating-zone (FZ) growth. Other physical growth methods, such as physical vapor transport (PVT), sputtering, and pulsed laser deposition (PLD), involve the transition of source materials from gas to solid phase.

2.2.2.1 Melt Growth Techniques

The CZ growth method was first demonstrated by Prof. Jan Czochralski in 1916. This technique grows crystals from a melt via a pulling mechanism (Figure 2.5a). Industrially, Si crystals with diameters exceeding 300 mm are produced using the CZ method. During Si CZ growth, polycrystalline silicon feedstock is loaded into a silica crucible. This crucible withstands the high temperatures required for melting silicon at 1413 °C within an argon atmosphere. Dipping a monocrystalline silicon seed crystal initiates growth. Gradual withdrawal of the seed and the nascent crystal from the melt induces continuous crystallization, forming a single-crystal Si ingot. Controlling the pulling speed governs crystal diameter. The diameter is first increased to a specific thin-neck value; maintaining this diameter for a certain crystal length facilitates dislocation reduction and defect minimization. Subsequently, the diameter increases to the nominal value, which remains constant during the main cylindrical body growth. Finally, the diameter decreases to form the tail cone, terminating crystal growth. While necessary to contain the molten silicon, the silica crucible inevitably introduces oxygen impurities into the crystal. Additional impurities such as carbon, boron (B), aluminum (Al), and iron (Fe) further degrade carrier mobility and carrier lifetime in the resulting semiconductor material.

Beyond CZ growth, the FZ method also produces high-quality single crystals (Figure 2.5b). This approach employs a zone-melting technique. A narrow zone heater—typically a contactless RF induction coil generating heat via its magnetic field—melts a confined radial segment of a polycrystalline semiconductor feedstock rod. Recrystallization occurs immediately after the molten zone passes. Typically, the induction heater remains stationary while the crystal feed rod moves downward during growth, causing the molten zone to traverse the semiconductor body. Impurities exhibit higher solubility in the liquid phase; thus, they concentrate and migrate with the moving molten zone, leaving purified crystalline material behind. This yields high-purity crystals due to impurity segregation. Maintaining the floating molten zone relies on the strong surface tension of the liquid silicon and support from the adjacent solid material, eliminating the need for a container and preventing contamination. The needle-eye technique enhances capillary stability, enabling larger-diameter FZ crystal growth. Nevertheless, inductive heater limitations restrict FZ silicon wafer diameters (200 mm) compared to CZ wafers (450 mm). FZ and CZ silicon possess complementary advantages: FZ silicon exhibits impurity concentrations two to three orders of magnitude lower than CZ silicon. This property is critical for high-voltage applications, rendering FZ silicon essential for high-power electronics.

2.2.2.2 Vapor Phase Physical Deposition

Synthesizing bulk single crystals of semiconductors with extremely high thermal stability and melting points—SiC, AlN, and GaN—proves challenging using the aforementioned melt-growth techniques. These materials typically dissociate before melting, and crucibles capable of containing their molten forms are lacking. Instead, these wide-bandgap semiconductors are synthesized via PVD or PVT (Figure 2.5c). These materials subsequently facilitate high-power electronics and other semiconductor applications.

The fundamental growth principle of PVT relies on the vaporization of a source material from a hot region and its subsequent condensation onto a colder target surface or seed crystal. Vaporization arises from purely physical processes, such as thermal/laser heating or sputtering via ion bombardment. A concentration gradient—typically created by a temperature gradient—transports vapor from the higher-temperature source zone to the lower-temperature condensation zone. Nucleation upon a crystal seed facilitates oriented growth with controlled polytype formation. Without seeding, saturated vapor nucleates randomly, yielding crystals of limited size. The vaporization rate at the source and condensation rate at the growth surface establish an equilibrium that governs the growth kinetics. PVT is currently the dominant method for producing bulk SiC and AlN single crystals.

Sputtering is another method for preparing thin films (Figure 2.6a). However, differing from the previously discussed methods, this PVD technique is predominantly utilized for depositing polycrystalline or amorphous films. A primary advantage of sputtering is the low deposition temperature. Large-area thin films exhibiting compositions closely resembling the target material can be deposited. Materials amenable to sputtering encompass various metals, semiconductors, oxides, nitrides, and so forth. The simplest sputtering configuration comprises a temperature-controlled cathode and anode within a vacuum chamber. For nonreactive sputtering, the chamber is filled with low-pressure inert gases, such as neon, argon, and krypton. Energetic ionized particles are typically generated by an RF signal, while a DC voltage is maintained between the cathode and anode. These energetic ions bombard the target positioned at the cathode, thereby sputtering atoms from the target material. The sputtered atoms subsequently transport to the anode and deposit onto the substrate. For reactive sputtering, a reactive gas, such as oxygen or nitrogen, is additionally introduced. Beyond generating bombardment ions, the reactive gas reacts with the target material to form the desired oxide or nitride compound, which is deposited as a thin film onto the substrate.

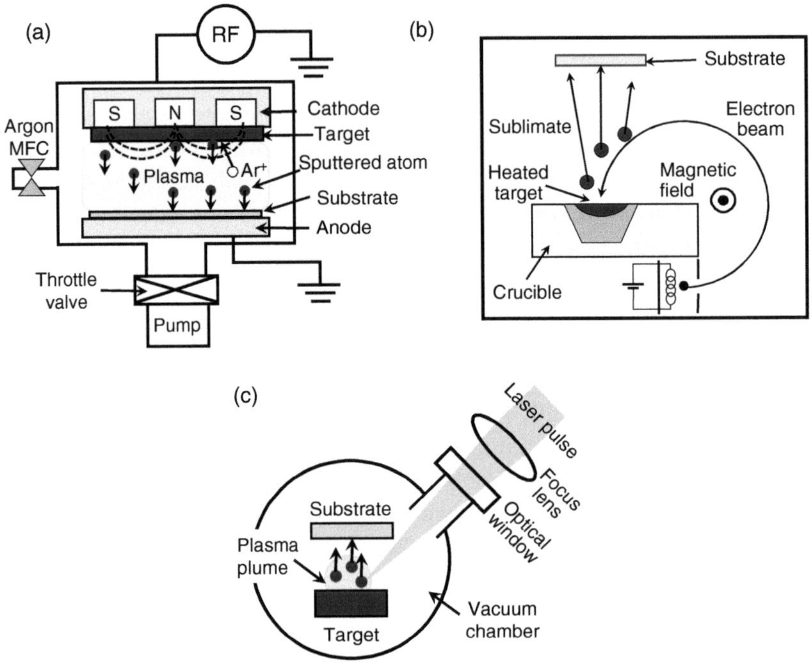

Figure 2.6 Physical methods used to deposition amorphous or polycrystalline films. (a) Schematic of magnetron sputtering. (b) Schematic of electron-beam physical vapor deposition. (c) Schematic of pulsed laser deposition.

In conventional sputtering, the RF signal, besides generating energetic ions, also creates electrons within the plasma. Electrons near the substrate at the anode region can cause substrate heating and bombardment, resulting in a lower deposition rate. Magnetron sputtering alleviates this issue by establishing a magnetic field near the target material at the cathode. The magnetic field confines electron movement, reduces electron bombardment of the substrate, and increases the ionization plasma density. Consequently, higher sputtering efficiency and deposition rates are achieved, even at lower operating pressures and voltages. Currently, the sputtering method is widely applied to fabricate various semiconductor devices, such as interlayer connection metals, dielectric layers in integrated circuits, antireflection coatings, and contact electrodes in solar cells.

Electron-beam physical vapor deposition (EBPVD), or E-beam evaporation, constitutes another physical deposition technique (Figure 2.6b). In contrast to the ion bombardment utilized in sputtering, the target anode is bombarded by a high-energy electron beam, generated via thermionic emission or field emission. The electrons are then accelerated to high kinetic energy and directed toward the evaporation target material. Most kinetic energy converts into thermal energy, with additional energy dissipated through X-ray production and secondary electron emission. The electron beam heats the target to an elevated temperature, vaporizing its atoms into the gas phase and emitting them into the chamber with thermal velocity. These vaporized atoms then transport freely and precipitate onto the substrate, forming a solid thin film. Typically, the chamber pressure in EBPVD systems must remain low to facilitate electron bombardment and unimpeded transport of vaporized atoms.

A high-energy pulsed laser beam can serve as an excitation source to vaporize the target material (Figure 2.6c). The associated deposition method is termed PLD. The pulsed laser beam enters the vacuum chamber through an optical window and focuses onto the target surface. To ablate the target material, the laser power density within the nanosecond pulse duration must exceed a specific threshold. This ablation produces plasma plumes containing various species, encompassing neutral atoms, electrons, and ions. These species flow toward the substrate and deposit a thin film. When reactive gases like oxygen or nitrogen are introduced, oxide or nitride thin films can also be deposited. Additional gases, such as argon and hydrogen, may also be introduced to enhance gas-phase reactions of the ablated species during PLD.

2.2.3 Epitaxy Growth

Epitaxy refers to the ordered growth or deposition of a crystalline material on a crystalline substrate, whereby the new material establishes a well-defined crystallographic relationship with the substrate across the interface. This newly formed material is termed an epitaxial film, epitaxial layer, or epilayer. Epitaxy is classified as homoepitaxy when the epilayer and substrate are composed of identical materials; otherwise, it is designated heteroepitaxy. Based on the source material phase, epitaxial methods are categorized into solid-phase epitaxy (SPE), liquid-phase epitaxy (LPE), and VPE.

Due to significantly slower atomic transport in solids compared to liquids or vapors, SPE is less prevalent than LPE or VPE. In SPE, amorphous silicon (Si) is deposited on a crystalline Si substrate. Subsequent heating facilitates the transformation of the amorphous phase into a crystalline structure, initiated at the interface, owing to the lower Gibbs free energy per atom in crystalline Si relative to amorphous Si. This phase transition occurs at approximately 500 °C for Si, while germanium (Ge) crystallizes at a lower temperature of about 350 °C. The crystallization velocity v is governed by classical nucleation and growth theory:

$$v = v_0 e^{-\frac{\Delta E_a}{kT}} \tag{2.2}$$

where ΔE_a is the activation energy, k is Boltzmann constant, and T is the temperature. For Si (001) surface, the activation energy is 2.7 eV, while for Ge (001) surface, the activation energy is 2.15 eV. According to this equation, the crystallization of Si is slower than Ge at the same temperature. The crystallization has a faster speed at a higher temperature.

LPE employs a liquid-phase source for epitaxial growth, enabling the fabrication of crystalline materials ranging from thin films to bulk crystals with specific orientations on a seeding substrate. LPE has been instrumental in advancing optoelectronic devices based on III–V binary or alloy semiconductors, including LEDs, laser diodes, solar cells, and heterojunction bipolar transistors. For instance, GaAs epitaxial layers are grown via LPE using a gallium-rich liquid melt saturated with arsenic. Controlled cooling induces supersaturation, and upon contact with a GaAs seeding substrate, a lattice-matched, high-crystallinity GaAs epilayer precipitates.

Ternary or quaternary III–V compounds, such as $Ga_xIn_{1-x}P_yAs_{1-y}$, can also be synthesized via LPE under lattice-matching constraints. This material exhibits a tunable bandgap and emits light at wavelengths near 1.55 μm, coinciding with the minimal absorption and dispersion window of silica optical fibers, rendering it crucial for optical telecommunications. Similarly, II–VI materials like $Cd_xHg_{1-x}Te$ are grown from tellurium or mercury solutions using LPE. Compositional tuning adjusts its bandgap for mid-wavelength (3–5 μm) and long-wavelength (8–10 μm) infrared detection, enabling applications in night vision and surveillance technologies.

LPE offers distinct advantages, including high growth rates, superior crystal quality, and highly anisotropic or faceted growth. It facilitates the growth of relatively thick (>100 μm) epitaxial layers. For example, thick SiC and GaN epilayers grown by LPE can serve as substrates for subsequent epitaxy via other techniques.

However, LPE exhibits critical limitations. Precise control over epilayer thickness, particularly for sub-0.1 μm layers, is challenging. Lattice mismatch and incompatibility between seed and epilayer also hinder the synthesis of complex heterostructures. Undesirable compositional and doping gradients may additionally arise during LPE growth.

VPE utilizes vapor-phase source materials for epilayer deposition and represents the most extensively employed epitaxial method. VPE classifications differ based on operating pressure: ballistic beam epitaxy, low-pressure VPE, and high-pressure VPE. Ballistic beam epitaxy operates under UHV, ensuring source vapor transport occurs ballistically toward the substrate. This category subdivides into molecular beam epitaxy (MBE) and chemical beam epitaxy (CBE), depending on the beam composition. Vapor transport in low-pressure VPE occurs in the viscous flow regime, encompassing methods such as metal–organic VPE (MOVPE), halide transport VPE (HTVPE), and hydride VPE (HVPE), among others.

MOVPE—also termed metal–organic chemical vapor deposition (MOCVD)—employs metalorganic precursors to deposit epilayers at moderate pressures. It is a principal method for manufacturing III–V optoelectronics. For example, indium phosphide (InP) growth utilizes $(CH_3)_3In$ as the Group III source and PH_3 as the Group V precursor; surface reactions of these precursors drive epitaxial growth on the substrate.

MBE grows epilayers using ballistic beams of pure elemental sources (Figure 2.7). Achieving long mean free paths necessitates a UHV chamber environment without carrier gases. Conversely, CBE employs molecular beams of reactive gases (e.g., hydrides or metal–organics) as reactants, also requiring UHV conditions. The terms CBE and metal–organic molecular beam epitaxy (MOMBE) are often synonymous. Both MBE and CBE exhibit lower growth rates compared to other VPE methods, necessitating stricter vacuum conditions to minimize impurity incorporation during layer growth. For GaAs MBE, beams of gallium and arsenic are generated by subliming pure elements in effusion cells or electron-beam evaporators; these beams impinge upon the substrate to form a single-crystal epilayer. MBE can also deposit organic semiconductors using molecular beams.

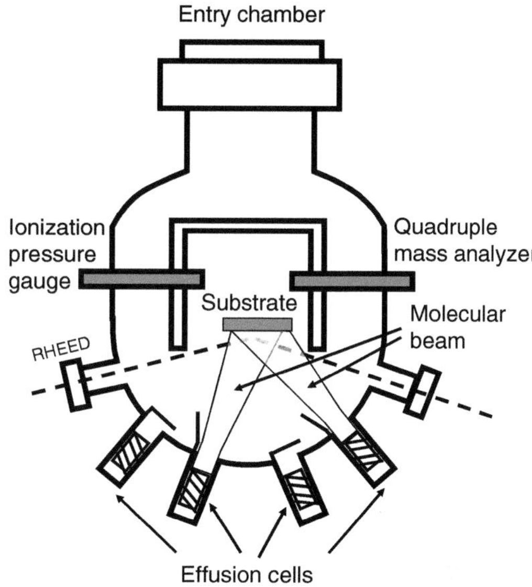

Figure 2.7 Schematic cross section of a typical MBE chamber.

Epilayer growth rate and quality are modulated by source and substrate temperature. Higher source temperatures increase beam flux and deposition rate, while substrate temperature governs atom condensation or desorption kinetics. *In-situ* monitoring during MBE, such as via reflection high-energy electron diffraction (RHEED), combined with effusion cell shutters, enables precise monolayer-thickness control. This facilitates the fabrication of intricate heterostructures (e.g., quantum wells) for semiconductor lasers and LEDs. Additional *in-situ* characterization tools like angle-resolved photoemission spectroscopy (ARPES), Auger electron spectroscopy, and EELS can be integrated with MBE systems.

The growth of epitaxial thin films on a single-crystal surface critically depends on the interaction strength between adatoms and the substrate surface (Figure 2.8). For either LPE or VPE, three distinct epitaxial growth modes exist: layer-by-layer deposition (Frank–van der Merwe mode or FM mode), island growth (Volmer–Weber mode or VM mode), and layer-by-layer growth followed by island growth (Stranski–Krastanov mode or SK mode). These modes are primarily determined by the interplay between surface energy and strain energy at the epilayer–substrate interface. Layer-by-layer deposition (FM mode) is favored when

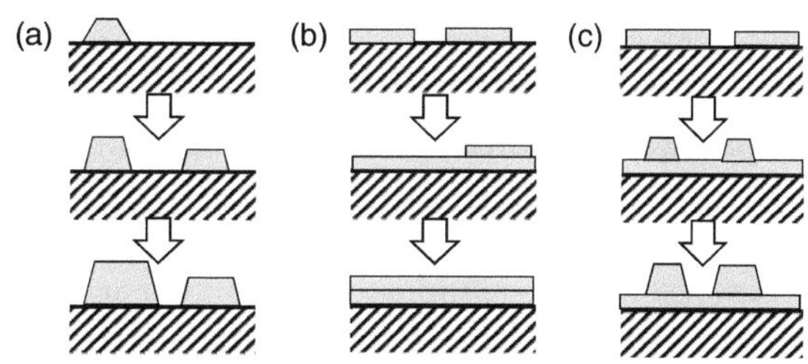

Figure 2.8 Three different growth modes of epilayers. (a) Island growth (VM mode), (b) layer-by-layer deposition (FM mode), and (c) layer-by-layer followed by island growth (SK mode).

both the epilayer's surface energy and its lattice mismatch with the substrate are low, resulting in continuous, uniform film growth. Conversely, higher surface energy or substantial lattice mismatch promotes island formation (VM mode) to minimize the combined surface and strain energy. The SK mode arises from competition between low surface energy and accumulating high strain energy, leading to an initial wetting layer followed by island nucleation. In the SK mode, lattice-mismatch-induced strain energy gradually builds up as the epilayer thickens. Eventually, upon exceeding a critical thickness, the growth transitions from layer-by-layer to island formation on the wetting layer to release the accumulated strain energy. This mode, also referred to as the island-on-wetting-layer growth mode, has been extensively studied and provides a pathway for fabricating various novel nanostructures, such as quantum dots, for applications in optoelectronics.

2.3 Characterization

2.3.1 Characterization of Electrical Properties

Semiconductors are widely utilized in electronic components, such as diodes and transistors. Consequently, the systematic characterization of their electrical properties is fundamentally important. This subsection introduces key techniques for this purpose, including four-probe measurement, Hall effect measurement, and I–V and C–V characterization.

2.3.1.1 Four Probe Measurement

The resistivity of semiconductors is closely related to crystal structure, defects, doping conditions, and device performance. Resistivity ρ is defined by

$$\rho = \frac{1}{q(n\mu_n + p\mu_p)} \tag{2.3}$$

where q is the elementary charge, n and p are the electron and hole densities, respectively, and μ_n and μ_p are the corresponding carrier mobilities. Typically, semiconductors are dominated by one carrier type, where the majority carrier exhibits significantly higher density than the minority carrier. Under this condition, resistivity is primarily determined by the properties of the majority carrier.

The four-probe method is extensively employed for semiconductor resistivity measurements, as it overcomes limitations inherent to the two-probe method. The necessity for this approach arises from the ubiquitous presence of wire resistance and the uncertain contact resistance at each probe, illustrated in Figure 2.9. The four-probe configuration resolves this issue by utilizing two probes exclusively for potential measurement, without establishing new current paths. This is enabled by the voltmeter's high input impedance, resulting in negligible current flow through these probes. Consequently, parasitic voltage drops from wiring and contact resistances are eliminated, permitting accurate measurement of the potential difference, which can subsequently be used to determine resistivity.

Although the four-probe method mitigates the influence of parasitic resistances, the interpretation of resistance $R = V/I$ is contingent upon probe arrangement, probe

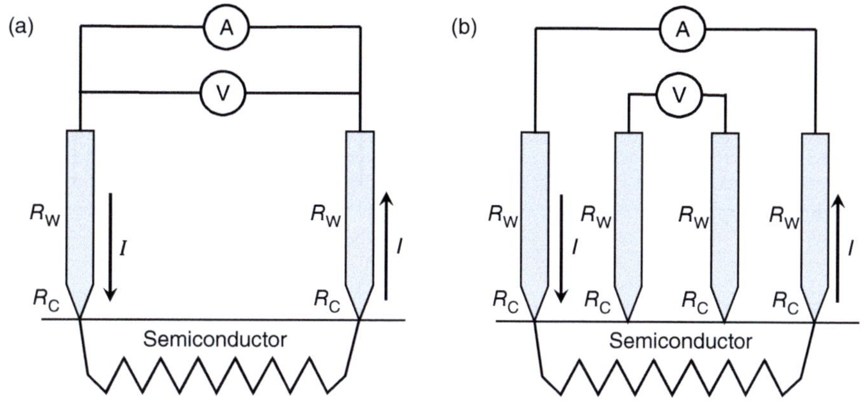

Figure 2.9 (a) Two-probe method and (b) four-probe method. The parasitic resistances are included in the two-probe method and cause measurement error. In the four-probe method, the additional probes of the voltmeter introduce no current flow; thus, they can measure voltage drop without the influences of wiring and contact resistances.

spacing, semiconductor thickness, and boundary conditions. Probes are typically arranged in-line with equal spacing, though alternative configurations are feasible. For a thick semiconductor sample, resistivity ρ is given by

$$\rho = 2\pi s \frac{V}{I} \tag{2.4}$$

where s denotes the probe spacing. This equation is derived under the assumption that the semiconductor is infinitely thick, ensuring current injection spreads radially within a semi-infinite bulk medium.

The derivation begins by examining current injection from a single-point probe, as shown in Figure 2.10a. The resulting electric field and current density exhibit radial symmetry. The current density magnitude within the semiconductor is

$$J = \frac{I}{2\pi r^2} \tag{2.5}$$

Given Ohm's law within the semiconductor, where current density $J = \sigma E = \frac{E}{\rho}$, the electric field E is

$$E = \rho J = \frac{\rho I}{2\pi r^2} \tag{2.6}$$

The potential difference between two points r_1 and r_2 is then

$$V = \int_{r_1}^{r_2} \frac{dV}{dr} dr = \int_{r_1}^{r_2} E dr = \frac{\rho I}{2\pi}\left(\frac{1}{r_1} - \frac{1}{r_2}\right) \tag{2.7}$$

For two probes in Figure 2.10b, with probe 1 injecting current and probe 2 collecting current, the electric field and current density become the vector superposition of the fields from each independent probe. The potential drop between two additional measurement points r_1 and r_2 relative to probe 1 and r_3 and r_4 relative to probe 2 is

$$V = \frac{\rho I}{2\pi}\left(\frac{1}{r_1} - \frac{1}{r_2}\right) + \frac{-\rho I}{2\pi}\left(\frac{1}{r_3} - \frac{1}{r_4}\right) \tag{2.8}$$

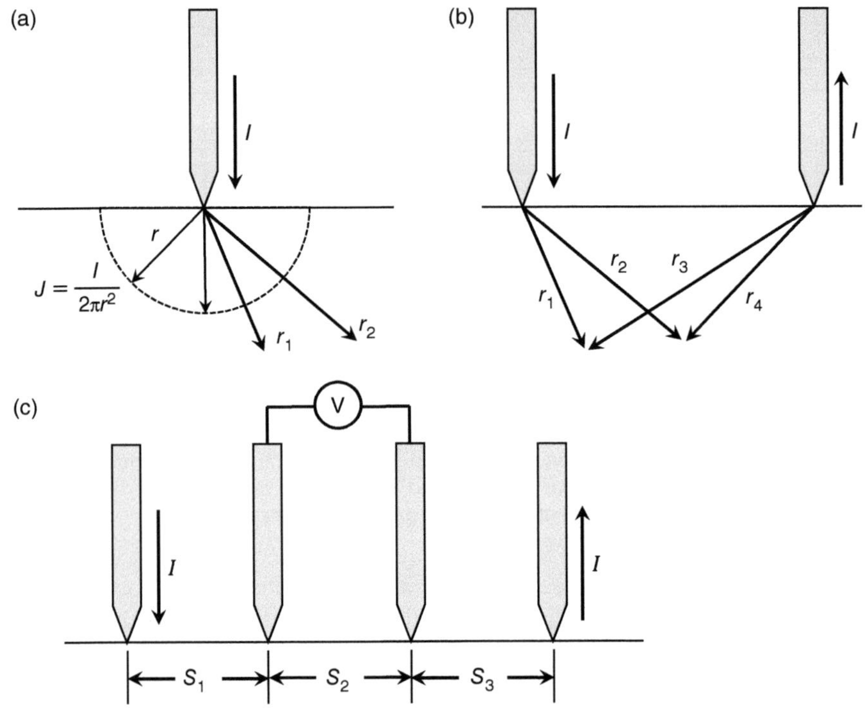

Figure 2.10 (a) Current injection from single-probe to semi-infinite semiconductor. (b) Current injection and collection of two probes. (c) Current injection and voltage measurement of four probes.

The negative sign in the second term arises because probe 2 acts as a current sink. Considering aligned voltage probes spaced s_1, s_2, and s_3 apart (Figure 2.10c), the measured voltage becomes

$$V = \frac{\rho I}{2\pi}\left(\frac{1}{s_1} - \frac{1}{s_1 + s_2}\right) + \frac{-\rho I}{2\pi}\left(\frac{1}{s_2 + s_3} - \frac{1}{s_3}\right) \tag{2.9}$$

Under the condition of equidistant probe spacing $s = s_1 = s_2 = s_3$, Eq. (2.9) simplifies to

$$V = \frac{\rho I}{2\pi s} \tag{2.10}$$

The resistivity can thus be extracted from Eq. (2.10). It is important to note that these derivations assume that the semiconductor extends infinitely in all dimensions. For finite geometries, such as samples of limited thickness or probes near the wafer edge, corrections are necessary to mitigate errors. For instance, in a semiconductor with thickness $t > 2s$ on an insulating substrate, the error is less than 9.4%, and for $t > 5s$, the error reduces to below 0.7% [1].

Measurement of materials with thicknesses significantly smaller than the probe spacing $t < s$ requires a different expression:

$$\rho = \frac{\pi}{\ln(2)} t \frac{V}{I} \tag{2.11}$$

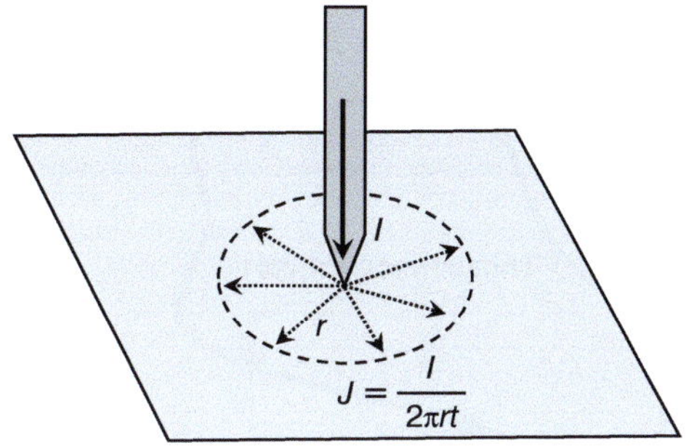

Figure 2.11 The distribution of current injection from one probe into a thin semiconductor film with thickness t.

Equation (2.11) is valid for $t < s/2$ and follows a derivation analogous to the bulk case. In this regime, as shown in Figure 2.11, the lateral current flow dominates, and the voltage drop between points r_1 and r_2 on the surface of a thin film of thickness t is

$$V = \int_{r_1}^{r_2} \frac{dV}{dr} dr = \int_{r_1}^{r_2} \frac{\rho I}{2\pi r t} dr = \frac{\rho I}{2\pi t} \ln\left(\frac{r_2}{r_1}\right) \tag{2.12}$$

Sheet resistance R_{sh} is a more frequently utilized parameter for characterizing thin films:

$$R_{sh} = \frac{\rho}{t} = \frac{\pi}{\ln(2)} \frac{V}{I} \tag{2.13}$$

which reflects the average resistivity of thin film over the thickness and has the unit of Ohms per square.

Although the collinear probe arrangement is prevalent for thin-film resistivity measurement, it requires large lateral dimensions and is sensitive to probe placement relative to edges. Alternative configurations include the square arrangement, where probes contact the perimeter of a square sample. A more versatile method, applicable to arbitrarily shaped samples, was established by van der Pauw and is supported by conformal mapping theory. As depicted in Figure 2.12, four probes contact the boundary of an arbitrary sample shape. The resistivity is calculated by [2]

$$\rho = \frac{\pi}{\ln(2)} t \frac{R_{12,34} + R_{23,41}}{2} f \tag{2.14}$$

where $R_{12,34}$ denotes the resistance measured with current injected through probes 1 and 2, and the voltage is measured between probes 3 and 4; $R_{23,41}$ is measured similarly; and f is a correction factor defined by

$$\frac{R_{12,34} - R_{23,41}}{R_{12,34} + R_{23,41}} = \frac{f}{\ln(2)} \operatorname{arccosh}\left[\frac{\exp\left(\frac{\ln(2)}{f}\right)}{2}\right] \tag{2.15}$$

A common variant of the van der Pauw structure is the Greek cross, as shown in Figure 2.13a. Its symmetry ensures $R_{12,34} = R_{23,41}$, simplifying the sheet resistance calculation to

$$R_{sh} = \frac{\pi}{\ln(2)} R_{12,34} \tag{2.16}$$

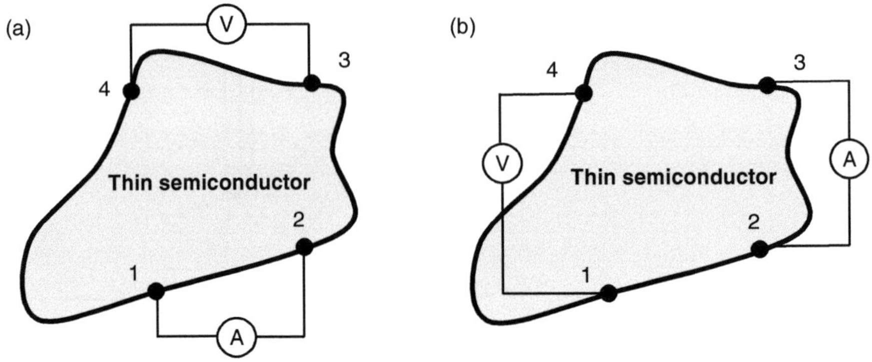

Figure 2.12 The four-probe contact of an arbitrarily shaped sample. (a) Resistance measured with current injected through probes 1 and 2, and the voltage is measured between probes 3 and 4. (b) Resistance measured with current injected through probes 2 and 3, and the voltage is measured between probes 4 and 1.

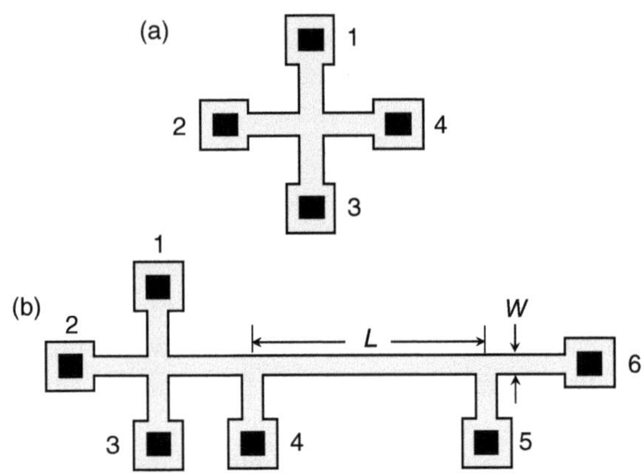

Figure 2.13 (a) The Greek cross structure. (b) Greek cross integrated with the bridge structure to measure both sheet resistance and line width.

where $R_{12,34}$ is measured with current injected through probes 1 and 2, while the voltage is measured between probes 3 and 4. Furthermore, when the Greek cross is integrated with the bridge structure and current I_{26} is injected through probes 2 and 6 as shown in Figure 2.13b, and the voltage V_{45} is measured between probes 4 and 5, then

$$R_{26,54} = \frac{V_{45}}{I_{26}} = R_{sh}\frac{L}{W} \tag{2.17}$$

and the line width can be calculated by

$$W = R_{sh}\frac{I_{26}}{V_{45}}L \tag{2.18}$$

2.3.1.2 Hall Effect Measurement

The resistivity measurement described previously was conducted without an applied magnetic field. When a magnetic field is applied to the conductor, the

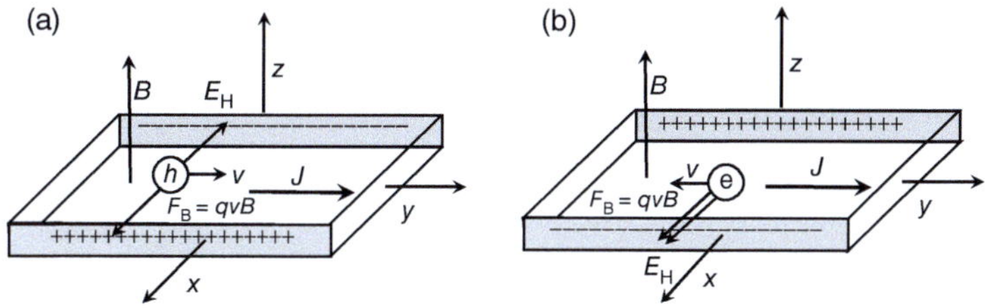

Figure 2.14 Hall effect of (a) *n*-type and (b) *n*-type semiconductors. The magnetic field causes the hole and electron to accumulate or dissipate on one side, resulting in an electric field E_H and Hall voltage V_H.

Hall effect is observed. This phenomenon, first discovered by Hall in 1879, occurs when the magnetic field is perpendicular to the current flow direction. Under this condition, charge accumulation develops on the sidewalls of the conductor, generating an electric field perpendicular to both the magnetic field and the current [3]. The polarity of the charge accumulations and the direction of this electric field depend on the carrier type, as illustrated in Figure 2.14, enabling carrier type identification. The resulting voltage difference between opposite sides of the conductor is termed the Hall voltage, V_H.

Experimentally, the Hall coefficient is defined by

$$R_H = \frac{tV_H}{BI} \tag{2.19}$$

where t is the sample thickness, V_H is the Hall voltage, B is the magnetic field strength, and I is the current. While sample thickness is typically well defined, caution is required for structures such as thin epitaxial or implanted layers on substrates of opposite conductivity type, or in cases involving strong depletion effects and band bending caused by Fermi level pinning at surfaces and interfaces. Under these conditions, the active layer thickness deviates from the physical thickness, necessitating corrections to the Hall coefficient to account for errors.

Hall measurements provide a powerful method for determining carrier density, carrier type, and mobility. Theoretically, the Hall coefficient is expressed as

$$R_H = \frac{r(p - b^2 n)}{q(p + bn)^2} \tag{2.20}$$

where $b = \mu_n/\mu_p$ and r is the scattering factor, ranging between 1 and 2 depending on the dominant scattering mechanism within the semiconductor. The scattering factor is defined by $r = \langle\tau^2\rangle/\langle\tau\rangle^2 = \mu_H/\mu$, where τ is the scattering time (which may depend on energy or momentum), μ_H is the Hall mobility, and μ is the drift mobility. For three-dimensional (3D) nondegenerate carriers, the scattering factor is $r = 3\pi/8 \approx 1.18$ for acoustic phonon scattering and $r = 315\pi/512 \approx 1.93$ for ionized impurity scattering. For degenerate carriers, r approaches unity regardless of dimensionality and scattering mechanism. r also varies with magnetic field and temperature, approaching 1 in the high magnetic field limit. However, magnetic fields typically employed in Hall measurements smaller than 1 T yield $r > 1$. As r is generally unknown, it is frequently assumed equal to 1 for simplicity.

For extrinsic n-type semiconductor with $p \gg n$, the Hall coefficient reduces to

$$R_{\mathrm{H}} = \frac{r}{qp} \tag{2.21}$$

and for extrinsic n-type semiconductor, it becomes

$$R_{\mathrm{H}} = -\frac{r}{qn} \tag{2.22}$$

Utilizing experimental Hall coefficient values and assuming $r = 1$, the carrier type and density can be determined. If the conductivity σ is known, the carrier mobility can also be derived. For example, in an n-type semiconductor where conductivity relates to carrier concentration and mobility by $\sigma = qp\mu$, the mobility is given by

$$\mu \approx \mu_H = \sigma R_{\mathrm{H}} \tag{2.23}$$

Plotting carrier density versus temperature provides additional information, including donor/acceptor type, density, and activation energy. Hall measurements typically employ two device structures. For the bridge-type Hall bar (Figure 2.15a), current is injected and collected via probes 1 and 4. The Hall voltage is measured between probes 2 and 6 or probes 3 and 5 in the presence of a magnetic field, while the resistivity is determined by the voltage difference between probes 2 and 3 or probes 5 and 6 in the absence of a magnetic field.

As previously introduced, the van der Pauw structure (Figure 2.15b) measures resistivity using the equation

$$\rho = \frac{\pi}{\ln(2)} t \frac{R_{12,34} + R_{23,41}}{2} f \tag{2.24}$$

where f is a geometric correction factor dependent on the ratio $R_{12,34}/R_{23,41}$. This structure can also measure the Hall coefficient. By applying current between probes 1 and 3 and measuring the change in voltage ΔV_{24} between probes 2 and 4 with and without the magnetic field applied, the Hall coefficient is calculated as

$$R_{\mathrm{H}} = \frac{t\Delta V_{24}}{BI_{13}} \tag{2.25}$$

Standard Hall measurements assume uniform doping. Nonuniformly doped samples yield averaged carrier properties. To obtain spatially resolved carrier density profiles, differential Hall effect measurements are required. This technique involves sequential anodic oxidation and oxide etching steps to strip the sample layer by layer. After each removal step, the sheet Hall coefficient $R_{\mathrm{Hsh}} = R_H/t = V_{\mathrm{H}}/BI$ and the sheet conductance $\sigma_{\mathrm{sh}} = \sigma t = t/\rho$ are measured. Subsequently, the depth-dependent carrier density and mobility profiles can be extracted using these measurements [4, 5].

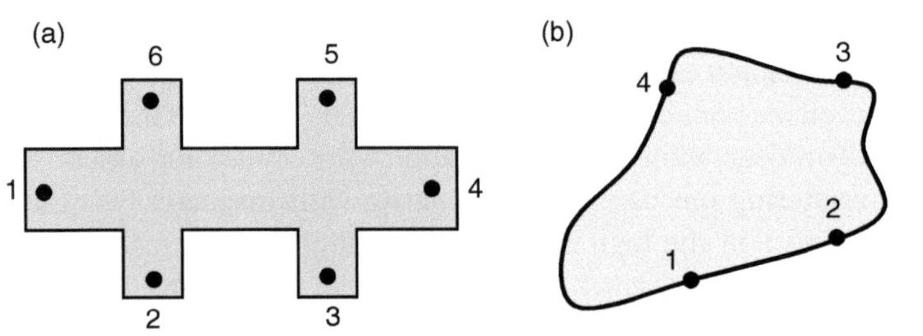

Figure 2.15 Two typical device structures to measure the Hall effect. (a) Bridge structure and (b) van der Pauw structure.

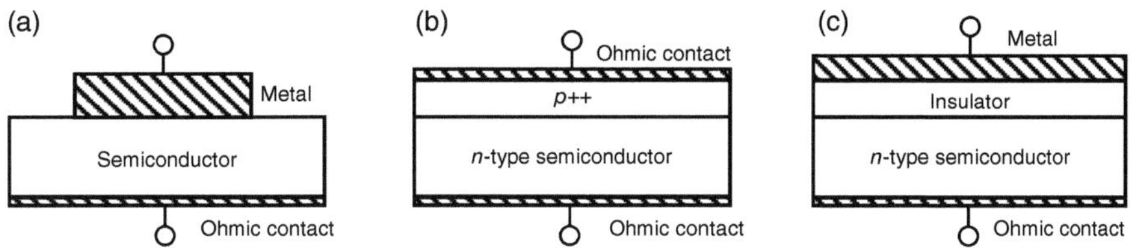

Figure 2.16 Typical structures that can use C–V measurement to extract doping profile. (a) Metal–semiconductor Schottky junction, (b) one-sided heavily doped PN junction, and (c) MOS capacitor.

$$\mu(x) = \frac{\dfrac{\mathrm{d}\left(R_{\mathrm{Hsh}}\sigma_{\mathrm{sh}}{}^2\right)}{\mathrm{d}x}}{r \cdot \dfrac{\mathrm{d}\sigma_{\mathrm{sh}}}{\mathrm{d}x}} \tag{2.26}$$

$$n(x) = \frac{r \cdot \left(\dfrac{\mathrm{d}\sigma_{\mathrm{sh}}}{\mathrm{d}x}\right)^2}{q \cdot \dfrac{\mathrm{d}\left(R_{\mathrm{Hsh}}\sigma_{\mathrm{sh}}{}^2\right)}{\mathrm{d}x}} \tag{2.27}$$

2.3.1.3 C–V and I–V Measurements

The performance of semiconductor devices is closely related to their doping conditions. Both C–V and I–V measurements are employed for semiconductor characterization. When a semiconductor is fabricated into structures such as Schottky junctions, PN junctions, metal–oxide–semiconductor (MOS) capacitors, or metal–oxide–semiconductor field-effect transistors (MOSFETs), the C–V method facilitates the extraction of the dopant profile (refer to Figure 2.16). This capability relies on the property that the width of the space-charge region in a reverse-biased semiconductor junction device is modulated by the applied voltage, consequently influencing the junction capacitance.

For a metal–semiconductor Schottky junction, a one-sided heavily doped PN junction, or an MOS capacitor, the measured capacitance per unit area C is given by

$$C = \frac{\mathrm{d}Q}{\mathrm{d}V} = \frac{\mathrm{d}}{\mathrm{d}V}\int_0^W qN_\mathrm{D}\mathrm{d}x = qN_\mathrm{D}(W)\frac{\mathrm{d}W}{\mathrm{d}V} \tag{2.28}$$

where Q represents the charge density per unit area, V is the bias voltage, N_D denotes the dopant density, and W is the depletion width within the semiconductor, as illustrated in Figure 2.17. This equation assumes an abrupt depletion region edge. For Schottky junctions or one-sided heavily doped abrupt PN junctions, the capacitance also relates to the depletion width as

$$C = \frac{\varepsilon_s}{W} \tag{2.29}$$

In contrast, for an MOS capacitor, the measured capacitance results from the series combination of the oxide capacitance and the depletion capacitance:

$$C = \frac{1}{\dfrac{W}{\varepsilon_s} + \dfrac{1}{C_{ox}}} \tag{2.30}$$

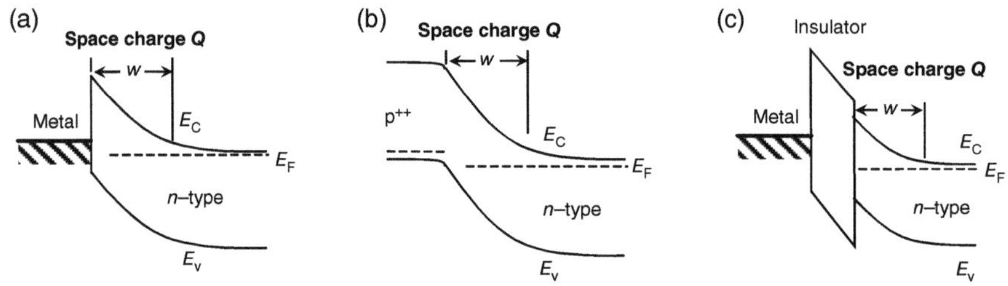

Figure 2.17 Band diagrams of (a) Schottky junction, (b) one-sided heavily doped PN junction, and (c) MOS capacitor. The voltages are biased to deplete the semiconductors with space-charge width of W.

Equations (2.29) and (2.30) enable the extraction of the depletion width W. For Schottky or PN junctions:

$$W = \frac{\varepsilon_s}{C} \tag{2.31}$$

For MOS capacitors:

$$W = \varepsilon_s \left(\frac{1}{C} - \frac{1}{C_{\text{ox}}} \right) \tag{2.32}$$

Differentiating the width yields:

$$\frac{\mathrm{d}W}{\mathrm{d}V} = -\frac{\varepsilon_s}{C^2} \frac{\mathrm{d}C}{\mathrm{d}V} \tag{2.33}$$

Combining these, the dopant density can be expressed as

$$N_{\mathrm{D}}(W) = \frac{C}{q} / \frac{\mathrm{d}W}{\mathrm{d}V} = -\frac{C^3}{q\varepsilon_s \mathrm{d}C/\mathrm{d}V} = \frac{2}{q\varepsilon_s \mathrm{d}(1/C^2)/\mathrm{d}V} \tag{2.34}$$

These equations form the basis of the C–V method for extracting the doping profile applicable to Schottky junctions, one-sided heavily doped abrupt PN junctions, and MOS capacitors. Notably, the depletion approximation is intrinsic to these derivations; it assumes complete depletion of majority carriers within the space-charge region while neglecting minority carriers. Consequently, the extracted density $N_{\mathrm{D}}(W)$ represents an effective carrier density rather than the true chemical doping density. Deviations between the majority carrier concentration and the doping density correlate with the Debye length L_{D}:

$$L_{\mathrm{D}} = \sqrt{\frac{kT\varepsilon_s}{q^2(p+n)}} \tag{2.35}$$

While C–V measurements provide dopant profiles, I–V characterization also offers extraction capabilities. Unlike the C–V method, which necessitates large-area devices to mitigate stray capacitance effects and enhance signal-to-noise ratios, the I–V technique overcomes these limitations, enabling doping profile analysis in small-area devices.

Methods have been proposed to characterize semiconductors configured as four-terminal surface-channel MOSFETs (see Figure 2.18). One method involves tuning both the substrate V_{SB} and gate V_{GS} voltages while actively maintaining a

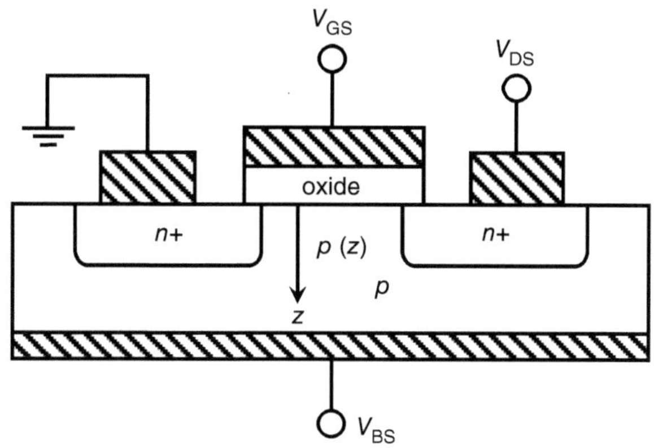

Figure 2.18 A four-terminal surface-channel *n*-MOSFET structure to deplete the *n*-type bulk substrate and measure the *n*-type dopant profile using I–V measurement.

constant channel carrier density [6]. Practical implementation typically involves first biasing the MOSFET in its linear region with a small drain-source voltage V_{DS}. Subsequently, both V_{GS} and V_{SB} are adjusted continuously via a feedback loop designed to sustain a constant channel current.

Under these conditions, the inversion charge density in the conduction channel remains effectively constant. Poisson's equation and Gauss's law establish a direct link between changes in V_{GS} and changes in the depletion charge density: $dQ_b = C_{ox}dV_{GS}$. The total gate charge Q_0 is

$$Q_0 = qn_s + Q_b \tag{2.36}$$

where n_s is the inversion layer carrier density per unit area at the oxide–semiconductor interface, and Q_b represents the charge density per unit area within the depletion region. With the channel current constant and n_s consequently fixed, changes in the depletion charge are directly measurable by

$$dQ_b = dQ_0 = C_{ox}dV_{GS} \tag{2.37}$$

where $C_{ox} = \varepsilon_{ox}/t_{ox}$ is the oxide capacitance per unit area and dV_{GS} denotes the change in gate-to-source voltage.

As shown in Figure 2.19, Poisson's equation and Gauss's law govern the depletion region:

$$Q_b = \int_0^w qp(z)dz \tag{2.38}$$

$$V_{SB} + 2\phi_b = \int_0^w \frac{qp(z)}{\varepsilon_s} zdz \tag{2.39}$$

where $p(x)$ represents the acceptor (p-type) dopant density, ε_s is the semiconductor permittivity, and w signifies the depletion region depth. Differentiation yields

$$dQ_b = qp(w)dw \tag{2.40}$$

$$dV_{SB} = \frac{qp(w)}{\varepsilon_s} wdw \tag{2.41}$$

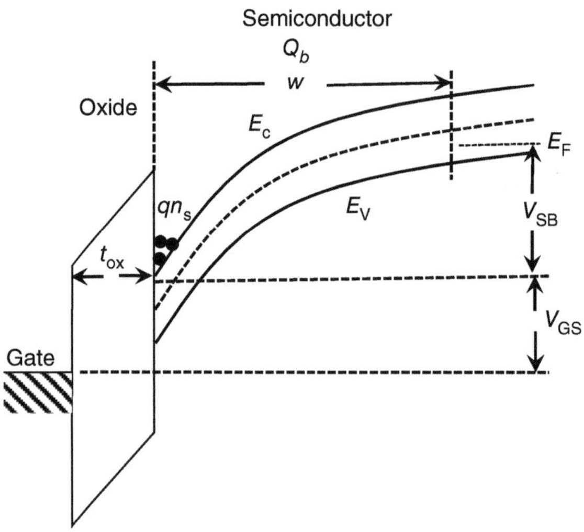

Figure 2.19 Energy bands of an *n*-type substrate fabricated in MOSFET structure to extract the substrate dopant profile using I–V measurement. Both gate and substrate are biased and continuously tuned, while the inversion carrier density is maintained constant.

Based on the aforementioned equations, we can get

$$w = \frac{\varepsilon_s \mathrm{d}V_{\mathrm{SB}}}{\mathrm{d}Q_b} = \frac{\varepsilon_s \mathrm{d}V_{\mathrm{SB}}}{\mathrm{C}_{ox}\mathrm{d}V_{\mathrm{GS}}} \tag{2.42}$$

$$p(w) = \frac{\mathrm{d}Q_b}{q\mathrm{d}w} = \frac{C_{ox}^{\ 2}}{q\varepsilon_s} / \frac{\mathrm{d}^2 V_{\mathrm{SB}}}{\mathrm{d}V_{\mathrm{GS}}^{\ 2}} \tag{2.43}$$

Thus, by measuring the V_{SB} versus V_{GS} relationship under constant inversion carrier density, the doping density profile $p(w)$ can be extracted using Eqs. (2.42) and (2.43).

2.3.2 Characterization of Optical Properties

2.3.2.1 Photoluminescence Spectrum

Following excitation by incident light, a semiconductor absorbs photons, generating electron–hole pairs. The subsequent radiative recombination process emits photons, resulting in PL. A typical PL measurement configuration is shown in Figure 2.20, comprising a laser source, sample stage, light-collection optics, monochromator, photodetector, and associated electronics. PL mapping necessitates additional components such as a scanning mirror or stage.

PL spectroscopy serves as a powerful, nondestructive technique for characterizing semiconductor properties, including bandgap energy, exciton binding energy, and impurity/defect densities. The laser photon energy must exceed the semiconductor bandgap $h\nu > E_{\mathrm{G}}$. Photon absorption creates electron–hole pairs, which rapidly thermalize to the conduction band minimum and valence band maximum, respectively, via phonon scattering. Radiative recombination efficiency and PL emission depend critically on the bandgap type. As illustrated in Figure 2.21, direct-bandgap semiconductors like GaAs (III–V Group) exhibit high PL efficiency, while indirect-bandgap semiconductors like Si exhibit low PL efficiency due to momentum conservation constraints requiring large-momentum phonon scattering.

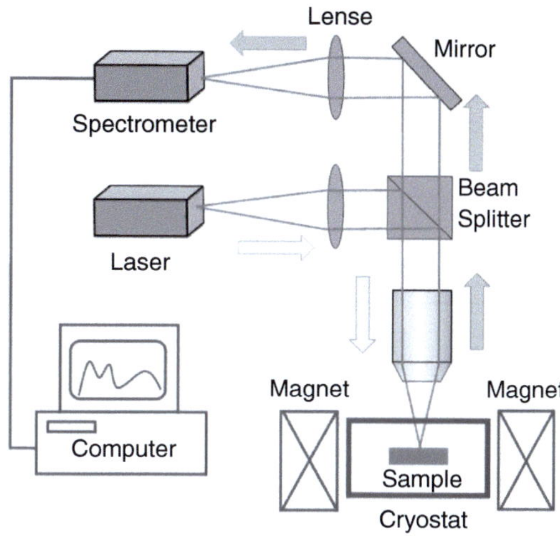

Figure 2.20 Typical PL measurement setup. Cryostat and magnet are used to tune the temperature and apply magnetic field during PL measurement.

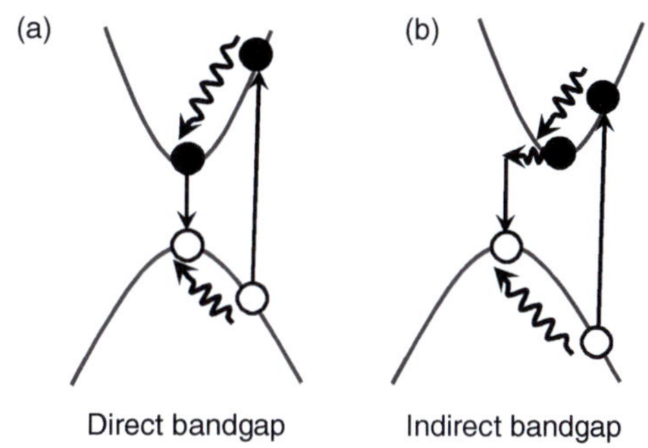

Figure 2.21 PL of (a) direct- and (b) indirect-bandgap semiconductors. Electron and hole cannot recombine directly without phonon scattering for indirect-bandgap semiconductor, due to momentum conservation, resulting in low PL intensity.

Beyond band-to-band recombination, excitonic recombination is significant, particularly at low temperatures. Coulomb attraction binds electrons and holes into hydrogen-like states called excitons (Figure 2.22a). In direct-bandgap semiconductors, excitons radiatively recombine, emitting photons with energy slightly below the bandgap.

$$h\nu = E_{\mathrm{G}} - E_{\mathrm{B}} \tag{2.44}$$

where E_{G} is the bandgap energy and E_{B} is the exciton binding energy. For indirect-bandgap semiconductors, exciton recombination requires scattering by a large-momentum phonon E_{ph}, leading to emitted photon energy further reduced by E_{ph}.

Electron-hole pair generation can deform the local lattice structure, inducing self-trapped excitons (Figure 2.22b). Self-trapping occurs predominantly in materials with soft lattices and strong electron-phonon coupling, producing broad-spectrum, low-energy emission. The resulting large Stokes shift minimizes self-absorption, benefiting applications like white-light emitters and X-ray scintillators. The photoluminescence excitation (PLE) spectrum reveals free exciton peaks, whereas self-trapped excitons appear in the PL spectrum.

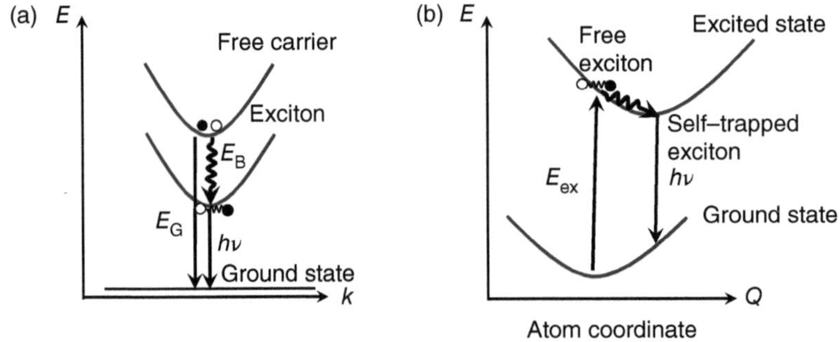

Figure 2.22 (a) Free electron and hole bind together due to Coulomb interaction, forming exciton with binding energy of E_B. Both free electron-hole pair and exciton can have non-zero total momentum and result in the energy dispersion. (b) The strong interaction between the excited electron–hole pair and phonon causes the lattice distortion, which can reduce the system energy and induce the self-trapped exciton. Stokes shift of $E_G - h\nu$ between excitation and emission energy can be observed.

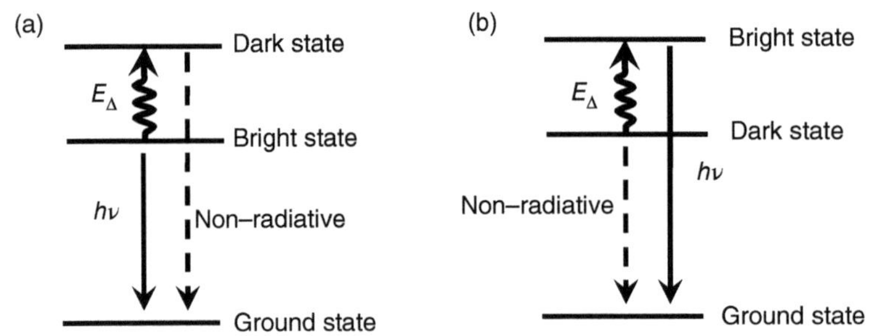

Figure 2.23 (a) Thermal activation of bright states to dark/lossy states, making the PL intensity decrease with increasing temperature. (b) Thermal activation of dark states to bright states, making the PL intensity increase with increasing temperature.

Supplementary instruments such as cryostats or electromagnetic coils expand PL capabilities, enabling investigations of temperature or magnetic field dependencies of PL intensity. Temperature-dependent PL intensity variations provide insights into emission mechanisms. Typically, PL intensity decreases with increasing temperature due to enhanced phonon scattering, non-radiative recombination, exciton dissociation, or excitation to dark state (Figure 2.23a). This is modeled by:

$$I(T) = I_0 \frac{1}{1 + Ae^{-\frac{E_\Delta}{kT}}} \tag{2.45}$$

where I_0 is low-temperature PL intensity, A is a fitting parameter, E_Δ is the activation energy, k is Boltzmann's constant, and T is temperature. Important physical parameters such as exciton binding energy can be extracted. According to this equation, high temperature lends more energy to exciton and dissociate the exciton, decreasing the PL intensity.

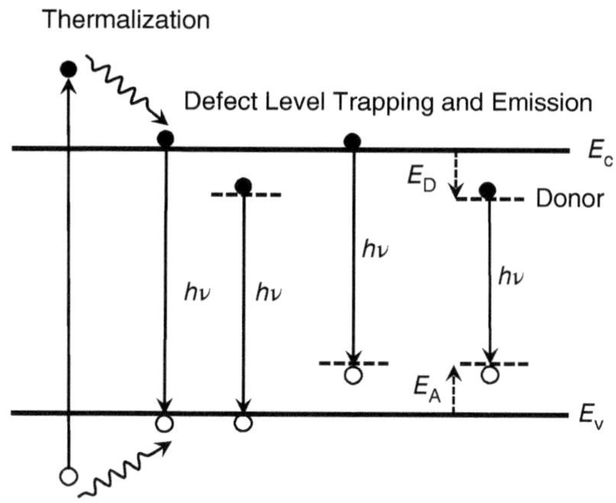

Figure 2.24 Photoluminescence caused by the trapping of excited electron–hole pairs in defect levels. After the thermalization of electron and hole toward the conduction band minimum and valence band maximum, respectively, electron or hole can be trapped in donor or acceptor defect level, resulting in lower emission energy.

Occasionally, PL intensity increases with temperature, indicating thermal activation from a low-lying dark state to a bright state with a higher radiative recombination rate (Figure 2.23b). This behavior is described by

$$I(T) = I_0 \frac{Ae^{-\frac{E_\Delta}{kT}}}{1 + Ae^{-\frac{E_\Delta}{kT}}} \tag{2.46}$$

PL intensity is exceptionally sensitive to semiconductor defects (Figure 2.24). For characterized defect types, PL can quantify defect density if calibrated; however, it only detects defects involved in radiative transitions. Despite this limitation, comparative PL intensity analysis remains a powerful tool for defect characterization. Reduced defect density typically enhances total PL intensity by minimizing Shockley–Read–Hall non-radiative recombination and serves as an indicator of crystal quality.

If electron is trapped by the donor, the emission energy will be

$$h\nu = E_G - E_D \tag{2.47}$$

where E_D is the donor energy level with respect to conduction band minimum. If hole is trapped by the acceptor, the emission energy will be

$$h\nu = E_G - E_A \tag{2.48}$$

where E_A is the acceptor energy level with respect to valence band maximum. If both electron and hole are trapped by donor and acceptor, respectively, forming charge-neutral donor and acceptor, the recombination will create the charged donor and acceptor. As a result, depending on the spatial distance r between donor and acceptor, the emission wavelength also needs to be modified by the Coulomb interaction:

$$h\nu = E_G - E_A - E_D + \frac{q^2}{4\pi\varepsilon_s r} \tag{2.49}$$

where ε_s is dielectric constant of the semiconductor.

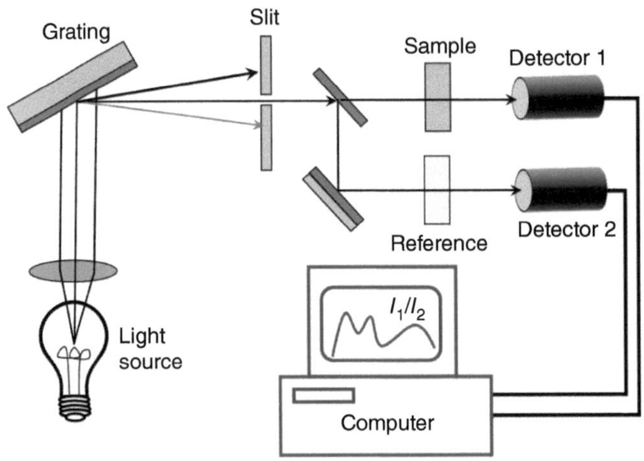

Figure 2.25 Schematics of the double-beam UV–vis spectrometer.

2.3.2.2 Ultraviolet–Visible Absorption Spectrum

UV–vis optical absorption spectroscopy is a fundamental technique for characterizing key semiconductor properties, including bandgap energy, exciton binding energy, and defect states. A typical experimental configuration for such measurements, a double-beam spectrometer, is illustrated in Figure 2.25. UV–vis radiation typically spans wavelengths from 190 to 900 nm, representing a specific portion of the electromagnetic spectrum. Light within this range possesses varying photon energies. When the photon energy is sufficiently high, it induces electronic transitions from the ground state to excited states, resulting in light absorption.

An absorption spectrum is experimentally acquired by measuring the intensity of transmitted light following normal incidence of monochromatic light, where the wavelength is continuously tuned. This requires a monochromator and a photodetector. A monochromator isolates a narrow bandwidth of wavelength from a broadband light source using various methods. A fundamental approach employs optical dispersion within a prism. However, modern instruments typically utilize diffraction gratings, which angularly disperse light by wavelength due to interference. Gratings consist of many closely spaced, parallel grooves inscribed on a substrate and are essential components in contemporary spectrometers.

Following dispersion by the grating, light of the desired wavelength is selected using a slit. A narrower slit provides higher spectral resolution but reduces transmitted light intensity, which can be problematic for achieving a high signal-to-noise ratio. Implementing intensity-modulated light coupled with a lock-in amplifier technique can significantly enhance measurement accuracy. To account for instrumental contributions, the absorption spectrum of a reference sample is typically measured first. Alternatively, a beam splitter can divide the monochromatic beam into two separate paths, enabling simultaneous measurement of the sample and reference. This double-beam configuration offers the advantage of mitigating time-dependent environmental perturbations, thereby improving the stability and accuracy of the results.

The absorption of light by a semiconductor is intimately linked to surface reflection, the complex refractive index, and sample thickness. For a semiconductor with complex refractive index $\widetilde{n} = n - i\kappa$, the electric field of a propagating wave inside the material is expressed as

$$E(x, t) = E_0 e^{i\omega t - i(n - i\kappa)k_0 x} \tag{2.50}$$

where $E(x, t)$ represents the electric field at position x and time t, ω is the angular frequency of oscillation, and $k_0 = 2\pi/\lambda$ is the wave vector in vacuum. Since the light intensity $I(x, t) \propto |E(x, t)|^2$, the absorption coefficient is defined as $\alpha = 4\pi\kappa/\lambda$.

The experimentally measured transmitted light intensity is influenced by surface reflection, thin-film interference effects, and bulk absorption. For a sufficiently thick film where interference effects are negligible, the reflection coefficient for normally incident light from air onto the semiconductor surface is given by the Fresnel equations:

$$R = \left| \frac{1 - (n - i\kappa)}{1 + (n - i\kappa)} \right|^2 \tag{2.51}$$

For samples with finite thickness d, interference must be considered. The reflectance R and transmittance T are then derived as

$$R = \left| \frac{r\left(1 - e^{-i2\widetilde{n}k_0 d}\right)}{1 - r^2 e^{-i2\widetilde{n}k_0 d}} \right|^2 \tag{2.52}$$

$$T = \left| \frac{(1 - r^2)e^{-i\widetilde{n}k_0 d}}{1 - r^2 e^{-i2\widetilde{n}k_0 d}} \right|^2 = \left| \frac{1 - r^2}{1 - r^2 e^{-i2\widetilde{n}k_0 d}} \right|^2 e^{-\alpha d} \tag{2.53}$$

where $r = (1 - \widetilde{n})/(1 + \widetilde{n})$ is the wave reflection coefficient. From the transmission equation, it can be seen that the transmission rate is roughly related to the film thickness d by $e^{-\alpha d}$. However, the denominator term includes d by $\left|1 - r^2 e^{-i2\widetilde{n}k_0 d}\right|^2$. If the resolution Δk of spectrometer is not enough or the sample is thick, i.e., $\Delta k \gg \frac{1}{2nd}$, the spectrometer will not be able to resolve the oscillations in the denominator term. As a result, the measured reflection and transmission rate will be

$$R = \frac{|r|^2(1 + e^{-2\alpha d})}{1 + |r|^4 e^{-2\alpha d}} \tag{2.54}$$

$$T = \frac{|1 - r^2|^2 e^{-\alpha d}}{1 + |r|^4 e^{-2\alpha d}} \tag{2.55}$$

When $\alpha d \gg 1$, the aforementioned equations can be simplified as

$$R \approx |r|^2 \tag{2.56}$$

If the imaginary part of surface reflectance r is ignorable and the transmission rate is further simplified as $T = |1 - r^2|^2 e^{-\alpha d}$, the absorption coefficient can be calculated by

$$\alpha = -\frac{\ln\left(\frac{T}{(1-R)^2}\right)}{d} \tag{2.57}$$

The absorption coefficient of semiconductor is closely related to the band structure. For light with energy below the bandgap, the absorption is weak. The light absorption is strong for light energy higher than the bandgap. From the absorption coefficient as a function of photon energy, the bandgap can be determined. For direct-bandgap semiconductor, the absorption coefficient α will be related to photon energy $h\upsilon$ by

$$\alpha \propto \sqrt{h\upsilon - E_G} \tag{2.58}$$

By plotting α^2 as a function of $h\upsilon$, the bandgap can be determined from the onset of α^2. This kind of plot is also called Tauc plot. For indirect-bandgap semiconductor, the absorption coefficient α relates to photon energy $h\upsilon$ by

$$\alpha \propto (h\upsilon \mp h\omega_p - E_G)^2 \tag{2.59}$$

where $h\omega_p$ is energy of the scattering phonon. The absorption or emission of scattering phonon can maintain momentum conservation. If the phonon energy can be ignored, by plotting $\sqrt{\alpha}$ against $h\upsilon$, the bandgap of the indirect semiconductor can also be determined from the onset of $\sqrt{\alpha}$. Special care should be taken for direct-bandgap semiconductors with forbidden transition, such as the centrosymmetric Cu_2O. In this case, the absorption coefficient α relates to photon energy $h\upsilon$ by

$$\alpha \propto (h\upsilon - E_G)^{\frac{3}{2}} \tag{2.60}$$

2.3.2.3 Fourier Transform Infrared Spectrum

FTIR spectroscopyis a significant technique for characterizing chemical bonds in semiconductors, particularly those with vibrational energies within the infrared range, such as oxygen or carbon impurities in silicon. It serves as a powerful analytical tool for investigating semiconductor properties. A typical experimental setup is illustrated in Figure 2.26. Unlike techniques utilizing monochromators to measure transmission and absorption spectra sequentially, FTIR employs an interferometer. Monochromators isolate a narrow bandwidth of light at each measurement step. This narrowband isolation inherently yields low spectral intensity, resulting in a weak signal that necessitates substantial time to scan across the entire wavelength range of interest. FTIR circumvents this limitation by utilizing a Michelson interferometer to measure the intensity of transmitted light.

An infrared broadband light source, such as a heated filament, arc lamp, or Glowbar, generates the incident radiation. This light is initially collimated and directed toward a 50:50 beam splitter, which divides the beam into two spatially separate optical paths. One path reflects off a fixed mirror while the other reflects off a movable mirror translating back and forth. The light returning from both mirrors recombines partially at the beam splitter and traverses the sample. The transmitted light intensity is subsequently detected and recorded electronically. Crucial to the measurement, the movable mirror must maintain parallelism during translation. Although the diverse wavelengths from the broad-spectrum source are mutually incoherent, the beam-splitting action creates path-dependent coherence for each

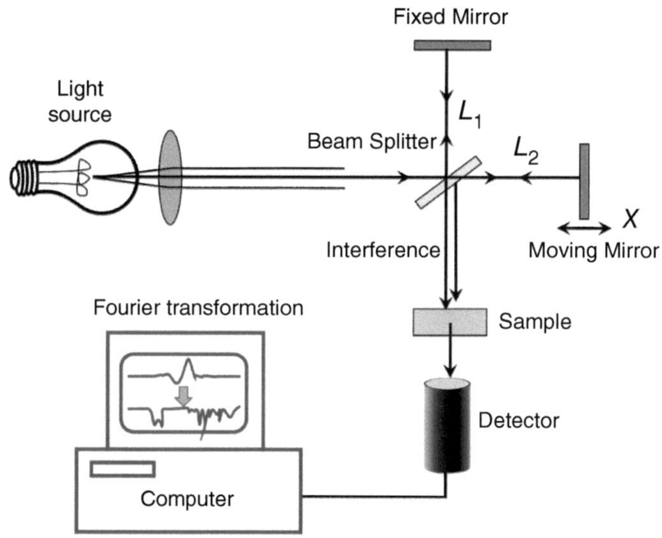

Figure 2.26 Schematic of Fourier transform infrared spectrum (FTIR) measurement setup.

specific wavelength; thus, light of the same wavelength from the two paths can interfere upon recombination. When the two path lengths are equal ($L_1 = L_2$), constructive interference occurs for all wavelengths, maximizing the detector output. When the movable mirror translates a distance x, an optical path difference $\delta = 2x$ is introduced. For light of wavelength λ, a mirror displacement of $x = \lambda/4$ yields a retardation $\delta = 2x = \lambda$. This introduces a 180° phase shift between the beams, leading to destructive interference and a minimal output. Conversely, a displacement $x = \lambda/2$ produces $\delta = 2x = \lambda$ and a 360° phase shift (0° equivalent), resulting in constructive interference and a maximum output. The detected light intensity $I(f, x)$ for light of wavenumber $f = 1/\lambda$ with oscillation amplitudes A_1 and A_2 can be quantified as

$$I(f,x) = |A_1(f) + A_2(f)e^{i4\pi fx}|^2$$
$$= A_1(f)^2 + A_2(f)^2 - 2A_1(f)A_2(f)\cos(4\pi fx) \tag{2.61}$$

The oscillation amplitudes A_1 and A_2 are generally wavelength-dependent functions. For a precisely 50:50 beam splitter $A_1 = A_2$, the modulation amplitude is proportional to $A_1(f)A_2(f)$ and thus to the incident light source intensity $I(f)$ at the corresponding wavelength $\lambda = 1/f$. The detector integrates intensities across all wavelengths simultaneously; therefore, the total measured intensity at displacement x is

$$I(x) = \int_0^{f_m} I(f,x)df \tag{2.62}$$

FTIR measures the interferogram, $I(x)$. Maximum intensity occurs at $L_1 = L_2$ (termed the center burst) where constructive interference maximizes across all wavelengths. By analyzing the measured intensity $I(x)$ as a function of mirror displacement x, the spectral amplitude information, $A_1(f)A_2(f)$, or equivalently the incident spectral intensity, $I(f)$, is recovered via the Fourier transform:

$$I(f) \propto A_1(f)A_2(f) \propto \left| \int_{-x_m}^{x_m} I(x)\cos(4\pi f x)\mathrm{d}x \right| \tag{2.63}$$

The spectral resolution Δf of the interferogram is fundamentally limited by the maximum mirror displacement x_m, specifically $\Delta f = \frac{1}{x_m} \geq \frac{1}{L}$, where L is the interferometer arm length.

Whereas FTIR provides light spectra, the absorption characteristics of the sample itself are often of primary interest. Absorption spectra are deduced by comparing FTIR measurements with and without the sample in place, as depicted in Figure 2.27. The measurement without the sample establishes the background spectrum, thereby eliminating artifacts associated with the source spectral distribution. In practice, a dry nitrogen purge eliminates atmospheric absorption caused by H_2O and CO_2. FTIR offers two principal advantages over monochromator-based methods. Firstly, the entire spectrum is acquired simultaneously ("multiplex advantage"), contrasting sharply with the sequential narrowband detection of monochromators. This significantly reduces acquisition time, often to mere seconds per scan. Secondly, FTIR utilizes a larger aperture without the restrictive slit required by monochromators ("throughput advantage"), allowing greater light flux onto the detector. Consequently, FTIR achieves significantly enhanced signal intensity and a superior signal-to-noise ratio.

2.3.2.4 X-Ray Photoelectron Energy Spectrum

When incident X-rays are absorbed by materials, photoelectrons acquire sufficient energy to escape the material surface and can be collected by detectors (Figure 2.28). Analysis of these photoelectrons provides critical information about the elemental composition and chemical states present on the material surface. XPS is a technique based on this mechanism. X-ray photoelectron emission constitutes a photoelectric effect initiated by high-energy X-ray photons, which interact predominantly with core-level electrons. When incident photon energies are low, typically less than 50 eV, the ejected electrons originate primarily from the valence band; the corresponding technique is ultraviolet photoelectron spectroscopy (UPS).

Photoelectric effect during XPS measurement. According to the photoelectric effect, the kinetic energy of a photoelectron, E_{kin}, is determined by the following equation:

$$E_{\mathrm{kin}} = h\upsilon - E_b - q\phi \tag{2.64}$$

where $h\upsilon$ is the photon energy and ϕ is the work function of the analyzed material, defined as the energy difference between the Fermi level and the vacuum level. E_b is the electron binding energy, referenced to the Fermi level. However, before reaching the electron energy analyzer, unknown electrostatic potentials arising from sample charging or within the measurement chamber can shift the measured electron kinetic energy. Consequently, E_{kin} is typically not measured directly. To mitigate

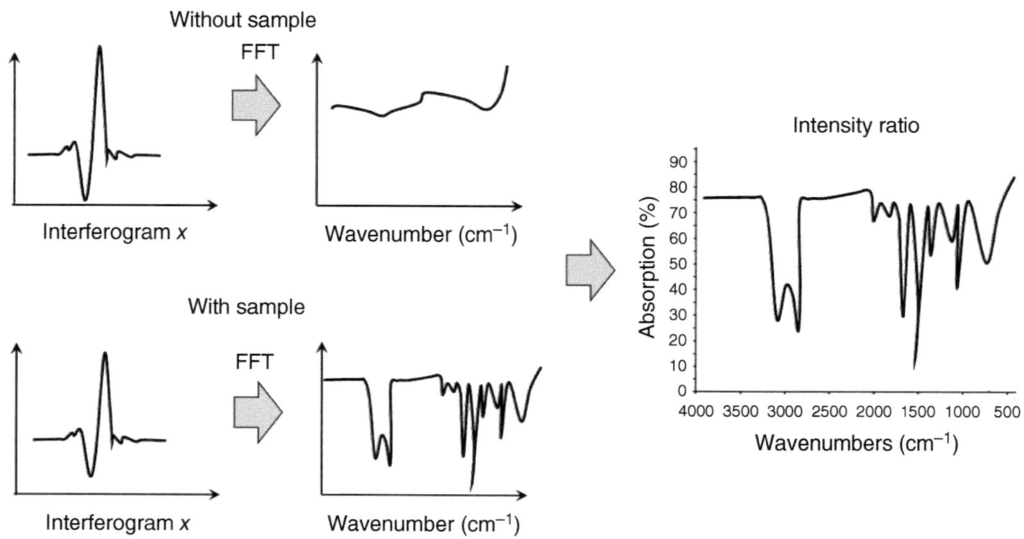

Figure 2.27 FFT (fast Fourier transform) of interferograms to obtain the infrared light intensities with and without placing sample in the arm. The sample absorption can be extracted by the intensity ratio with and without sample.

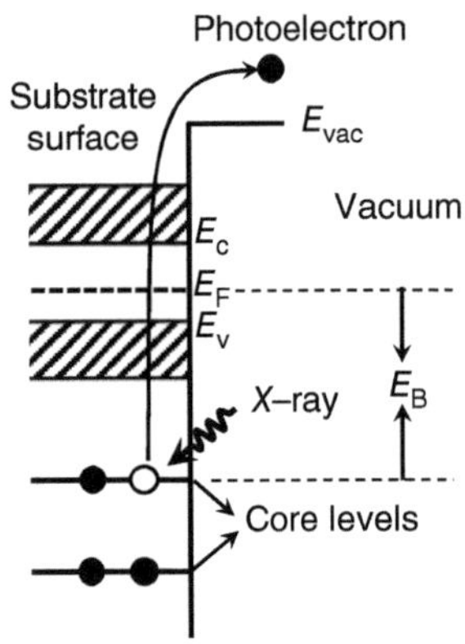

Figure 2.28 Photoelectric effect during XPS measurement.

this, the sample is electrically connected to the spectrometer to establish Fermi level alignment. By precisely adjusting the voltage difference between the sample and the spectrometer, the electron energy reaching the analyzer can be regulated. Subsequent calibration using reference materials, such as a gold substrate, enables elimination of the charging potential influence and accurate determination of the electron binding energy E_b.

The equipment for XPS measurements is illustrated schematically in Figure 2.29. An X-ray source employing high-intensity and narrow–line width radiation is essential. Mg Kα (1253.6 eV) and Al Kα (1486.6 eV) lines are commonly utilized. While the typical X-ray beam diameter ranges from millimeters to centimeters, micro-XPS with micrometer-scale spot sizes is now available. Samples are irradiated with X-rays under UHV conditions. The generated photoelectrons are focused by electron optics and their energy distributions analyzed using a hemispherical energy analyzer, with detection often accomplished by electron multipliers. Photoelectron

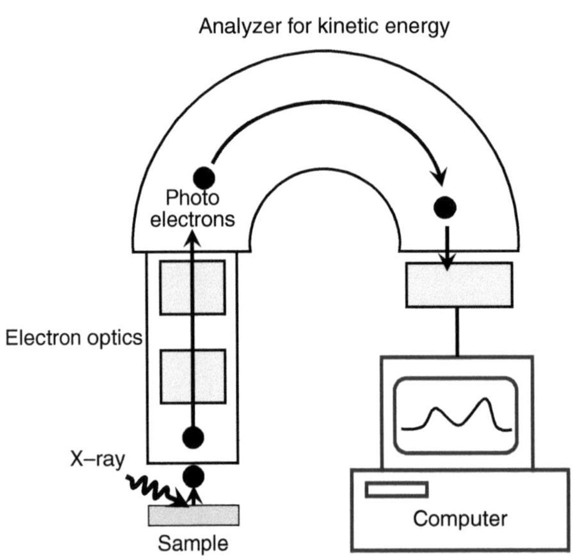

Figure 2.29 Schematic setup of XPS measurement.

emission induces positive sample charging. For insulating or poorly conductive samples, an electron flood gun is frequently employed to neutralize accumulated charge.

Following energy calibration and charge corrections, the XPS spectrum is plotted as intensity versus binding energy. Core-level binding energies, referenced to the Fermi level, act as elemental fingerprints. Consequently, XPS peaks are conventionally labeled with the element symbol and the corresponding electronic shell designation. Crucially, subtle variations (<1 eV) in core-level binding energies reflect the local chemical environment of the emitting atom. For example, the binding energy of a metal differs from its oxide phase. Therefore, beyond elemental identification, XPS can determine the chemical states of elements. Since the binding energy reference is the Fermi level, and the Fermi level position depends on doping concentration, peak shifts, especially in UPS valence band spectra, can be utilized to characterize material doping levels. XPS is intrinsically surface-sensitive, probing elemental composition and chemistry within the uppermost several nanometers.

This sensitivity arises from the short inelastic mean free path (escape depth) of photoelectrons, typically <10 nm, while the incident X-rays possess a penetration depth of several micrometers. Depth profiling of element distributions is achievable by combining XPS with ion-beam sputtering. However, caution must be exercised during interpretation, as sputtering can alter surface chemistry. Quantitative analysis of elemental ratios (sensitivity ~0.1%) and distribution information requires consideration of relative sensitivity factors and photoelectron escape probabilities.

References

1 Weller, R.A. (2001). *Rev. Sci. Instrum.* 72: 3580–3586.
2 van der Pauw, L.J. (1958). *Philips Res. Rep.* 13: 1–9.
3 Hall, E.H. (1879). *Am. J. Math.* 2: 287–292.
4 Petritz, R.L. (1958). *Phys. Rev.* 110: 1254–1262.
5 Baron, R., Shifrin, G.A., Marsh, O.J., and Mayer, J.W. (1969). *J. Appl. Phys.* 40: 3702–3719.
6 Buehler, M.G. (1980). *J. Electrochem. Soc.* 127: 701–704.

3

EUV Lithography Process of Semiconductor Devices

With the continuous shrinking of feature size, developing advanced patterning methods to fabricate semiconductor devices with high resolution and high volume is increasingly critical. To address this challenge, extreme ultraviolet (EUV) lithography has emerged as a leading candidate for sub-7 nm manufacturing, offering significant potential to meet the semiconductor industry's current requirements. This chapter introduces the EUV lithography system used in semiconductor device processing and describes its working mechanisms. Additionally, recent progresses in EUV photoresists and masks are summarized. Finally, the challenges and future directions for EUV lithography are discussed.

3.1 Introduction

The rapid development of the semiconductor industry necessitates advanced fabrication technologies to meet the challenges presented by Moore's law (Figure 3.1) [1]. This law states that the number of transistors per integrated circuit doubles approximately every two years. Over past decades, this projection has consistently guided the semiconductor industry. Presently, EUV lithography is widely regarded as one of the most cornerstone technologies for sustaining the sub-10 nm era and enabling the continued advancement of Moore's law.

Lithography is a patterning technology used to fabricate desired micro/nanoscale patterns on substrates by irradiating photosensitive or electron-sensitive materials with ultraviolet light (photolithography) or electron/ion beams (e-beam lithography or focused ion beam [FIB] lithography), respectively [2]. The schematic of a typical photolithographic process (Figure 3.2), primarily comprises spin-coating, exposure, development, etching, and liftoff steps.

To date, five generations of lithography equipment have been developed (Table 3.1), each distinguished by its unique light source. The first-generation lithography equipment, introduced in the 1980s, employed a mercury g-line blue-light source at 436 nm and achieved critical dimensions of 1 μm. Subsequently, the second-generation equipment utilized a mercury i-line blue-light source at 365 nm, enabling feature sizes of 220 nm. During the mid-1980s, deep ultraviolet (DUV) excimer lasers emerged as novel light sources. The third-generation equipment

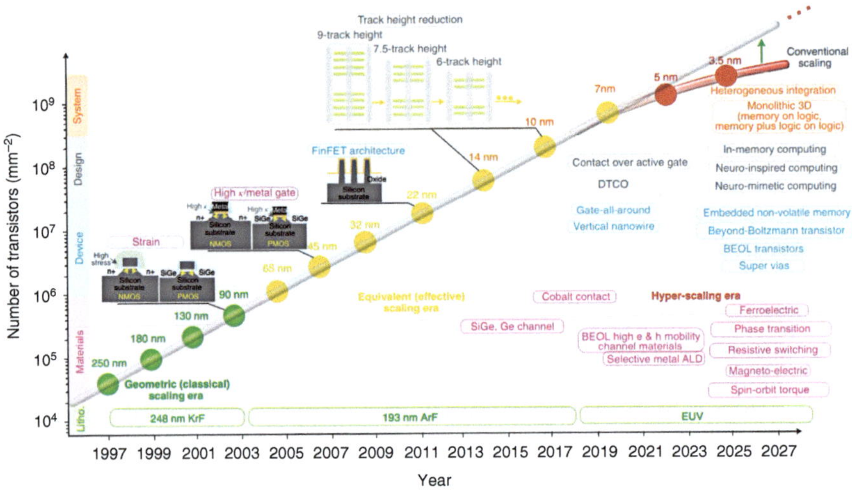

Figure 3.1 Moore's law. *Source:* [1] / with permission of Springer Nature.

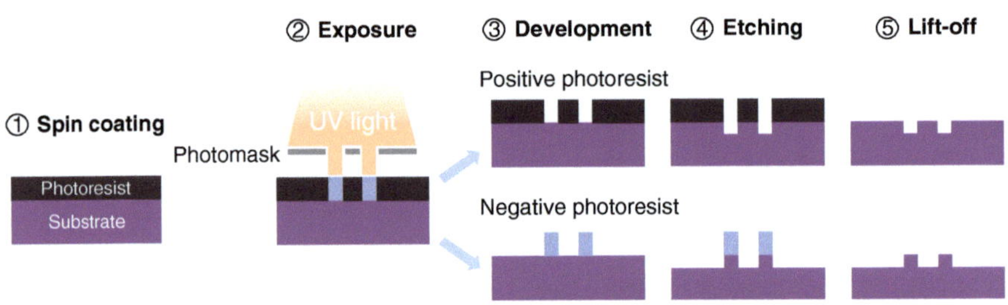

Figure 3.2 Schematic diagram of photolithography process, including spin-coating, exposure, development, etching, and liftoff.

Table 3.1 The five generations of lithography equipment.

Generation	Light source	Wavelength (nm)	Lithography system
1	Mercury arc lamp (g-line)	436	Contact/proximity aligner
2	Mercury arc lamp (i-line)	365	Contact/proximity aligner
3	Excimer laser (KrF)	248	Scanning projection aligner
4	Excimer laser (ArF)	193	Step-and-projection aligner
5	Laser-produced plasmon (EUV)	13.5	EUV system

adopted krypton fluoride (KrF) laser technology with a wavelength of 248 nm, which reduced feature sizes to 80 nm. This was followed by the fourth-generation equipment, which utilized an argon fluoride (ArF) excimer laser at 193 nm and realized feature sizes of 38 nm. More recently, the fifth-generation equipment employs laser-produced plasma (LPP) as its light source to generate EUV radiation at 13.5 nm, with the feature size anticipated to attain sub-7 nm dimensions.

Lithography techniques, particularly EUV lithography, represent pivotal solutions for scaling feature sizes to sub-7 nm nodes by overcoming diffraction limitations inherent in conventional optical systems. This advancement enables precise pattern transfer through innovative light sources and resist materials, facilitating the fabrication of high-density interconnects essential for next-generation integrated circuits.

Research into EUV lithography commenced in the 1980s, with industrial development efforts initiated in 1994. Numerous research institutions and companies have contributed to its advancement. Among these, Advanced Semiconductor Materials Lithography Holding N.V. (ASML) has played a leading role. In 2006, ASML developed the world's first EUV lithography demonstration tool. Subsequently, in 2010, the company manufactured the first pre-production EUV system, the TWINSCAN NXE:3100. By 2013, ASML had produced the first EUV production system, the TWINSCAN NXE:3300. Recently, ASML announced that its latest-generation system, the TWINSCAN NXE:3600D, supports volume EUV production at the 5 and 3 nm logic nodes, alongside leading-edge dynamic random-access memory (DRAM) nodes.

Despite rapid progress, EUV lithography faces substantial challenges and technical obstacles requiring near-term resolution. For example, resolution is severely constrained by resist and mask limitations, highlighting an urgent need for superior materials possessing high sensitivity, enhanced resolution, and low line-edge roughness (LER), in conjunction with novel mask designs and materials. Ongoing innovations in mask technology and computational lithography further refine resolution enhancement techniques, underpinning efforts to achieve atomic-scale precision while addressing power consumption and thermal management constraints in nanoscale devices.

3.2 EUV Lithography System and Working Principles

3.2.1 EUV Light Source and Exposure System

The realization of EUV lithography necessitates the integration of an EUV light source within the lithography system. Initially, synchrotron radiation sources were employed to generate the required EUV radiation. However, given their substantial spatial requirements and high operational costs, tin-based discharge-produced plasma (DPP) and LPP sources have been adopted as alternative EUV generators. Schematic diagrams of DPP and LPP light sources are presented in Figure 3.3 [3]. Tin (Sn) and xenon (Xe) represent the most promising target materials for these

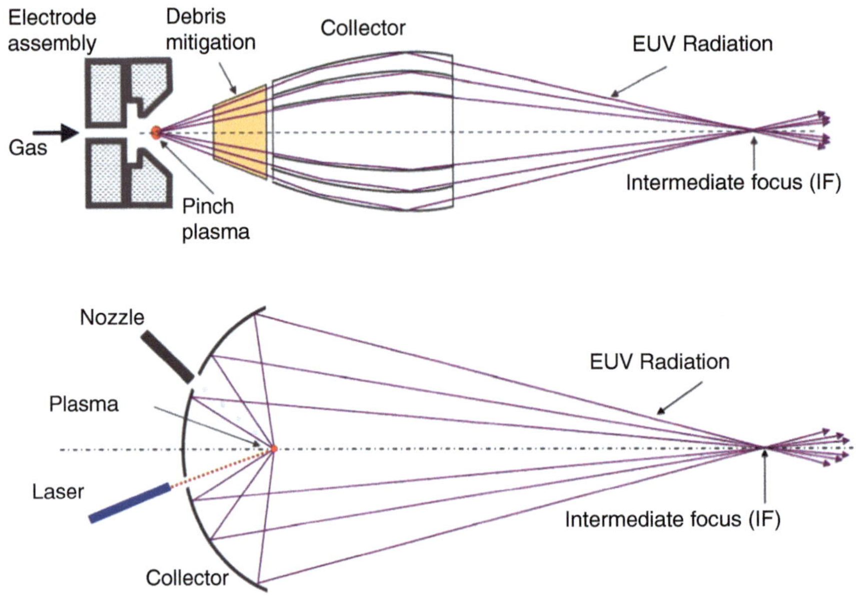

Figure 3.3 Schematic diagrams of DPP (top) and LPP (down) sources. *Source:* [3] / with permission of AIP Publishing.

plasma sources. Nevertheless, the significantly stronger emission intensity of tin compared to xenon has led to its predominant utilization in contemporary EUV light sources.

The DPP light source exhibits high conversion efficiency; however, its collection efficiency is low due to the limited solid angle over which EUV radiation can be collected. Additionally, issues related to debris mitigation and a short operational lifetime restrict its further application. In the LPP light source, molten Sn droplets (~25 μm diameter) are ejected from a generator at ~70 m/s. Subsequent low-intensity CO_2 laser pulses strike these droplets, flattening them into planar discs. A subsequent, high-intensity laser pulse then vaporizes the flattened droplets, generating plasmas that emit EUV light. Consequently, LPP has become the leading candidate for EUV light source technology.

Figure 3.4 illustrates the schematic of the EUV lithography exposure system [4]. Typically, 13.5 nm EUV light is first generated by the LPP source. The light is then collected and focused by the collector before entering the illuminator. Subsequently, the light illuminates the reticle stage (i.e., mask), with the reflected light directed into the projection optics. Finally, the projection optics focuses the light onto the wafer stage to fabricate the designed patterns on the substrate. This entire exposure process occurs within a low-hydrocarbon, high-vacuum environment.

3.2.2 EUV Lithography Equipment

The EUV lithography machine represents one of the most sophisticated technologies within the semiconductor industry, enabling the fabrication of sub-7 nm process

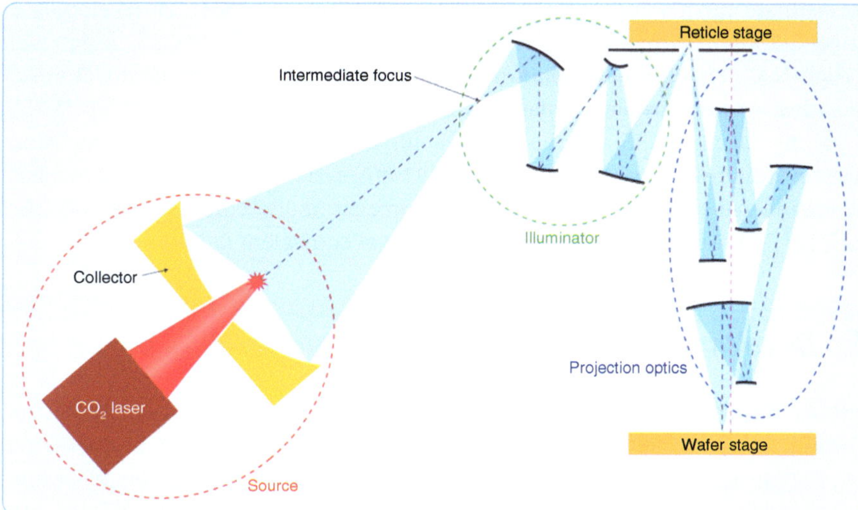

Figure 3.4 Schematic diagram of the EUV lithography exposure system. Wagner and Harned [4].

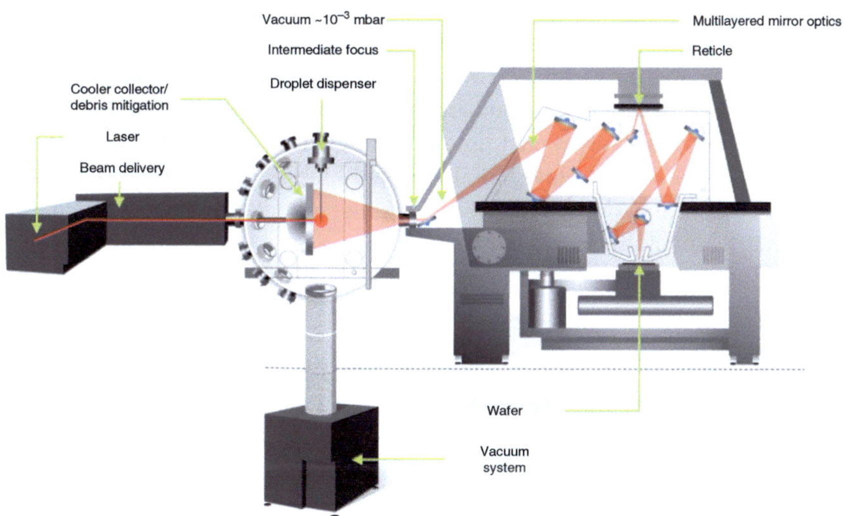

Figure 3.5 Schematic diagram of EUV lithography machine. *Source:* [5] / with permission of SPIE.

node chips, as illustrated in Figure 3.5 [5]. To ensure maximum efficiency by minimizing EUV light absorption, the system is maintained at high vacuum. In 2013, ASML introduced the world's first production EUV system, the TWINSCAN NXE:3300. Subsequently, the TWINSCAN NXE:3600D emerged as ASML's most advanced EUV lithography machine, capable of producing logic nodes at 5 and 3 nm, as well as leading-edge DRAM nodes, achieving a resolution of 13 nm with a numerical aperture (NA) of 0.33.

To facilitate further innovation in chip manufacturing, ASML is developing a next-generation EUV platform that increases the NA from 0.33 to 0.55, designated as "High-NA". This high-NA platform, also known as EXE, incorporates a novel optical design and substantially faster wafer and reticle stage speeds. The 0.55 NA lithography machine offers enhanced resolution, enabling a critical dimension below 8 nm half-pitch, compared to the 0.33 NA system. The EXE platform has been engineered to support multiple future process nodes, commencing with the 2 nm logic node and followed by memory nodes of equivalent density.

3.3 Development of Resist for EUV Lithography

Resist is indispensable for EUV lithography to transfer patterns onto a chip. Resists are broadly categorized into positive-tone and negative-tone types. In positive-tone resist, regions exposed to light become soluble in the photoresist developer, while unexposed regions remain insoluble. Conversely, in negative-tone resist, exposed regions become insoluble to the developer, whereas unexposed regions are dissolved by it. The mask design primarily dictates the selection of resist tone.

To fabricate high-quality patterns, resists must meet demanding criteria, including high transparency, high etching resistance, high resolution, high sensitivity, high environmental stability, low absorbance, low exposure dose, low gas production, low LER, low line width roughness (LWR), and excellent reproducibility. Notably, a fundamental interdependency exists among resolution (R), LER/LWR (L), and sensitivity (D), forming a significant limitation known as the RLS trade-off triangle. Achieving low sensitivity, low LER/LWR, and high resolution simultaneously is extremely challenging. Consequently, optimizing resist performance necessitates careful management of this RLS relationship.

To simplify resist evaluation, the Z-factor is introduced, where the Z value is defined as follows [6]:

$$Z = R^3 \times L^2 \times \mathrm{D} \tag{3.1}$$

where R represents resolution (nm), L is LWR or LER (nm), and D is dose or sensitivity (mJ/cm^2). Generally, a smaller Z value indicates better overall resist performance. However, it is critical to note that the Z-factor does not account for the imaging performance of the exposure tool or illumination quality. Therefore, the Z-factor is only suitable for comparing lithography results obtained using the same or highly similar exposure setups. Despite advancements in EUV resist technology, the synthesis of novel resists continues to confront critical challenges, including resolution limitations, LWR control, photospeed optimization, pattern collapse resistance, outgassing mitigation, tolerance to out-of-band radiation, defect minimization, and stringent quality control requirements.

3.3.1 Polymer-Based Chemically Amplified Resist

Polymer-based chemically amplified resists (CARs) are a type of resist system employed for EUV lithography. These resists typically comprise a backbone matrix polymer, a photoacid generator (PAG), and a dissolution inhibitor. During light exposure, the PAG produces acid, which subsequently diffuses into the matrix polymer. A post-exposure bake (PEB) then induces the acid to catalyze the deprotection of the polymer's pendant groups, altering the resist's solubility in the developer. Importantly, the photogenerated acid is not consumed during the deprotection reaction and can repeatedly act as a catalyst. Based on the interconnection between the PAG and the polymer matrix, polymer-based CARs are categorized as either PAG-blend type or PAG-bound type. The PAG-bound type CAR often exhibits superior resolution and sensitivity compared to the blend type, making it widely used for EUV resists in sub-10 nm patterning applications.

Polymer-based CARs have been extensively investigated for EUV lithography [7–10]. For instance, Rao et al. fabricated 70 nm line/space (L/S) patterns using a PAG-bound type CAR for EUV exposure [11]. Ongayi et al. incorporated novel EUV-absorbing chromophores into the polymer structure, achieving a resolution of 17 nm L/S patterns with a dose sensitivity of 14.5 mJ/cm^2 [12]. Fujii et al. reported that adding di-*p*-tolyl sulfone enhanced acid-generation efficiency in the CAR, leading to increased sensitivity and reduced LER [13]. Kruger et al. demonstrated that incorporating acid amplifiers into the EUV resist formulation improved resolution, LER, and sensitivity [14].

Due to the relatively small EUV absorbance cross sections of carbon and oxygen, conventional polymer-based CARs typically exhibit low photon absorption. To overcome this limitation, nonmetallic elements possessing high EUV absorption cross sections can be introduced into the polymer to increase the resist's absorption coefficient. Yamamoto et al. observed that incorporating fluorine atoms into poly(4-hydroxystyrene) (PHS) significantly increased the backbone polymer's absorption coefficient [15], consequently enhancing both acid generation and resist sensitivity. Furthermore, Christianson et al. synthesized a backbone-fluorinated polymer via the free radical polymerization of 3,3,3-trifluoromethacrylate and vinyl ethers [16]. Their findings indicated that fluorinated polymers not only exhibited increased EUV absorption but also demonstrated considerable potential for high-sensitivity EUV lithography.

3.3.2 Molecular Glass Resist

As feature sizes continue to diminish, researchers have sought to reduce molecular resist dimensions to enhance resolution and minimize LWR. Consequently, molecular glass (MG) resists have been developed for EUV lithography [17–22]. As illustrated in Figure 3.6, an MG resist typically comprises three components: an MG core moiety, acidic pendant groups, and protecting groups [21]. The MG core material necessitates a high glass transition temperature (T_g) to ensure stable thin-film formation for patterning applications. Simultaneously, high etch resistance and optical transparency at relevant wavelengths constitute critical criteria for MG core selection. Acidic groups within the MG resist must facilitate hydrogen bonding, demonstrate robust adhesion to the core framework, and exhibit sufficient solubility in development solvents. Protecting groups modulate resist solubility while modulating sensitivity, polarity, dimensional characteristics, and etch resistance properties.

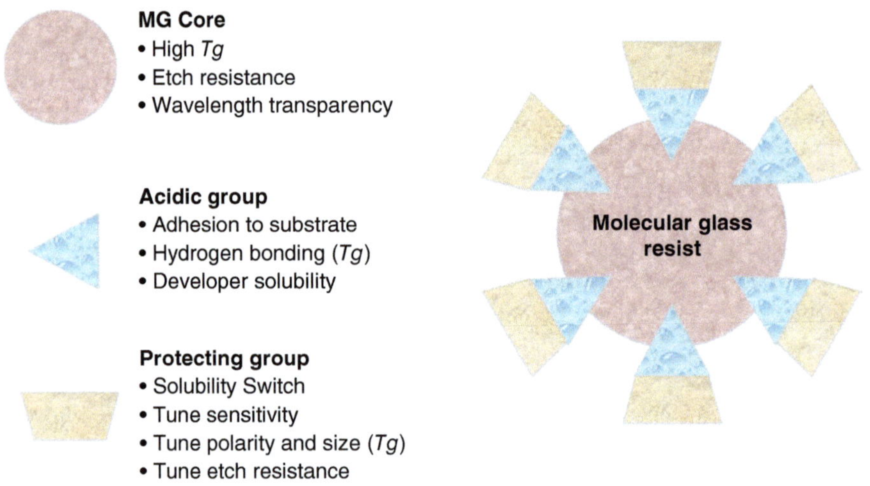

Figure 3.6 Model of modular molecular glass resist. *Source:* [21] / with permission of John Wiley & Sons.

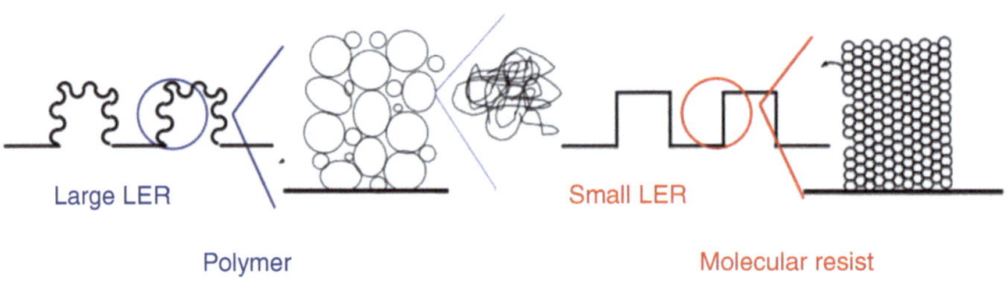

Figure 3.7 Comparison of LER between polymer-based resist and molecular glass–based resist. Adapted from [2].

Compared to CAR, MG resists are synthesized from monodisperse molecular building blocks possessing inherently smaller dimensions. This molecular architecture facilitates higher-resolution patterning with reduced LWR/LER, as demonstrated in Figure 3.7 [2]. Furthermore, the constrained free volume inherent to MG resists effectively restricts photogenerated acid diffusion, thereby enhancing local critical dimension uniformity.

Based on structural topology, MG resists can be classified as branched MGs or ring-structured MGs. Branched polyphenolic compounds synthesized from PHS derivatives serve as promising MG candidates for EUV lithography, typically exhibiting excellent glass transition temperatures (T_g), high etch resistance, and favorable solubility. Figure 3.8 illustrates three distinct branched phenolic MG resists capable of patterning sub-50 nm features via EUV lithography [21]. Ring-structured MG resists similarly demonstrate desirable characteristics including high T_g and thermal stability. Calix[*n*]arenes and calix[4]resorcinarenes are archetypal ring MG resist materials. For instance, tert-butoxycarbonyl (*t*-BOC)-protected calix[4]resorcinarene has been shown to resolve sub-50 nm patterns using EUV lithography [23]. De Silva and Ober reported the synthesis of a chemically

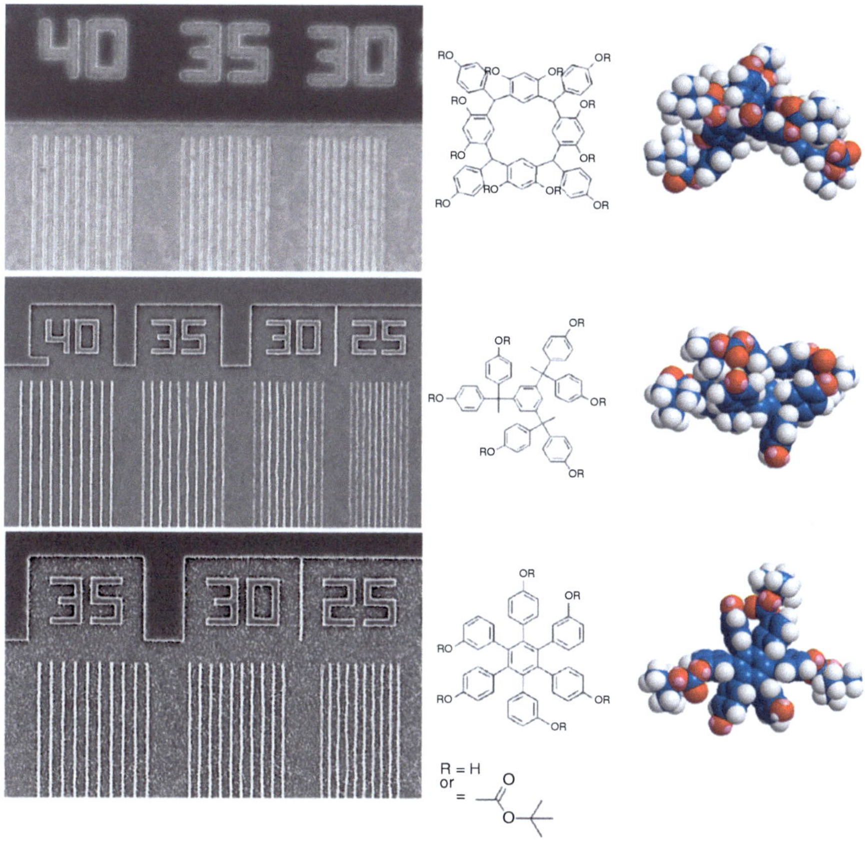

Figure 3.8 High-resolution EUV patterning results with three different types of molecular glass cores. Adapted from [21].

amplifiable derivative, *c*-4-hydroxyphenyl-calix[4]resorcinarene MG resist [22]. This resist successfully achieved sub-30 nm patterning resolution. Building upon the calix[4]resorcinarene scaffold, a double-cyclic oligomer featuring ladder-like linkages between the rings are synthesized, exhibiting a waterwheel-like configuration (Figure 3.9) [24]. This novel molecule, termed "Noria" (MD resist), demonstrates considerable potential as a ring-MG resist capable of resolving features below 30 nm using EUV lithography.

3.3.3 Metal-Based Nanoparticles Resist

Metal-based nanoparticle resists have gained prominence for EUV lithography due to the dual function of metal elements: they act as sensitizers for EUV photon absorption and enhance interaction efficiency between the resist material and secondary electrons generated by EUV radiation [25–27]. Compared to organic resists, metal-based alternatives exhibit superior EUV absorption, enhanced etching resistance, and greater thermostability. As demonstrated by the atomic absorption

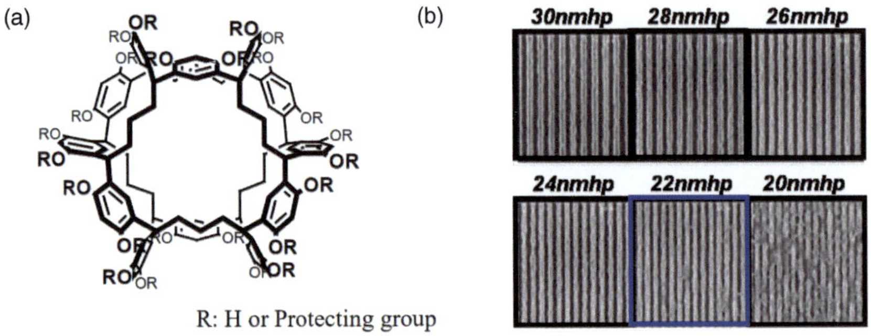

Figure 3.9 (a) Structure of protected Noria resist. (b) EUV exposure result of Noria MD resist. Shioy et al. [24] / with permission of The Society of Photopolymer Science and Technology (SPST).

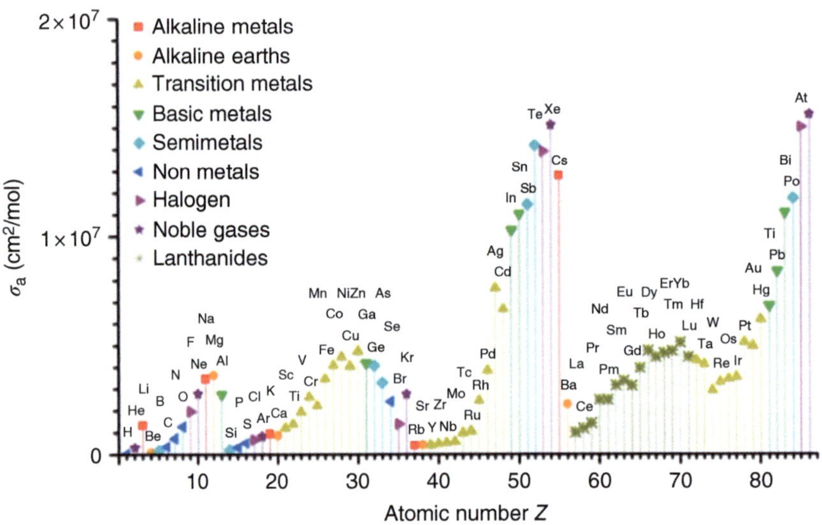

Figure 3.10 Atomic absorption cross section of EUV light source versus atomic number Z from 1 to 86. Fallica et al. [28] / SPIE / CC BY.

cross section versus atomic number (Z = 1–86) in Figure 3.10 [28], the primary metal elements incorporated into these nanoparticle resists include hafnium (Hf), zirconium (Zr), zinc (Zn), tin (Sn), bismuth (Bi), antimony (Sb), platinum (Pt), palladium (Pd), and cobalt (Co) [29–31].

For instance, the Ober and Giannelis group at Cornell University developed EUV resists based on hafnium oxide (HfO_2) and zirconium oxide (ZrO_2) core nanoparticles functionalized with photosensitive organic ligand shells (Figure 3.11) [32]. The metal-oxide core provides high mechanical strength and etch resistance, while simultaneously enhancing EUV absorption to supply sufficient energy for activating photochemical reactions and preventing pattern deformation induced by absorption of excess photons. The organic ligand shell facilitates dispersion in solvent and undergoes a solubility switch upon EUV exposure.

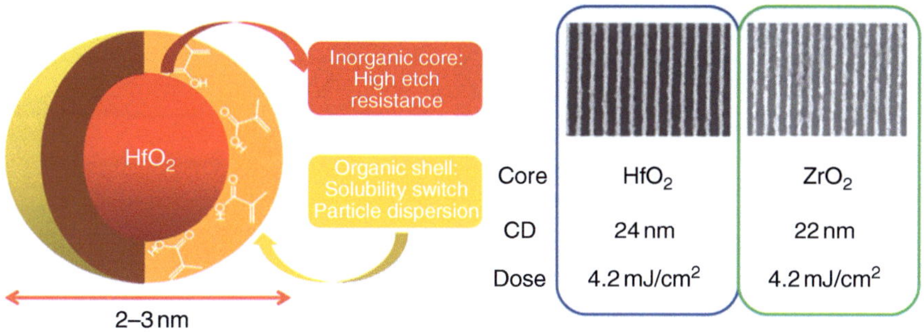

Figure 3.11 Schematic diagram of HfO_2-based EUV resist, accompanied by SEM images of patterns fabricated using HfO_2- and ZrO_2-based EUV resists. *Source:* [32] / with permission of The Royal Society of Chemistry.

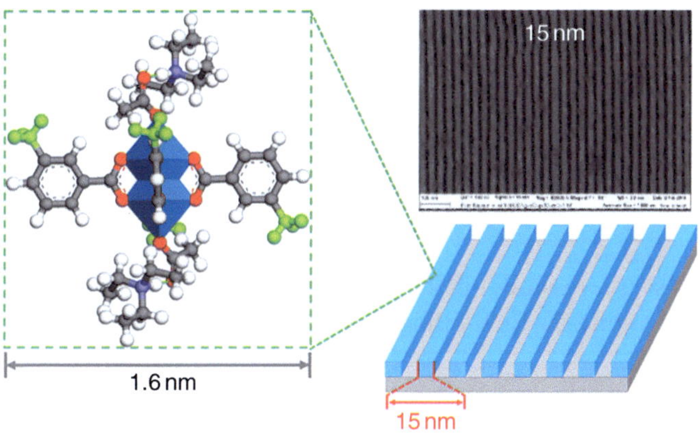

Figure 3.12 Molecular structure of Zn-mTA formed by $Zn_2(CO_2)_4$ modified with a 3-methyl-phenyl group. Schematic diagram and scanning electron microscope (SEM) image of 15 nm line-and-space dense patterns fabricated with Zn-mTA resist. *Source:* [32] / with permission of The Royal Society of Chemistry.

Utilizing trans-2,3-dimethylacrylic acid (DMA) as the ligand, HfO_2- and ZrO_2-based resists achieved 20 nm line/space patterns at doses of 2.4 and 1.6 mJ/cm^2, respectively. Additionally, incorporating photoactive compounds improved resist sensitivity and patterning resolution by modulating the photochemical pathways.

Inspired by the metal–organic framework 2 (MOF-2) structure, Xu et al. synthesized a 3-methylphenyl-modified $Zn_2(CO_2)_4$ resist (denoted Zn-mTA) (Figure 3.12) [33]. Resist Zn-mTA resolved 16, 15, 14, and 13 nm dense 1:1 line/space patterns at EUV doses of 45, 47, 36, and 35 mJ/cm^2, respectively. Its reduced particle size and enhanced structural uniformity significantly improved LER at high resolutions compared to ZrO_2- and HfO_2-based resists.

Cardineau et al. investigated tin-oxo clusters with the formula $[(RSn)_{12}O_{14}(OH)_6]X_2$ as EUV resist materials, where R represents organic ligands and X denotes carboxylic counter-anions (Figure 3.13) [34]. Their study revealed that resist sensitivity depends on molecular weight and organic-ligand bond energy, indicating that photolysis of the ligands mediates the patterning mechanism. Consequently, the

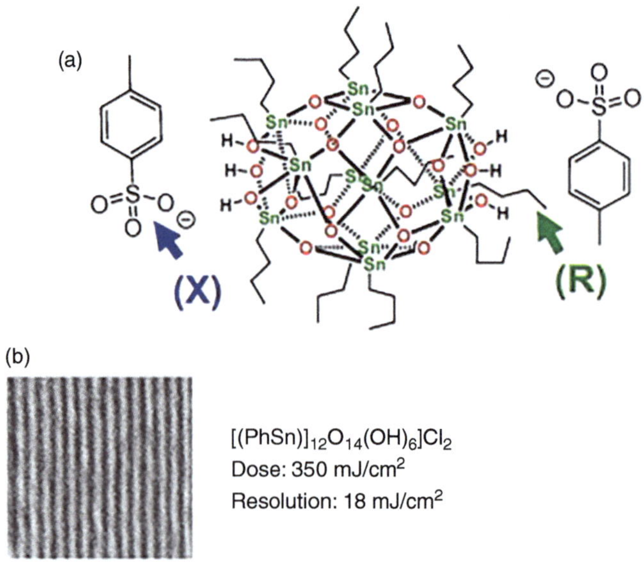

Figure 3.13 (a) Molecular structure of $[(RSn)_{12}O_{14}(OH)_6]X_2$ resist. (b) SEM image of dense patterns fabricated using the $[(RSn)_{12}O_{14}(OH)_6]X_2$ resist. *Source:* [34] / with permission of Elsevier.

$[(RSn)_{12}O_{14}(OH)_6]X_2$ resist successfully patterned 18 nm dense-line features, demonstrating significant promise for EUV lithography applications.

3.4 Mask Materials and Designs for EUV Lithography

To date, the fabrication of EUV masks has emerged as one of the most urgent issues for EUV lithography. Since EUV light is strongly absorbed by most materials, the design of an EUV mask differs significantly from that of a conventional optical photomask. As shown in Figure 3.14 [35], an EUV mask typically consists of a conductive layer, a substrate, a multilayer (ML) mirror layer, a capping layer, and an absorber layer. The ML mirror layer comprises multiple alternating layers with

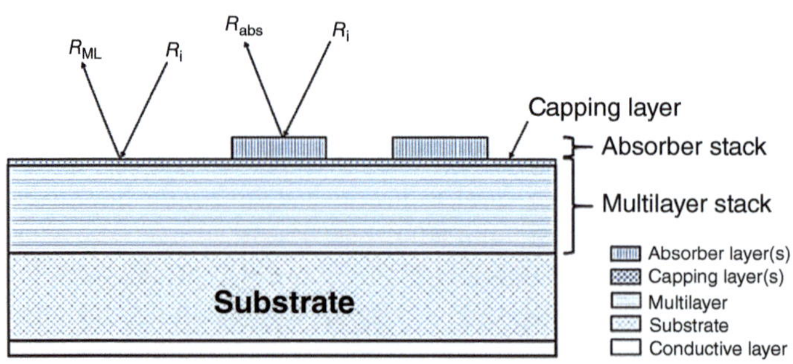

Figure 3.14 Structure of the patterned EUV mask. The image contrast is obtained by the reflectivity difference between the ML mirror layer and the absorber layer. *Source:* [35] / with permission of Springer Nature.

distinct optical constants and serves the vital function of reflecting EUV light. To achieve optical contrast for imaging, an absorber layer with high EUV absorptivity is patterned atop the ML mirror layer. The resulting image contrast depends on the difference in reflectivity between the ML mirror layer and the absorber layer, given by

$$\text{Image contrast} = \frac{\mathrm{R_{ML}} - \mathrm{R_{abs}}}{\mathrm{R_{ML}} + \mathrm{R_{abs}}} \times 100\% \tag{3.2}$$

where R_{ML} and R_{abs} represent the reflectivities of the ML mirror layer and the absorber layer, respectively. Note that the term "ML mirror layer" in the figure and equation encompasses both the ML stack and the capping layer.

3.4.1 EUV Mask Fabrication Process

EUV mask fabrication involves two primary steps: mask blank fabrication and mask patterning, illustrated in Figure 3.15 [35]. Fabrication begins with a defect-free substrate of low-thermal-expansion material (LTEM). This substrate is carefully polished and then coated with the ML mirror and protective capping layers. Ru is the most common capping layer material due to its effectiveness in preventing mask degradation. Subsequently, the absorber layer is deposited. Since defect inspection occurs using DUV wavelengths and metallic absorbers typically exhibit high reflectivity in this range, an antireflective coating (ARC) is applied to the absorber layer to achieve sufficient optical contrast during DUV inspection. Finally, a conductive layer is added to the substrate's backside to function as an electrostatic chuck. The patterning process comprises electron-beam writing, dry etching, and defect inspection and repair. Imperfections inherent in the substrate or introduced during

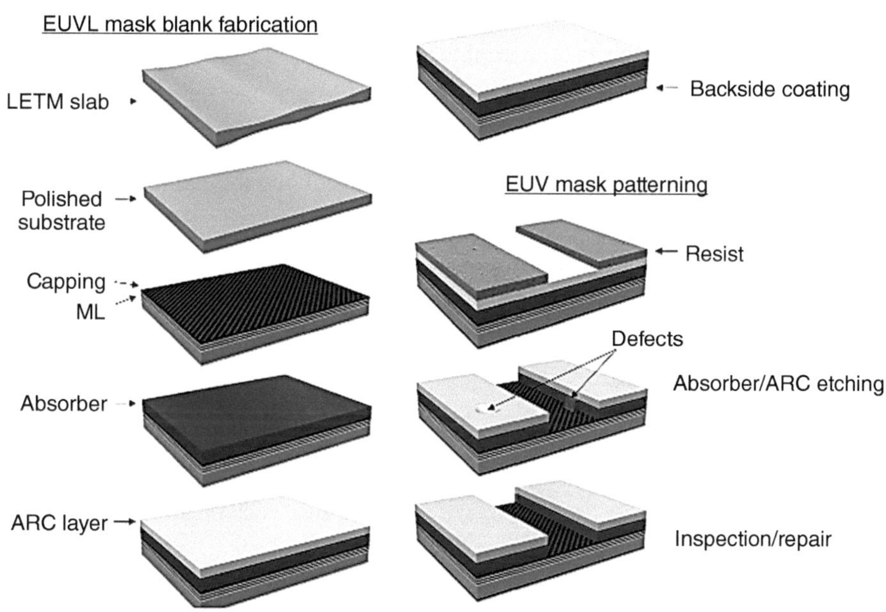

Figure 3.15 Schematic diagram of EUV mask fabrication process flow. *Source:* [35] / with permission of Springer Nature.

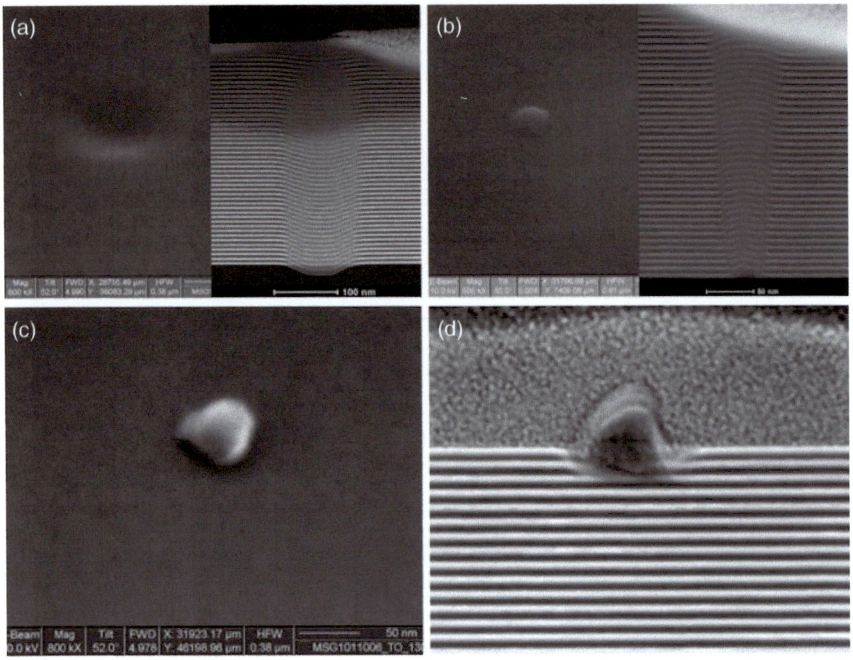

Figure 3.16 SEM images of EUV lithography mask blank defect. (a) A substrate pit-generated phase defect. (b) A substrate bump-generated phase defect. (c) A surface particle defect. (d) An embedded particle defect. *Source:* [36] / with permission of Optica Publishing Group.

fabrication can result in defects on the EUV mask, as depicted in Figure 3.16 [36]. Pre-exposure detection and repair of these defects are critically important, as their presence significantly impacts lithographic results.

3.4.2 Mask Substrate

The selection of an appropriate mask substrate is crucial for fabricating high-quality EUV masks. As shown in Figure 3.17, the SEMI P37-1102 standard specifies

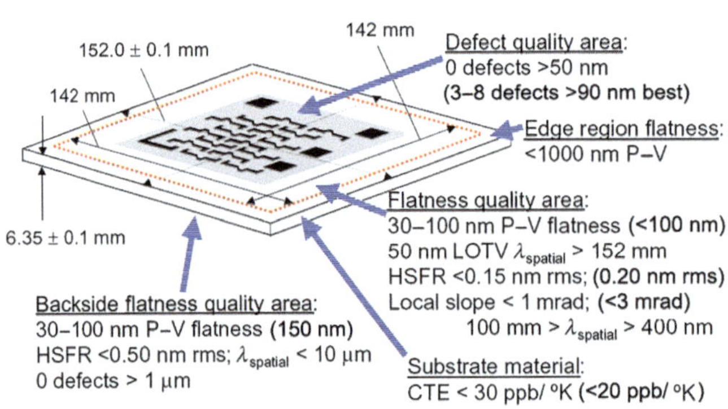

Figure 3.17 SEMI P37-1102 substrate requirements. *Source:* [37] / with permission of SPIE.

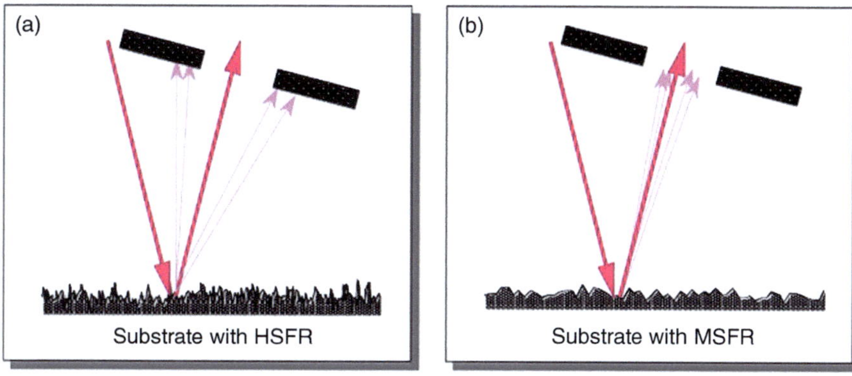

Figure 3.18 Schematic diagrams of substrates with (a) HSFR and (b) MSFR. *Source:* [38] / with permission of SPIE.

requirements for substrate size, material, flatness, defect density, surface roughness, and other properties [37].

Since temperature variations during exposure can induce mask distortion and cause image placement errors on the wafer, the substrate material must exhibit low thermal expansion. Moreover, the substrate distortion is influenced by the average coefficient of thermal expansion (CTE) of this material. Consequently, the SEMI P37-1102 standard requires the CTE to be controlled within specific ranges, typically from 0 ± 5 to 0 ± 30 ppb/K. The EUV reflectivity of the substrate is also significantly affected by its surface roughness, particularly the high-spatial-frequency roughness (HSFR) [38]. As presented in Figure 3.18a, substrates with HSFR cause large-angle scattering, leading to loss of EUV light from the projection lens. Conversely, Figure 3.18b illustrates how middle-spatial-frequency roughness (MSFR) induces wavefront errors and speckle through small-angle scattering. To ensure high-quality substrates, the surface roughness frequency should be constrained to specific bands: MSFR should fall within approximately 10^{-6}/nm $< f <$ 0.004/nm, and HSFR within approximately 0.004/nm $< f <$ 0.02/nm. According to SEMI P37-1102, the HSFR should be less than 0.15 nm rms, and the local slope angle of the front surface must be smaller than 1 mrad.

The non-telecentric nature of EUV lithography illumination systems further necessitates strict substrate flatness control. Non-flatness of the patterned mask surface, as shown in Figure 3.19 [39], contributes to image placement errors and overlay errors on the wafer [40, 41]. This relationship is quantified by the equation $\Delta x = \Delta z \times \tan(\theta)/M$, where M is the reduction ratio, and θ is the illumination angle (approximately 6°). Under this condition, the wafer image placement error (Δx) is approximately 0.026 times the mask non-flatness (Δz).

3.4.3 ML Mirror Layer

The deposition of the ML mirror layer is one of the greatest challenges in fabricating an EUV mask. Since defects in the ML mirror layer cause printable defects on the

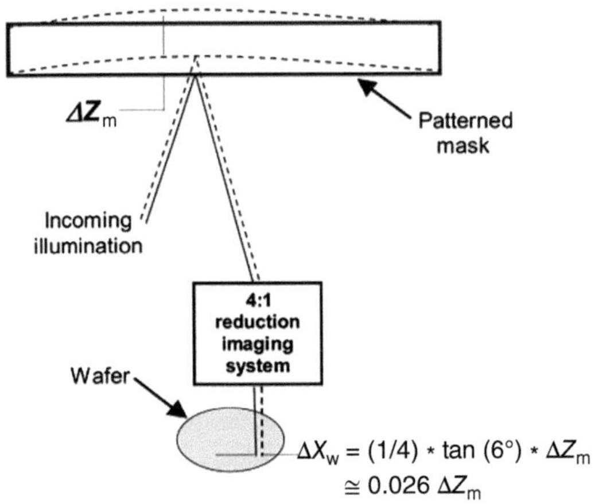

Figure 3.19 Image placement error of the wafer resulting from the non-flatness of mask in the non-telecentric illumination system. *Source:* [39] / with permission of SPIE.

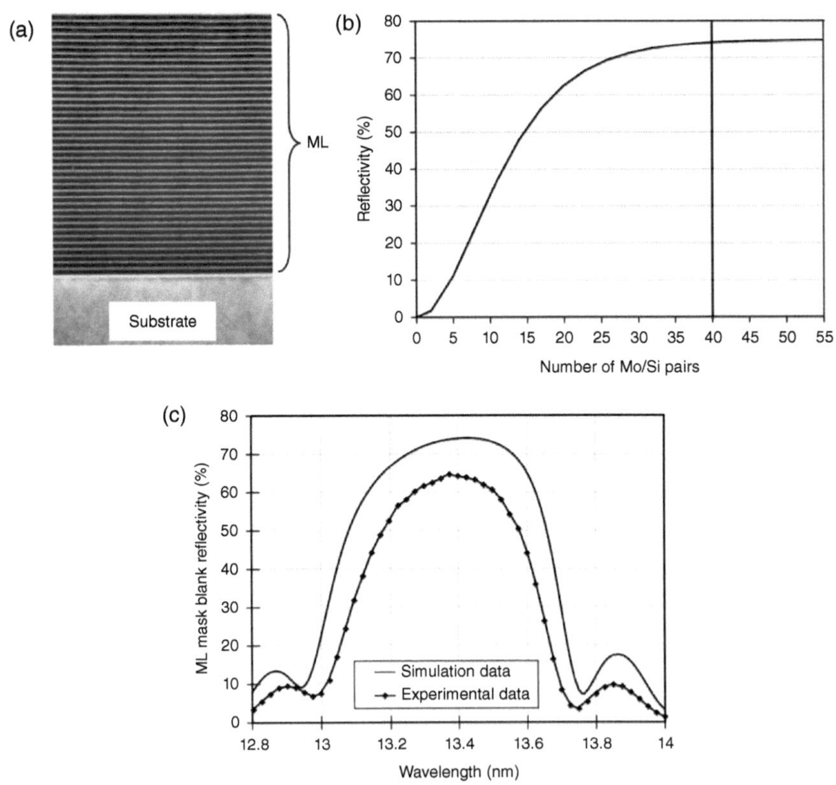

Figure 3.20 (a) Transmission electron microscope (TEM) image of the cross section of the Mo/Si ML mirror layer. (b) Calculated peak reflectivity as a function of the number of Mo/Si pairs. (c) Wavelength-dependent reflectance spectra comparing theoretical and experimental results. Adapted from [42].

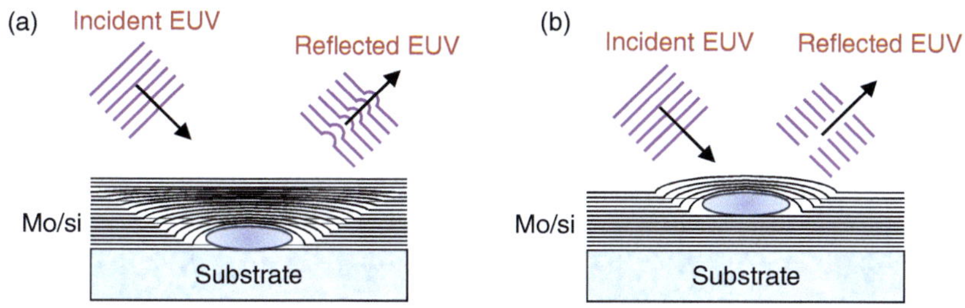

Figure 3.21 Schematic diagrams of the ML mirror layer with (a) a phase defect and (b) an amplitude defect.

wafer, the deposition process must produce essentially defect-free layers. This layer normally consists of multiple alternating material layers with distinct refractive indices, resulting in high reflectance for EUV illumination. Alternating layers of Mo and Si are one of the most commonly used ML mirror configurations for EUV masks [42–44], as shown in Figure 3.20a. To attain maximum reflectance at the EUV wavelength (λ), the period (d) of the alternating layer should satisfy Bragg's law [39]:

$$n\lambda = 2d\cos\theta\sqrt{1 - \frac{2\delta}{1 - \cos^2\theta}} \tag{3.3}$$

where n is an integer, θ is the incident angle to the mask, and δ is the bilayer-weighted δn. Typically, δn is defined as $1-n$, where n is the real part of the refractive index of the material. Optimization shows that the calculated peak reflectivity of the ML mirror layer reaches a maximum of ~75% when composed of 40 pairs of alternating Mo/Si bilayers (Figure 3.20b). The thickness of the Mo layer is ~2.8 nm and the Si layer is ~4.1 nm. Due to interfacial intermixing of Mo and Si, the measured reflectance is slightly lower than the theoretical value (Figure 3.20c). The measured peak reflectance at the EUV wavelength (13.5 nm) is approximately 70%.

Defects readily form in the ML mirror layer due to substrate defects (bumps, pits, or particles) or the introduction of particles during deposition [45]. ML mirror-layer defects are typically classified into two types: phase defects and amplitude defects. As shown in Figure 3.21, phase defects appear as protrusions or depressions on the ML surface, while amplitude defects consist of particles near the top surface or flaws within the layer. Consequently, phase defects disrupt the constructive interference of reflections from the Mo/Si interfaces, whereas amplitude defects directly absorb EUV radiation.

3.4.4 Absorber Layer

In contrast to the ML mirror layer with high reflectivity, the absorber layer must possess high absorption for EUV light. Additionally, eligible absorber layers should exhibit good stability, conductivity, etch selectivity, and DUV reflectivity to enable surface defect inspection. To date, a variety of materials have emerged as absorber layers, including titanium (Ti), tantalum (Ta), chromium (Cr), nickel (Ni), and their compounds (such as TiN, TaSiN, TaSi, TaN, and TaBN) [46–50]. Among these materials, Ta-based compounds have been demonstrated as promising candidates

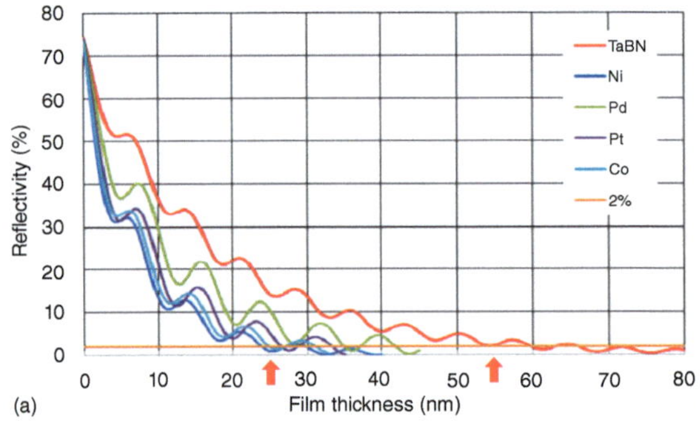

(b)

Film	R (%)	T (nm)
TaBN	2.10	54.2
Ni	0.94	25.4
Pd	1.00	36.0
Pt	1.10	27.6
Co	1.26	26.0

Figure 3.22 (a) Dependence of calculated net reflectivity on the film thickness of TaBN, Ni, Pd, Pt, and Co. The ML mirror layer consists of 40 alternating Mo/Si bilayers and the capping layer is Ru. (b) Thicknesses (T) of absorber layers where net reflectivities (R) fall below 2%. *Source:* [56] / with permission of SPIE.

for the absorber layer in EUV mask fabrication [51, 52]. Investigations into using transition metals such as Ni and Pd have also been conducted recently [53–55]. Compared to the popular absorber material TaBN, these transition metal-based absorber layers exhibit a higher extinction coefficient at EUV wavelengths, a smaller required thickness, and lower EUV light reflectivity (Figure 3.22) [56]. Since thinner absorber layers reduce the mask shadowing effect, they enable patterning with smaller dimensional deviations and facilitate high-resolution lithography. Note that the mask shadowing effect, caused by the surface topography of the patterned absorber layer combined with oblique illumination (Figure 3.23) [57], results in a printed space width slightly smaller and a printed line width slightly larger than the designed parameters. Furthermore, this detrimental effect amplifies as the absorber layer thickness increases.

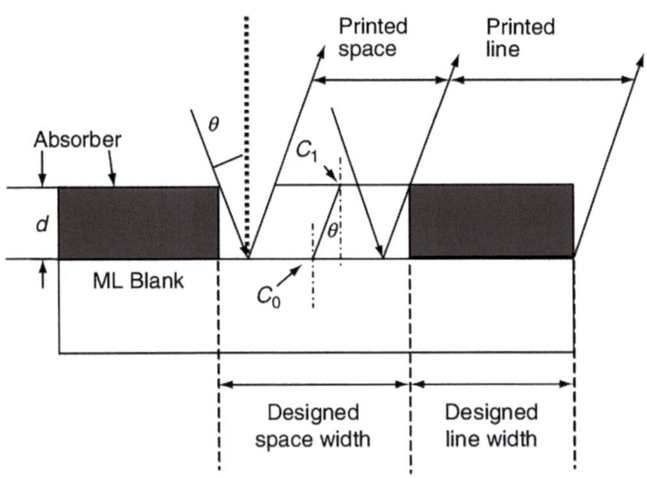

Figure 3.23 Geometrical optics illustration of the EUV mask shadowing effect. *Source:* [57] / with permission of SPIE.

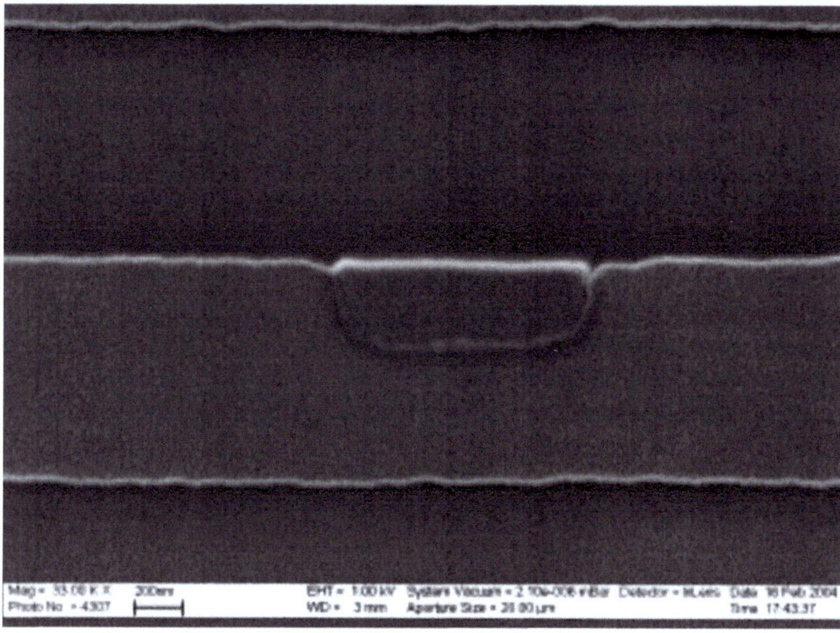

Figure 3.24 Example of e-beam deposition repair for mask defects. A mouse-bite-type clear defect is repaired on quartz mask. After e-beam deposition, defects are repaired via the deposited layer, preventing their transfer to the wafer. *Source:* [61] / with permission of AIP Publishing.

Since defects are sometimes generated on the patterned absorber layer during the electron-beam lithography process, it is necessary to repair them promptly. Several techniques exist to repair these defects, including electron-beam (e-beam), FIB, and scanning-probe-based mechanical nano-machining [58–60]. Figure 3.24 shows scanning electron microscope (SEM) images of an EUV mask after e-beam deposition repair illustrating a repaired mouse-bite-type defect [61].

3.5 Conclusions and Perspectives

Since EUV lithography offers a more affordable scaling path for chipmakers and enables the continuation of Moore's Law, it has been widely adopted in the semiconductor industry. Over the past decades, significant improvements have been achieved in the EUV source, exposure system, resist, and mask, driving the technology toward high-volume manufacturing. ASML reports that its latest-generation lithography system, the TWINSCAN NXE:3600D, supports volume production at the 5 and 3 nm logic nodes and leading-edge DRAM nodes. Despite these advances, the future development of EUV lithography presents both challenges and opportunities, summarized as follows.

Achieving EUV light source with higher power and longer lifetime: The light source remains a major challenge for EUV lithography. High source power is fundamentally required for high-volume manufacturing. To meet this need, the LPP source employs master oscillator power amplifier (MOPA) and pre-pulse technologies, with modified configurations demonstrating substantial power increases (~200 W) at the intermediate focus. Another critical challenge is contamination of collector optics by Sn debris, which significantly shortens source and collector lifetimes. This issue is mitigated using debris management systems. Collector longevity can be further increased by enhanced collector protection techniques and *in situ* cleaning methods, which show significant potential.

Developing high-quality EUV resist: Accelerating EUV lithography advancement necessitates high-quality resists featuring high sensitivity, high resolution, and low LER, even at low exposure power. Suitable resists must also overcome the RLS trade-off and address homogeneity, etch resistance, resist outgassing, high-volume manufacturability, and shelf life. Current progress indicates metal-based resists hold significant potential for sub-7 nm patterning. However, inventing novel resists combining all required properties remains crucial for advancing resolution limits.

Advancing mask defect inspection and repair techniques: Mask defects significantly impact EUV light transmittance during exposure, leading to patterning errors; consequently, improving defect inspection and repair techniques is paramount. Defects are categorized as phase or amplitude defects. Amplitude defects can be detected using optical or electron-beam inspection tools. In contrast, phase defects require expensive actinic inspection tools, which incorporate a stable EUV source, specialized optics, and signal capture/processing systems. Phase defects typically reside on the mask substrate surface or within the lower ML structure, distorting the overlaying layers and locally altering reflectivity [2]. To date, no successful phase defect repair method has been reported. Conversely, amplitude defects are located on the ML top surface and are repairable using techniques like gallium-FIB (Ga-FIB), mechanical milling, or electron-beam-assisted chemical processes. The critical size for amplitude defects is typically smaller than 80% of the technology node feature size.

Mitigating pattern collapse: Pattern collapse poses a significant challenge for high-resolution patterned resists. As collapse causes structural defects on the wafer, minimizing the number of collapsed patterns is critical. Proposed mitigation strategies include developing mechanically robust resists less susceptible to the capillary forces dominant during wet processing. Another effective approach involves optimizing resist film thickness, as thinner films reduce collapse risks during development and rinsing.

Mitigating the mask shadowing effect: Reducing the mask shadowing effect is essential to minimize its detrimental impact on printed feature size and resolution. This effect arises from the interaction between oblique EUV illumination and the mask absorber topography, resulting in printed spaces smaller than designed and lines wider than designed. Furthermore, it leads to horizontal–vertical critical dimension bias and resolution limitations. Shadowing intensifies with increased absorber thickness. However, using a very thin absorber layer degrades imaging quality by reducing mask contrast. To overcome this dilemma, strategies utilizing a thinner absorber with a high extinction coefficient or employing phase-shifting materials are effective. For example, Philipsen et al. demonstrated that absorber materials like pure nickel or cobalt reduce shadowing effects [54]. They also concluded that the optimal path forward involves developing metal alloys or compounds possessing high extinction coefficients.

References

1 Salahuddin, S., Ni, K., and Datta, S. (2018). *Nat Electron.* 1: 442–450.

2 Traub, M.C., Longsine, W., and Truskett, V.N. (2016). *Annu. Rev. Chem. Biomol. Eng.* 7: 583–604.

3 Wu, B. and Kumar, A. (2014). *Appl. Phys. Rev.* 1: 011104.

4 Wagner, C. and Harned, N. (2010). *Nat. Photonics* 4: 24–26.

5 Abhari, R.S., Rollinger, B., Giovannini, A.Z. et al. (2012). *J. Micro/Nanolithogr. MEMS MOEMS* 11: 021114.

6 Higgins, C.D., Szmanda, C.R., Antohe, A. et al. (2011). *Jpn. J. Appl. Phys.* 50: 036504.

7 Wang, M., Yueh, W., and Gonsalves, K.E. (2007). *Macromolecules* 40: 8220–8224.

8 Thackeray, J.W., Jain, V.J., Coley, S. et al. (2011). *J. Photopolym. Sci. Technol.* 24: 179–183.

9 Wang, M., Gonsalves, K.E., Rabinovich, M. et al. (2007). *J. Mater. Chem.* 17: 1699–1706.

10 Thiyagarajan, M., Dean, K., and Consalves, K.E. (2005). *J. Photopolym. Sci. Technol.* 18: 737–741.

11 Rao, V., Cobb, J.L., Henderson, C.C. et al. (1999). *Proc. SPIE* 3676: 615–626.

12 Ongayi, O., Christianson, M., Meyer, M. et al. (2012). *Proc. SPIE* 8322: 242–253.

13 Fujii, S., Okamoto, K., Yamamoto, H. et al. (2017). *Jpn. J. Appl. Phys.* 56: 06GD01.

14 Kruger, S., Revuru, S., Higgins, C. et al. (2009). *J. Am. Chem. Soc.* 131: 9862–9863.

15 Yamamoto, H., Kozawa, T., Tagawa, S. et al. (2008). *Appl. Phys. Express* 1: 047001.

16 Christianson, M.D., Meyer, M.M., Ongayi, O. et al. (2013). *Proc. SPIE* 8682: 306–316.

17 Chang, S.W., Ayothi, R., Bratton, D. et al. (2006). *J. Mater. Chem.* 16: 1470–1474.

18 Wang, Y., Chen, J., Zeng, Y. et al. (2022). *ACS Omega* 7: 29266–29273.

19 Chen, J., Hao, Q., Wang, S. et al. (2019). *ACS Appl. Polym. Mater.* 1: 526–534.

20 Green, D.P., Jain, V., Bailey, B. et al. (2013). *Proc. SPIE* 8679: 269–281.

21 De Silva, A., Felix, N.M., and Ober, C.K. (2008). *Adv. Mater.* 20: 3355–3361.

22 De Silva, A. and Ober, C.K. (2008). *J. Mater. Chem.* 18: 1903–1910.

23 Dai, J., Chang, S.W., Hamad, A. et al. (2006). *Chem. Mater.* 18: 3404–3411.

24 Shioya, T., Maruyama, K., and Kimura, T. (2011). *J. Photopolym. Sci. Technol.* 24: 217–222.

25 De Simone, D., Sayan, S., Dei, S. et al. (2016). *Proc. SPIE* 9776: 31–39.
26 De Simone, D., Mao, M., Lazzarino, F. et al. (2016). *J. Photopolym. Sci. Technol.* 29: 501–507.
27 Toriumi, M., Sato, Y., Koshino, M. et al. (2016). *Appl. Phys. Express* 9: 031601.
28 Fallica, R., Stowers, J.K., Grenville, A. et al. (2016). *J. Micro/Nanolithogr. MEMS MOEMS* 15: 033506.
29 Passarelli, J., Sortland, M., Del Re, R. et al. (2014). *J. Photopolym. Sci. Technol.* 27: 655–661.
30 Sortland, M., Hotalen, J., Del Re, R. et al. (2015). *J. Micro/Nanolithogr. MEMS MOEMS* 14: 043511.
31 Anderson, G.K., Lumetta, G.J., and Siria, J.W. (1992). *J. Organomet. Chem.* 434: 253–259.
32 Li, L., Liu, X., Pal, S. et al. (2017). *Chem. Soc. Rev.* 46: 4855–4866.
33 Xu, H., Sakai, K., Kasahara, K. et al. (2018). *Chem. Mater.* 30: 4124–4133.
34 Cardineau, B., Del Re, R., Marnell, M. et al. (2014). *Microelectron. Eng.* 127: 44–50.
35 Kim, J.S. and Ahn, J. (2018). *Electron. Mater. Lett.* 14: 533–547.
36 Goldstein, M. and Naulleau, P. (2012). *Optics Express* 20: 15752.
37 Hector, S.D. (2002). *Proc. SPIE* 4688: 134–149.
38 Hector, S.D. (2004). Standards for EUV Masks. Proc. EUV Mask Workshop, SEMATECH, Miyazaki.
39 Blaedel, K.L., Taylor, J.S., Hector, S.D. et al. (2002). *Proc. SPIE* 4688: 767–778.
40 Ballman, K., Lee, C., Dunn, T. et al. (2016). *Proc. SPIE* 9984: 188–199.
41 Turley, C., Rankin, J., Chen, X. et al. (2017). *Proc. SPIE* 10450: 104500A.
42 Rizvi, S. (ed.) (2005). Handbook of Photomask Manufacturing Technology. In: *Chapter 11. Masks for Extreme Ultraviolet Lithography*, 1ste. CRC Press. https://doi.org/10.1201/9781420028782.
43 Smith, B.W., Venkataraman, P., Kurinec, S.K. et al. (1998). *Proc. SPIE* 3331: 544–554.
44 Slaughter, J.M., Schulze, D.W., Hills, C.R. et al. (1994). *J. Appl. Phys.* 76: 2144–2156.
45 Barty, A., Mirkarimi, P.B., Stearns, D.G. et al. (2002). *Proc. SPIE* 4688: 385–394.
46 Yan, P., Zhang, G., Kofron, P. et al. (2000). *Proc. SPIE* 4066: 116–123.
47 Niibe, M.N.M., Watanabe, T.W.T., Nii, H.N.H. et al. (2000). *Jpn. J. Appl. Phys.* 39: 6815.
48 Philipsen, V., Luong, K.V., Souriau, L. et al. (2017). *Proc. SPIE* 10450: 104500G.
49 Mangat, P.J.S., Hector, S.D., Rose, S. et al. (2000). *Proc. SPIE* 3997: 76–82.
50 Hoshino, E., Ogawa, T., Hirano, N. et al. (2001). *Proc. SPIE* 4186: 749–755.
51 Yan, P., Zhang, G., Ma, A. et al. (2001). *Proc. SPIE* 4343: 409–414.
52 Green, M., Choi, Y., Ham, Y. et al. (2017). *Proc. SPIE* 10450: 18–27.
53 Lee, J.U., Hong, S., and Ahn, J. (2013). *Appl. Phys. Express* 6: 076502.
54 Philipsen, V., Luong, K.V., Souriau, L. et al. (2017). *Proc. SPIE* 10143: 174–188.
55 Burkhardt, M. (2017). *Proc. SPIE* 10143: 195–208.
56 Wood, O. II, Raghunathan, S., Mangat, P. et al. (2015). *Proc. SPIE* 9422: 122–133.
57 Yan, P. (2002). *Proc. SPIE* 4688: 150–160.

58 Liang, T., Stivers, A., Livengood, R. et al. (2000). *J. Vac. Sci. Technol. B* 18: 3216–3220.
59 Lee, S.Y., Kim, G.B., Sim, H.S. et al. (2008). *Proc. SPIE* 7122: 813–822.
60 Jonckheere, R., Bret, T., Van den Heuvel, D. et al. (2011). *Proc. SPIE* 8166: 388–398.
61 Edinger, K., Becht, H., Bihr, J. et al. (2004). *J. Vac. Sci. Technol. B* 22: 2902.

4
Transistors

4.1 Introduction

Transistor, derived from "trans-resistor", is a three-terminal device wherein the resistance between two terminals is controlled by the third terminal. Among semiconductor devices, the transistor represents one of the most significant innovations, originating from research conducted by a team at Bell Laboratories in 1947. Its introduction marked a watershed moment for the electronics industry and solid-state research. Before 1947, semiconductors were primarily utilized as thermistors, photodiodes, and rectifiers, all of which were two-terminal devices. In 1947, Bardeen and Brattain announced groundbreaking experimental observations concerning the point-contact transistor [1]. The following year witnessed the publication of Shockley's seminal paper elaborating the theory of junction diodes and transistors. The principle of minority-carrier injection in PN junctions underpinned the subsequent development of the junction transistor, with the first successful demonstration of a bipolar junction transistor (BJT) occurring in 1950 [2].

Since their inception, transistor theories have undergone substantial development to model high-frequency, high-power, and switching behaviors. Numerous breakthroughs have subsequently occurred in transistor manufacturing technology, particularly in crystal growth, epitaxy, diffusion, ion implantation, lithography, dry etching, surface passivation, planarization, and multilevel metallization [3]. These advancements have significantly enhanced transistor power handling, operational frequency capabilities, and reliability. This chapter provides knowledge on various transistor structures, principles governing current amplification, voltage–ampere characteristic curves, and their applications within electronic systems.

4.2 Bipolar Transistors

4.2.1 Introduction

Bipolar transistors, also referred to as BJTs, represent a milestone innovation originating from Bell Laboratories in 1947. This pivotal invention revolutionized electronic systems, aiming to supplant the cumbersome vacuum tubes prevalent in early-1950s data-processing systems [4]. With their compact size and efficient operation, BJTs rapidly gained prominence, heralding a new era of compact, reliable, and versatile electronic devices.

The advent of BJTs catalyzed a fundamental transformation in electronics, culminating in their large-scale commercialization during the 1960s. This era witnessed the introduction of the first television and radio components utilizing solid-state mechanisms into consumer markets, thereby replacing devices reliant on vacuum tubes. BJTs were engineered not merely as replacements for vacuum tubes, but also to provide significant advantages: lower power dissipation, enhanced reliability through reduced system failure rates, and a substantially reduced form factor [4]. Subsequently, BJTs have become essential components within a diverse array of electronic equipment, including both discrete circuits and integrated circuits (ICs) [5]. Their applications encompass, but are not limited to, wide-bandwidth systems, high-frequency circuits, and operations requiring substantial voltage gain [6]. Fabricated primarily from semiconductor materials, silicon crystals doped with specific impurities constitute the most prevalent type of BJTs globally.

4.2.2 Device Structures and Working Principles

BJTs comprise two back-to-back PN junctions within a three-layer semiconductor structure, manifesting in two configurations: NPN and PNP. Figure 4.1 presents their schematic diagrams and circuit symbols. Each BJT possesses three terminals designated as the emitter (E), base (B), and collector (C). The arrow incorporated within the circuit symbol consistently indicates the direction of conventional current flow between the emitter and base terminals. Specifically, for both transistor types, this arrow signifies current flow from the *p*-type region to the *n*-type region, analogous to the directional marking in a standard diode symbol.

A thorough understanding of BJT operation requires prior knowledge of PN junction principles. When an *n*-type semiconductor contacts a *p*-type semiconductor, typically possessing identical chemical composition but opposite doping

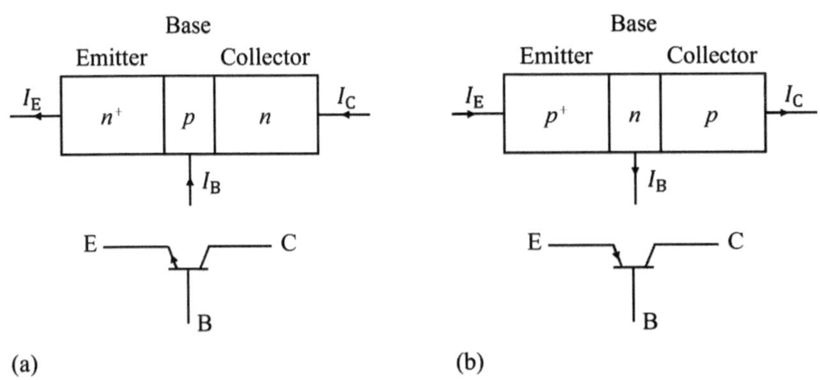

Figure 4.1 Schematic diagrams of (a) NPN and (b) PNP BJTs.

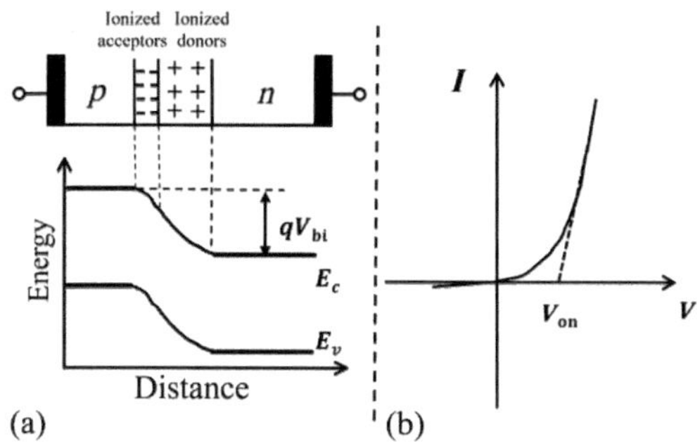

Figure 4.2 (a) A schematic illustration of a typical structure and the band diagram of a PN junction. (b) Qualitative current–voltage characteristics of a PN junction diode.

types, a PN junction forms. In *n*-type semiconductors, electrons constitute the majority charge carriers; in *p*-type semiconductors, holes serve as the majority carriers. Upon formation of an ideal PN junction, holes diffuse from the *p*-type region to the *n*-type region due to concentration gradients, while electrons diffuse from the *n*-type region to the *p*-type region. This carrier diffusion leaves immobile, negatively charged acceptor ions near the junction within the *p*-type material and immobile, positively charged donor ions near the junction within the *n*-type material, establishing an electric field and a space-charge region (depletion region).

Figure 4.2a illustrates the resulting charge distribution in the PN junction along with the potential energy profiles for electrons (*Ec*) and holes (*Ev*) as functions of position. It also depicts the charge densities associated with the ionized acceptors and donors. These fixed charges form the depletion region, characterized by near-zero concentrations of free electrons in the *n*-type region and holes in the *p*-type region. This region creates a potential barrier that opposes further electron diffusion into the *p*-region and hole diffusion into the *n*-region, resulting in zero net carrier flux under thermal equilibrium conditions.

The built-in potential (V_{bi}) in Figure 4.2a typically ranges between 0.5 E_g/q and E_g/q, where q denotes the elementary charge and E_g represents the bandgap energy. Applying an external voltage across the junction modulates this potential barrier. Forward bias (a positive voltage applied to the *p*-region relative to the *n*-region) injects holes from the *p*-region into the *n*-region and electrons from the *n*-region into the *p*-region, reducing the barrier height reduction to $V_{bi} - |V|$. Conversely, reverse bias (a negative voltage applied to the *p*-region relative to the *n*-region) sweeps carriers away from the junction, increasing the barrier height to $V_{bi} + |V|$. Consequently, as depicted in Figure 4.2b, the diode exhibits highly asymmetric resistance characteristics. This asymmetry defines three key operational regions: high resistance under reverse bias; exponential current growth in forward bias near or below the turn-on voltage (V_{on}); and, at high forward-bias voltages, current limitation dominated by the parasitic series resistance.

Figure 4.3a illustrates the structure of a fabricated silicon-planar NPN BJT. The base region, composed of *p*-type material, is sandwiched between two *n*-type layers representing the emitter and collector. The NPN BJT is typically biased according to Figure 4.3b, wherein the base–emitter (BE) PN junction is forward-biased, and the base–collector (BC) junction is reverse-biased. This configuration enables

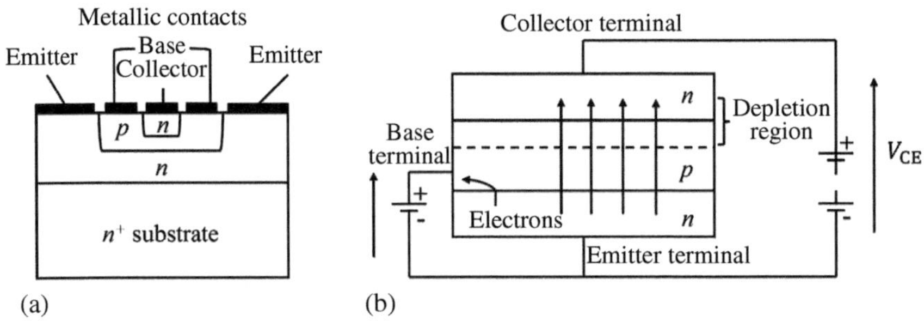

Figure 4.3 Schematic diagrams of a silicon-planar BJT. (a) Cross-section and (b) enlarged view of the active area.

hole flow from base to emitter and electron flow from emitter to base. Due to the emitter's heavy doping relative to the base, most charge-carrier injections are electrons from the emitter to the base. Under positive collector bias, an electric field gradient establishes across the reverse-biased collector–base junction. This widens the depletion region, which normally conducts negligible current. When electrons are injected into the base from the emitter, the collector–base electric field gradient attracts them toward the collector, with the majority reaching the collector terminal. This constitutes the primary current drawn from the V_{CE} supply. Conversely, the significantly smaller hole current from base to emitter creates a minor base terminal current. The current gain β is defined as the ratio of collector current to the base current:

$$\beta = \frac{I_c}{I_B} \tag{4.1}$$

Crucially, the BJT design makes I_C highly sensitive to changes in I_B, meaning small variations in base current induce significantly larger variations in collector current. For most general-purpose transistors, β typically ranges between 20 and 200.

Simultaneously, the collector-to-emitter current gain (α) is defined by

$$\alpha = \frac{I_c}{I_E} \tag{4.2}$$

Kirchhoff's current law dictates that the total current flowing into the transistor must equal the current flowing out. Since I_E equals the sum of I_B and the considerably larger I_c, α is nearly unity (typically 0.950–0.999 for low-power signal transistors). The relationships between α and β are expressed by the following equations:

$$\alpha = \frac{\beta}{\beta + 1} \tag{4.3}$$

$$\beta = \frac{\alpha}{\alpha - 1} \tag{4.4}$$

For high-current power transistors, β ranges from ~20 to >1000, with manufacturer datasheets typically specifying 50–200 for standard NPN devices. This demonstrates a fundamental BJT property: minimal base current controls substantially larger collector current. BJTs, often characterized as current-regulating devices, control output current proportionally to base current, functioning as current-controlled switches. The fundamental operating principle remains identical for PNP and NPN BJTs, differing only in power supply polarity and biasing configurations.

4.2.3 Biasing Modes and Circuit Configurations

4.2.3.1 Biasing Modes

The operation of a BJT is governed by the biasing conditions of its junctions, resulting in four distinct operational modes. The NPN BJT will be used as an example to elucidate these biasing modes. The biasing conditions for PNP BJT will be similar, except that the voltage polarities are reversed.

Forward-Active Mode The emitter–base junction is forward-biased, while the collector–base junction is reverse-biased. This configuration enables efficient collector current flow relative to the base current, thereby establishing this mode as predominant for amplification applications.

Reverse-Active Mode The emitter–base junction is reverse-biased, and the collector–base junction is forward-biased. Current flow direction is inverted relative to the forward-active mode. Due to the typically heavier doping concentration in the emitter compared with the collector, hole injection into the collector in reverse-active mode substantially exceeds hole injection into the emitter in forward-active mode, resulting in significantly diminished current gain compared with the forward-active mode.

Saturation Mode Both emitter–base and collector–base junctions are forward-biased, permitting substantial current flow between emitter and collector. However, electron injection originating from the collector opposes the direction of emitter injection. Consequently, the resultant collector current is lower than observed under forward-active biasing.

Cutoff Mode Both junctions are reverse-biased, effectively obstructing current flow through the device. Transistor currents in this state are negligible.

These four modes characterize standard BJT operation. Significantly, saturation mode exhibits less efficient current transfer than forward-active mode due to opposing injection mechanisms, while reverse-active mode yields substantially lower current gain owing to different base-to-collector doping asymmetry. To achieve high current gain, the BJT is most typically biased in the forward-active mode.

4.2.3.2 Circuit Configurations

BJTs comprise three terminals, supporting three fundamental circuit configurations. In each configuration, one terminal is common to both input and output ports. These configurations include the common-base configuration (CBC),

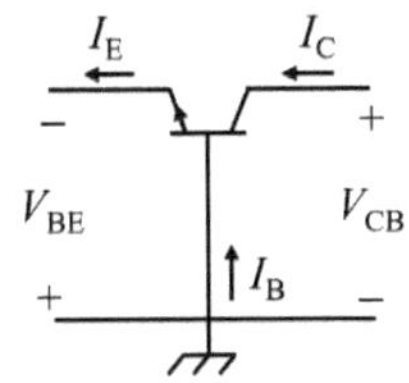

Figure 4.4 A schematic illustration of CBC.

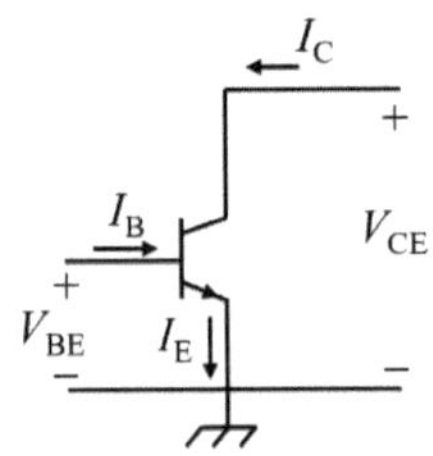

Figure 4.5 A schematic illustration of CEC.

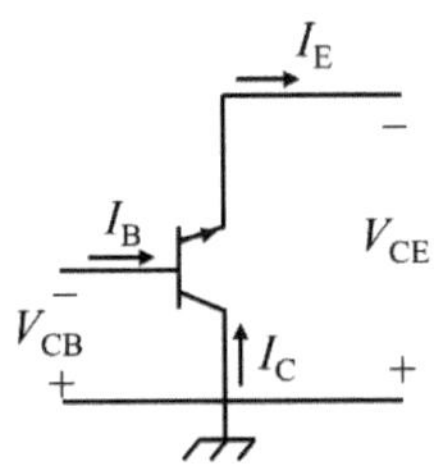

Figure 4.6 A schematic illustration of CCC.

providing voltage gain without current gain; the common-emitter configuration (CEC), delivering both voltage and current gain; and the common-collector configuration (CCC), exhibiting current gain without voltage gain. The distinct inherent characteristics of each configuration yield unique signal response behaviors. Corresponding circuitry is illustrated in Figures 4.4–4.6 using NPN BJT exemplars, with detailed analysis provided.

Common-Base Configuration (CBC) The CBC employs the base terminal as the common terminal for both input and output signals. As illustrated in Figure 4.4, with the base grounded or at a fixed reference potential, the input signal is applied between the emitter and base, while the output is obtained by voltage between the collector and base. Within this configuration, the emitter current equals the sum of the collector and base currents. Since the collector current is much larger than the base current, there is almost no current gain. According to the voltage polarity notation in Figure 4.4, the CBC operates as a non-inverting voltage amplifier, exhibiting in-phase input (V_{in}) and output (V_{out}) voltages. Although less prevalent with no current gain, it is characterized by high voltage gain. Additionally, a substantial resistance gain is achieved, characterized by a high ratio of output resistance (load resistance, R_L) to input resistance (R_{in}). The input characteristics are analogous to those of a forward-biased diode, while output characteristics of the CBC resemble those of an illuminated photodiode.

Due to its excellent high-frequency performance, the common-base circuit is extensively utilized in single-stage amplifiers, particularly within radio-frequency (RF) amplifiers and microphone preamplifiers.

Common-Emitter Configuration (CEC) In the CEC, as depicted in Figure 4.5, the input signal is applied between the base and emitter, while the output signal is developed across the collector and emitter terminals.

This configuration represents the standard method for connecting a bipolar transistor and is one of the most common configurations used for transistor-based amplifiers. It offers the highest power gain and current gain among the three fundamental transistor configurations. This high gain arises because the input terminal connects to a forward-biased PN junction, resulting in low input impedance, while the output is sourced from a reverse-biased PN junction, leading to high output impedance.

Compared to the CBC, the CEC offers higher input impedance, higher current gain, and higher power gain. As an inverting amplifier circuit, the CEC produces an output voltage signal that is 180° out of phase with the input voltage signal.

Common-Collector Configuration (CCC) In the CCC illustrated in Figure 4.6, the collector serves as the common terminal to the supply voltage. The output signal is obtained across the emitter load, while the input signal is applied directly to the base. Often referred to as an emitter-follower or voltage-follower, this configuration is ideal for impedance-matching applications, as it provides an extremely high input impedance (typically hundreds of kilohms) and a very low output impedance. In the CCC, the load resistance (R_L) is in series with the emitter; therefore, the emitter current equals the load current. Since the emitter current is the sum of the base current and collector current, the current gain of the circuit can be calculated mathematically, as follows:

$$A_i = \frac{I_E}{I_B} = \frac{I_C + I_B}{I_B} = \beta + 1 \tag{4.5}$$

This CCC configuration exhibits a non-inverting characteristic, where input and output voltages are in phase. The voltage gain is always less than unity, typically approaching one, providing substantial current amplification due to its high current gain, despite minimal voltage gain.

4.2.4 NPN and PNP BJTs

Having understood the fundamental configurations of BJTs, we now examine the two principal types: NPN and PNP.

The NPN transistor comprises two n-type–doped regions separated by a thin p-type–doped semiconductor layer, the base, resembling two diodes sharing a common anode region. Figure 4.7 depicts the schematic device structure and symbol of the NPN BJT.

In the standard forward-active operating mode for NPN BJTs, the BC junction is reverse-biased while the emitter–base junction is forward-biased. Applying a

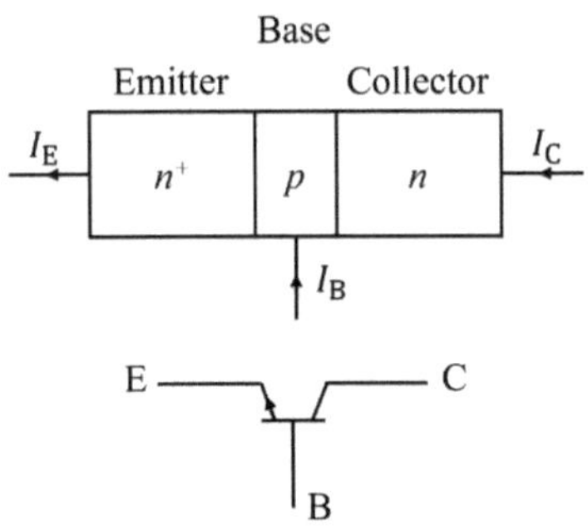

Figure 4.7 Schematic device structure and symbol of NPN BJT.

positive voltage across the BE junction disrupts the equilibrium between thermally generated carriers and the repulsive electric field within the depletion region. Consequently, thermally excited electrons diffuse from the high-concentration zone near the emitter into the base region, migrating toward the low-concentration region adjacent to the collector. These minority-carrier electrons traverse the *p*-type–doped base, where holes are the majority carriers. For correct transistor function, the collector terminal must sustain a positive voltage relative to both the base and emitter terminals. The collector connects to the supply voltage through a load resistor, which limits the maximum circuit current. Simultaneously, the base connects via a base resistor to confine the maximum base current.

Operated as a current-controlled device, the transistor conducts substantial collector current only when a smaller bias current flows into the base terminal, permitting base current regulation. To initiate the base current flow, the base voltage must be approximately 0.3 V higher than the emitter for germanium devices or 0.7 V higher for silicon devices. Within an NPN BJT operating in the active mode, the base current originates from three primary mechanisms: (i) The dominant contributor involves the recombination of electrons, injected from the forward-biased emitter, within the quasi-neutral base region. To preserve charge neutrality, the base terminal supplies holes to replace those lost through recombination. (ii) Hole injection from the *p*-type base into the *n*-type emitter occurs due to the forward bias across the emitter–base junction. This component, however, is substantially smaller than electron injection due to asymmetric doping concentrations. (iii) Thermally generated leakage currents within the reverse-biased collector–base depletion region constitute a minor, often negligible, component. Collectively, these processes establish the requisite small base current for transistor operation.

The CEC, extensively utilized with NPN transistors, provides both voltage amplification and current gain. In this topology, the emitter acts as the common terminal between the input and output circuits (base input, collector output), typically connected to ground or a shared reference node. Application of a minor input voltage across the BE junction controls a larger collector–emitter current, yielding significant voltage amplification at the collector. While offering voltage gain, moderate output impedance, and high input impedance, this configuration induces a 180° phase inversion between input and output. Appropriate DC base biasing establishes linear operation within the active region, forming an inverting amplifier circuit

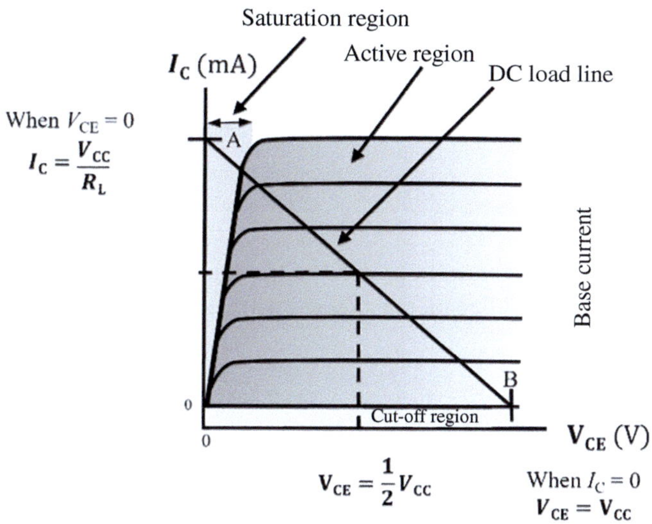

Figure 4.8 Typical output characteristics of a bipolar transistor.

(single-stage common-emitter amplifier). Practical implementations include Class A amplifiers, wherein a forward-biased BE junction permits precise replication of both positive and negative AC input signal components superimposed on the DC bias voltage.

In the CEC, output characteristics relate the collector current (I_C) to the collector–emitter voltage (V_{CE}) for differing values of base current (I_B), particularly for transistors with comparable current gain (β). Figure 4.8 displays typical NPN BJT output characteristics, offering valuable insights into device behavior and performance across diverse operating conditions. These curves are indispensable for understanding critical parameters such as current gain, saturation voltage, cutoff voltage, and maximum power dissipation ratings. Analysis of these curves enables circuit designers to optimize transistor circuits and ensure proper operation. However, obtaining precise characteristic curves for a specific transistor necessitates consultation of its datasheet or specifications, guaranteeing accurate performance assessment and valid application design.

When a load resistor is connected in series with the collector, according to the output characteristics of an NPN BJT, a DC load line plotted on these characteristics signifies all possible operating points for varying base current values. Establishing the initial collector voltage is crucial for amplifying AC input signals, as it enables symmetric positive and negative variations in the output voltage. This process, known as setting the operating point or quiescent point (Q-point), ensures proper transistor functionality. Furthermore, it is important to note that when V_{CE} is large enough, I_C is predominantly controlled by I_B and remains relatively insensitive to variations in collector voltage above this threshold. This characteristic causes the output circuit to behave as a constant-current source.

The Q-point or operating point of the transistor can be determined by drawing a straight line, known as the load line, directly onto the graph shown in Figure 4.8.

The slope of this load line is the reciprocal of the load resistance, which can be expressed as $-1/R_L$. This line extends from the saturation point (A) when $V_{CE} = 0$ to the cutoff point (B) when $I_C = 0$. The transistor's active region is represented by any point along the straight line connecting the saturation and cutoff points. Applying Ohm's Law, I_C can be expressed as

$$I_C = \frac{V_{CC} - V_{CE}}{R_L} \tag{4.6}$$

where R_L represents the load resistance and V_{CC} denotes the supply voltage.

As a result, the position of the load line on the characteristic curves can be calculated. Specifically,

$$I_C = \frac{V_{CC}}{R_L} \quad \text{when } V_{CE} = 0 \tag{4.7}$$

$$V_{CC} = V_{CE} \quad \text{when } I_C = 0 \tag{4.8}$$

The output characteristics of a BJT can be utilized to predict I_C when provided with V_{CE} and I_B. To establish an appropriate Q-point, I_B can be adjusted.

Similar to NPN BJTs, PNP BJTs are widely employed in amplification and switching applications. A PNP BJT consists of three semiconductor layers, with an n-type region sandwiched between two p-type regions, as depicted in Figure 4.9. Similar to the NPN BJT, the PNP BJT regulates current flow between its emitter and collector terminals by controlling a small current at the base terminal. When a negative voltage is applied to the base (relative to the emitter), current flows from the emitter to the collector. Conversely, reducing or eliminating the base current turns off the transistor, ceasing the current flow.

While sharing fundamental operational similarities with NPN BJTs, PNP BJTs exhibit a key distinction in the reversed polarities of applied voltages and the directions of currents. In contrast to NPN BJTs, the PNP emitter terminal exhibits a higher potential than the base terminal, as the base is consistently biased negatively relative to the emitter; consequently, the emitter voltage supply is positive relative to the base potential. For proper conduction in the forward-active mode, the emitter potential must be lower than those of both the collector and the base. The base, negatively biased relative to the emitter, connects to the base resistor, which governs the maximum base current.

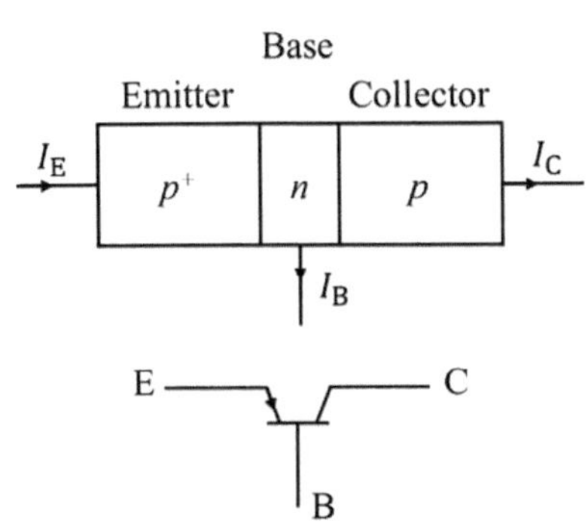

Figure 4.9 A schematic diagram showing a PNP transistor.

Similar to the NPN BJT, base current generation in a PNP transistor involves three primary mechanisms. First, hole injection occurs as holes injected from the emitter recombine with electrons in the base; the consequent reduction in base electron concentration necessitates base current injection to sustain charge neutrality. Second, electron injection arises from electrons being injected from the base into the emitter via the forward-biased emitter–base junction, requiring a corresponding base current. Third, thermal generation occurs at the reverse-biased collector–base junction: thermally generated electrons are swept into the base region, while thermally generated holes are swept into the collector, contributing a minor and often negligible component to the base current. Collectively, these processes establish the requisite small base current for transistor operation.

PNP BJTs provide versatile alternatives to NPN types in electrical and electronic circuits, characterized primarily by opposite current flow direction and voltage polarity. Functionally, PNP BJTs perform identical switching and amplification tasks as their NPN counterparts. The output characteristics of a typical PNP BJT in CEC resemble those of an NPN BJT, but are inverted by 180° to reflect the reversal in current and voltage polarities. This inversion facilitates the effective incorporation of PNP BJTs into circuits requiring reversed polarity configurations, thereby enhancing design flexibility and adaptability.

4.2.5 Heterojunction Bipolar Transistors

Heterojunction bipolar transistors (HBTs) represent a significant advancement over conventional homojunction BJTs, featuring different semiconductor materials for the emitter and base regions. The term "heterojunction" refers to the interface between two dissimilar semiconductor materials, where the emitter typically possesses a wider-bandgap energy than the base. This heterojunction structure enhances performance characteristics by optimizing carrier injection and reducing recombination losses. HBTs are prevalent in high-speed and high-frequency applications, including wireless communication systems and microwave devices. They offer significant advantages such as increased current gain, reduced noise, and improved power efficiency, rendering them essential components in modern electronic systems. Furthermore, their efficient high-frequency operation continues to propel advancements in semiconductor technology.

A typical HBT structure is depicted in Figure 4.10a. The collector region comprises a heavily doped n^+ GaAs layer, ensuring low-resistance Ohmic contacts and low series resistance, followed by a lightly doped n-GaAs layer adjacent to the base region. This configuration reduces the risk of punch-through. The base region consists of a heavily doped p^+ GaAs layer, selected to minimize base and base-contact resistance. This design stabilizes the base width under reverse-biased BC conditions, further reduces susceptibility to punch-through, allows for a thinner base region, and enables faster transistor operation.

A key distinguishing feature of HBTs is the emitter region, which incorporates a heterojunction formed by an n^+ GaAs layer and an n^+ AlGaAs layer. Compositionally graded layers between these materials facilitate efficient electron

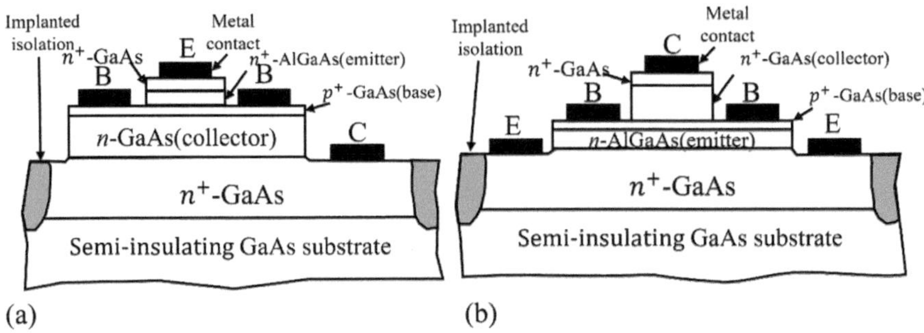

Figure 4.10 (a) A typical structure of an HBT. (b) A special structure using collector-up to minimize the collector capacitance.

transport from the emitter to the base, reducing the turn-on voltage and suppressing space-charge recombination currents at the heterojunction interface. Moreover, the potential barriers for hole injection at the graded emitter–base junction are dictated by the bandgap difference between the GaAs and AlGaAs layers. This bandgap engineering enhances key performance parameters such as current gain, noise reduction, and power efficiency, underscoring the value of HBTs in various high-frequency applications.

HBTs function analogously to conventional BJTs, with their operation governed by the biasing conditions applied to the constituent junctions. Under standard operating conditions, the collector–base junction is reverse-biased, whereas the BE junction is forward-biased. Forward-biasing the BE junction injects electrons from the heavily n^+-doped AlGaAs emitter into the lightly p^+-doped GaAs base, creating a significant perturbation of charge-carrier equilibrium within the base region. These injected electrons subsequently diffuse across the neutral base until encountering the high-field depletion region at the reverse-biased BC junction, where they are swept into the collector to constitute the dominant component of the collector current. Simultaneously, holes generated within the base due to recombination processes contribute primarily to the base current. Additional contributions to the base current arise from the injection of holes from the base into the emitter, although the heterojunction substantially suppresses this injection mechanism relative to homojunction devices. Depletion region recombination currents occurring at both the BE and BC junctions can also contribute to the base current.

The relative injection rates of electrons and holes in HBTs are primarily controlled by the heterojunction barrier height. Heavy emitter doping reduces parasitic resistance associated with the emitter region. Increased emitter doping concentration generally results in lower emitter resistance and can be compatible with maintaining high emitter injection efficiency because of the heterojunction's inherent ability to block hole back-injection. Alternative design strategies, such as the collector-up configuration exemplified in Figure 4.10b, offer distinct advantages, particularly through the minimization of specific parasitic capacitances relative to conventional emitter-up HBTs; this primarily facilitates enhanced optimization for high-frequency performance.

4.3 Field-Effect Transistors

The field-effect transistor (FET) is a semiconductor device that utilizes an electric field to control current flow. Each FET comprises three terminals: gate, source, and drain. Applying a voltage to the gate modulates the conductivity of the channel between the source and drain, enabling precise current control. As illustrated in Figure 4.11, applying a voltage to the gate electrode terminal induces an electric field across the insulating layer, leading to the formation of a depletion region or accumulation region within the channel. This depletion or accumulation region reduces or increases the concentration of free charge carriers, consequently modulating the channel's conductivity. This fundamental mechanism, known as the field effect, underpins FET operation.

FETs are designated as unipolar transistors because their operation involves only one type of charge carrier—electrons in n-channel devices or holes in p-channel devices. Unlike BJTs, which utilize both charge carriers, FETs are voltage-controlled devices offering significant advantages, including high input resistance, low-noise characteristics, low power consumption, a large dynamic range, and ease of integration into electronic circuits.

The conceptual origins of FET technology extend to the early twentieth century. In 1925, Julius Edgar Lilienfeld patented a device concept resembling a modern FET, describing a semiconductor layer situated between two metal electrodes. However, practical implementation was precluded by limitations in contemporary semiconductor technology. Transistor development accelerated significantly in the late 1940s, culminating in the invention of the BJT at Bell Laboratories in 1947 by William Shockley, John Bardeen, and Walter Brattain; they received the 1956 Nobel Prize in Physics for this achievement.

Although BJTs initially dominated semiconductor technology, FETs emerged as strong alternatives due to their superior input impedance, lower power dissipation, and greater suitability for large-scale integration in microelectronics. The breakthrough in metal–oxide–semiconductor FET (MOSFET) technology around 1959 further established FETs as the cornerstone of modern digital circuitry. Advancements in MOSFETs accelerated during the 1960s, driven by progress in semiconductor fabrication techniques. In 1960, Mohamed Atalla and Dawon Kahng at Bell Laboratories demonstrated the first functional MOSFET, a pivotal innovation that revolutionized transistor technology. This development facilitated MOSFET-based integrated circuits, enabling the miniaturization and high-scale integration of electronic components.

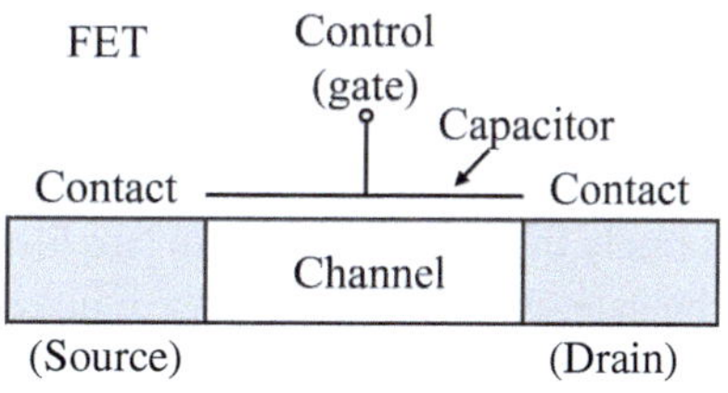

Figure 4.11 A schematic illustration of FETs.

Since their inception, FETs have undergone continuous refinement, fueling progress in telecommunications, computing, and consumer electronics. Currently, MOSFETs represent the most ubiquitous type of FET, forming the basis of contemporary semiconductor devices and enabling the realization of high-performance electronic systems.

FETs provide several advantages over BJTs, rendering them ideal for applications such as analog switches, high-input-impedance amplifiers, and microwave circuitry. A key benefit is their substantially higher input impedance, which simplifies integration into standard microwave systems. Additionally, FETs exhibit a negative temperature coefficient at elevated currents, whereby current diminishes as temperature increases. This characteristic promotes uniform thermal distribution across the device, mitigating the risk of thermal runaway—a common failure mode in BJTs. Another advantage is superior thermal stability, even with extensive active areas or when multiple devices operate in parallel. Unlike BJTs, FETs are immune to minority carrier-storage effects, permitting faster switching speeds in high-frequency applications. Furthermore, FETs function as essentially square-law or linear devices, resulting in reduced intermodulation and cross-modulation distortion compared to BJTs. This characteristic enhances their suitability for precision signal-processing and RF applications.

FETs can be categorized into three primary types based on the gate capacitance formation mechanism: insulated-gate FETs (IGFETs), junction FETs (JFETs), and metal–semiconductor FETs (MESFETs). In an IGFET, the gate capacitance arises from an insulator, whereas in JFETs or MESFETs, the capacitor is formed by the depletion region of a *p*–*n* junction or a Schottky barrier, respectively. IGFETs can be subdivided into MOSFETs/metal–insulator–semiconductor FETs (MOSFETs/MISFETs) and heterojunction FETs (HFETs). In a MOSFET, the insulator typically consists of a thermally grown oxide layer (e.g., SiO_2), whereas in a MISFET, the insulator involves a deposited dielectric material. In HFETs, the gate structure incorporates a high-bandgap semiconductor layer grown as a heterojunction, which acts functionally as an insulator. Although MOSFETs can be fabricated using various semiconductors such as Ge, Si, and GaAs, utilizing oxides and insulators like SiO_2, Si_3N_4, and Al_2O_3, the SiO_2–Si system remains predominant. This subchapter discusses the fundamentals of FETs, with emphasis on MOSFETs, JFETs, and MESFETs.

4.3.1 Metal–Oxide–Semiconductor FETs

Renowned for their versatility, MOSFETs are widely used in diverse electronic applications owing to their numerous advantages. A perspective view of an *n*-channel MOSFET is depicted in Figure 4.12. This MOSFET is a four-terminal device comprising a *p*-type semiconductor substrate in which two n^+ regions, namely the source and drain, are formed. The metal plate atop the oxide constitutes the gate, typically consisting of heavily doped polysilicon or a silicide combination. The fourth terminal provides an Ohmic contact to the substrate.

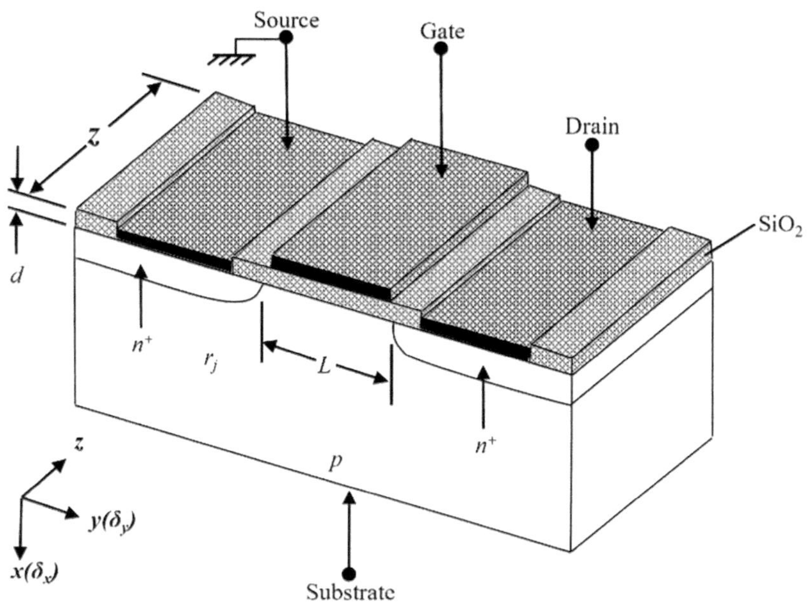

Figure 4.12 Schematic diagrams of a MOSFET.

FETs are primarily classified according to two criteria: the channel carrier type and their operational state under zero gate-bias conditions. *N*-channel FETs employ electrons as charge carriers, exhibiting increased conductivity under a positive gate bias. Conversely, *p*-channel FETs utilize holes as charge carriers, demonstrating enhanced conductivity when a negative gate bias is applied. FETs can also be distinguished based on conduction characteristics at zero gate bias. Depletion-mode (normally on) FETs possess a conductive channel at zero gate bias; gate voltage application is required to reduce or terminate current flow. In contrast, enhancement-mode (normally off) FETs exhibit negligible conduction at zero gate bias; application of a gate voltage is necessary to induce a conductive channel. Figure 4.13 depicts the four possible classification combinations arising from these criteria, along with their representative current–voltage (I–V) characteristics.

Figure 4.14 illustrates the symbols for both *n*-channel and *p*-channel MOSFETs, encompassing enhancement and depletion modes. For *n*-channel MOSFETs, the gate resembles one plate of a capacitor, while the opposite line signifies the channel. During turning on, electron current can flow between the source and the drain. A dashed channel line denotes enhancement mode, while a solid line signifies depletion mode. The arrow, pointing from the substrate toward the channel for *n*-channel MOSFET, can be understood as the direction of electron flow for turning on the MOSFET. In devices with only three terminals, the substrate terminal is internally connected to the source. The principles for *p*-channel MOSFETs are analogous, except that the conductive channel forms between the *p*-type source and drain regions. Consequently, hole current can flow between the source and the drain during turning on. The arrow, pointing from the channel toward the substrate

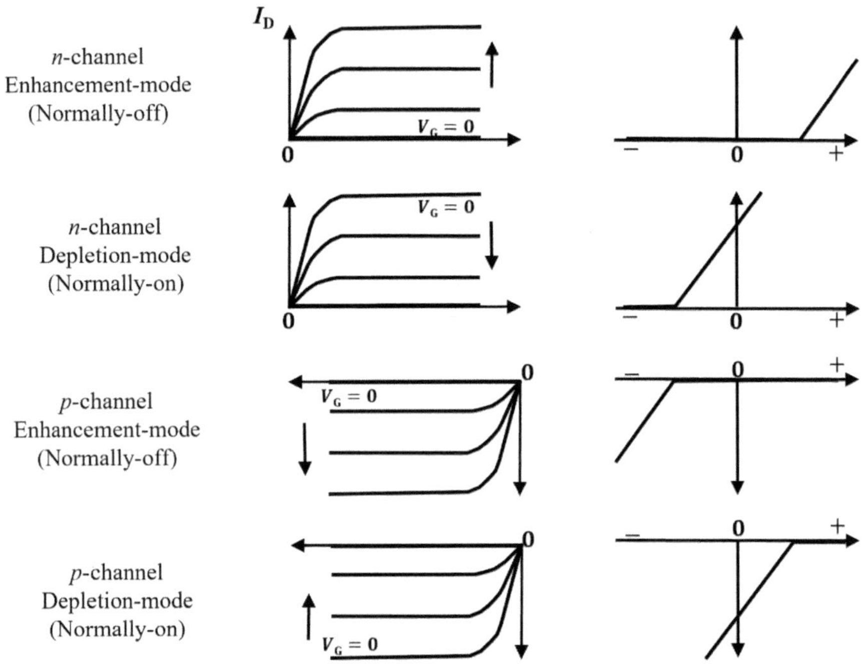

Figure 4.13 Output and transfer characteristics of MOSFETs with different modes.

MOSFET type \ Mode of operation	Enhancement (normally off)	Depletion (normally on)
n-type channel	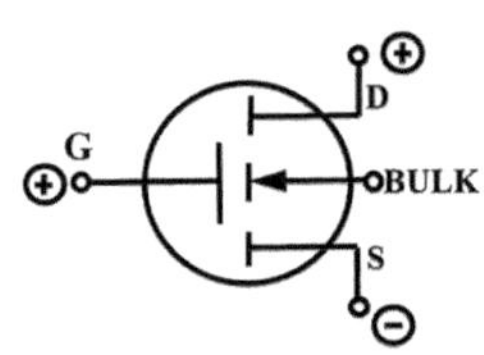	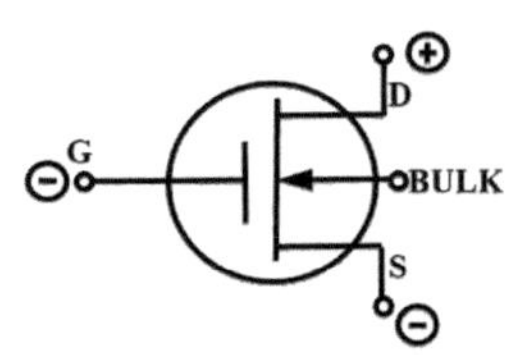
p-type channel	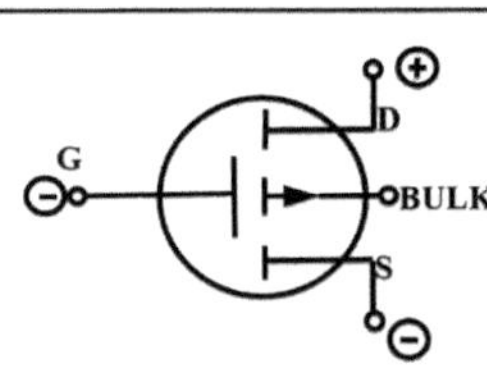	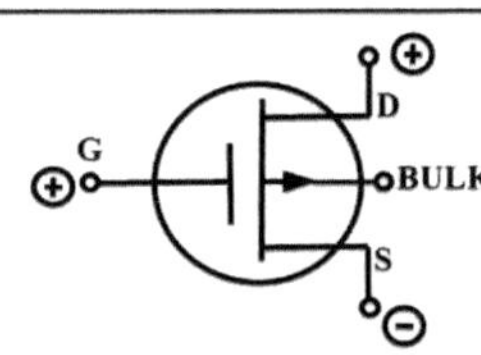

Figure 4.14 MOSFET symbols.

for p-channel MOSFET, can also be understood as the direction of electron flow for turning on the MOSFET.

In a typical MOSFET, the n-channel and gate function as the plates of a capacitor, with the metal–oxide layer serving as the dielectric. Variation of the gate voltage modulates the electric field across this capacitor, thereby influencing the channel carrier type, concentration, and conductivity. Since the gate is insulated from the n-channel, it can accommodate both positive and negative bias voltages. Enhancement mode is established when a positive gate voltage is applied, while depletion mode is initiated by a negative gate voltage.

4.3.1.1 Enhancement-Mode Operation

Figure 4.15 illustrates the circuit used for enhancement-mode operation. As previously described, the gate and n-channel act as the capacitor plates, with the oxide layer serving as the dielectric. When a positive voltage is applied to the gate, as shown in Figure 4.15, it induces an electric field that attracts free electrons into the n-channel, increasing its conductivity. This accumulation of charge carriers effectively reduces the n-channel resistance, allowing greater current flow from the source to the drain. As the positive gate voltage increases, the n-channel conductivity further improves, leading to an increase in drain current. Thus, in enhancement mode, the gate voltage directly controls the channel conductivity, enabling precise current regulation.

4.3.1.2 Depletion-Mode Operation

Figure 4.16 illustrates the circuit configuration utilized for depletion-mode operation. When the gate is negatively biased, electrons accumulate on the gate electrode, functioning as one capacitor plate. These accumulated electrons repel free electrons within the n-channel, resulting in the formation of a layer of positive ions near the oxide interface, as indicated in Figure 4.16. This process depletes the n-channel of charge carriers, effectively increasing its resistance and reducing current conduction. As the negative gate voltage magnitude increases, fewer electrons remain in the n-channel, further restricting the source-to-drain current.

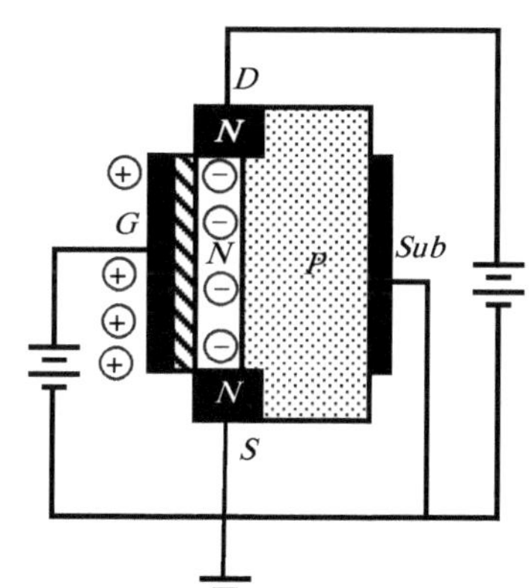

Figure 4.15 Schematics showing the circuit diagrams and the gate/channel charges for an enhancement-mode n-channel MOSFET.

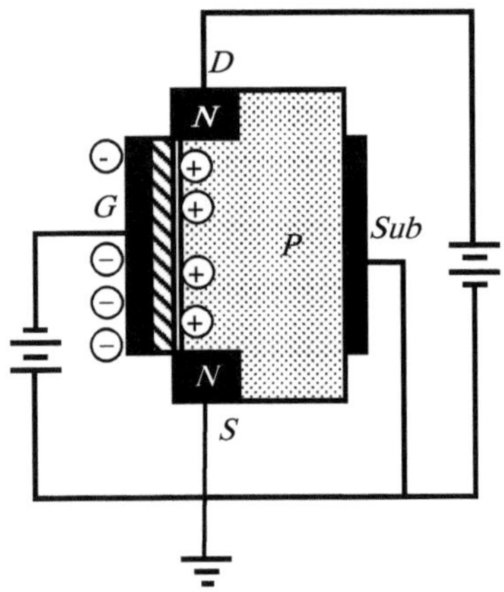

Figure 4.16 Schematics showing the circuit diagrams and the gate/channel charges for a depletion-mode *n*-channel MOSFET.

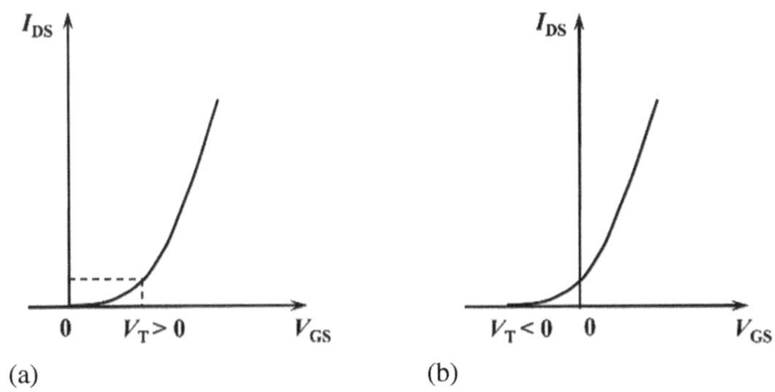

Figure 4.17 Typical transfer characteristics of (a) enhancement-mode and (b) depletion-mode *n*-channel MOSFETs.

Figure 4.17 illustrates the transfer characteristics of both depletion-mode and enhancement-mode *n*-channel MOSFETs. In enhancement-mode MOSFETs (Figure 4.17a), conduction begins only when V_{GS} exceeds a positive V_T. Conversely, in depletion-mode MOSFETs (Figure 4.17b), a negative V_{GS} reduces the already-existing channel conductivity, and the device is normally on at $V_{GS} = 0$ V.

Figure 4.18 illustrates the typical output characteristics of an *n*-channel MOSFET. When the gate-to-source voltage, V_{GS}, is below the threshold voltage, V_T, no drain current, I_{DS}, flows, and the MOSFET remains in the cutoff region, operating as an open switch. As V_{GS} exceeds V_T, the MOSFET enters the linear (Ohmic) region, functioning as a voltage-controlled resistor. In this region, I_{DS} depends in part on both V_{GS} and V_{DS}, enabling the transistor to act as an amplifier.

As V_{DS} increases, the MOSFET transitions into the saturation region, where the drain current reaches a saturated value. In this regime, V_{DS} is sufficiently high to establish the conductive channel fully, rendering I_{DS} relatively independent of further changes in V_{DS}. Consequently, increases in V_{DS} yield minimal variation in I_{DS}, as the channel formation is near completion; here, the device serves as a constant current source for high-gain amplification. The flat characteristic of the saturation region implies relative insensitivity of I_{DS} to V_{DS}, but this ideal behavior is altered by

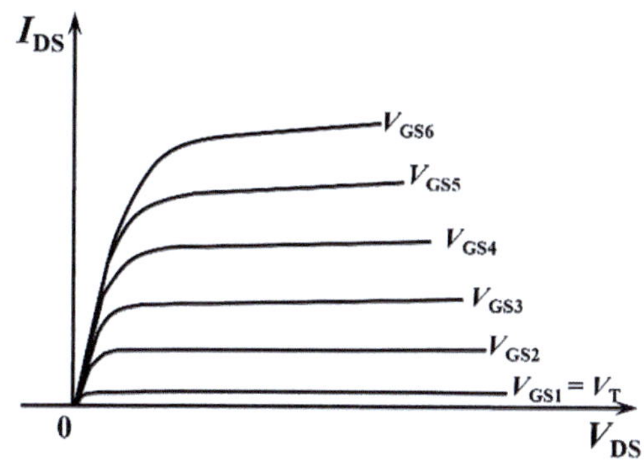

Figure 4.18 Typical output characteristics of *n*-channel MOSFETs.

channel length modulation. When V_{DS} exceeds the pinch-off voltage, the pinch-off point shifts toward the source, reducing the effective channel length, which results in a gradual rise of I_{DS} with V_{DS}. This effect manifests as a slight positive slope in the output characteristics.

A single-stage Class A amplifier can be constructed using MOSFETs, with the *n*-channel enhancement-mode common-source amplifier configuration being among the most prevalent, as illustrated in Figure 4.19. A primary advantage of MOSFETs over JFETs is their substantially higher input impedance, which minimizes loading effects on preceding stages. This high input impedance is predominantly governed by the resistive gate-biasing network, formed by R_1 and R_2, which establishes a stable DC operating point.

Additionally, the output signal of the common-source MOSFET amplifier exhibits an inversion relative to the input. This phase inversion arises because a low V_{GS} places the MOSFET in the cutoff region (turned off), resulting in a high drain voltage.

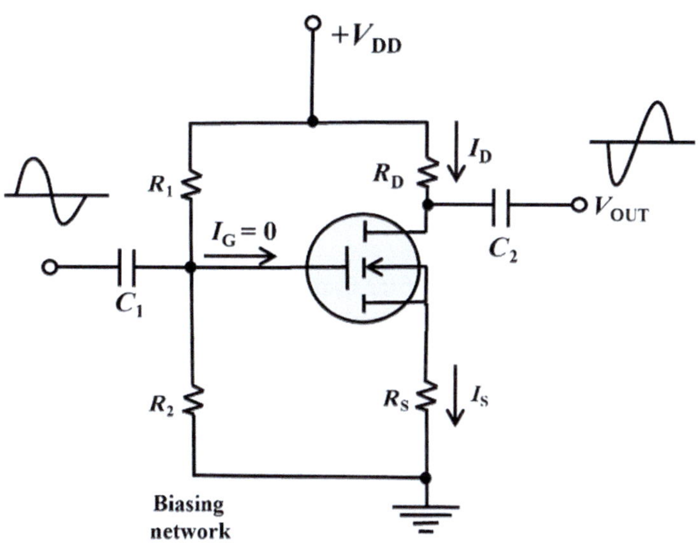

Figure 4.19 A schematic illustration of *n*-channel enhancement-mode MOSFET amplifier.

Conversely, a high V_{GS} drives the MOSFET into saturation, enabling current flow which subsequently reduces V_D due to the voltage drop across R_D. This behavior, depicted in Figure 4.19, explains the 180° phase shift between input and output voltages. The gate voltage and drain current can be mathematically expressed by the following equations:

$$V_{GS} = \frac{R_2}{R_1 + R_2} V_{DD} - V_S \tag{4.9}$$

$$V_S = I_{DS} R_S \tag{4.10}$$

MOSFETs are indispensable in modern electronics, acting as fundamental components in applications ranging from digital circuits to power management systems. Their efficacy in controlling and amplifying electrical signals, coupled with high speed, low power consumption, and excellent scalability, renders them the backbone of integrated circuits. MOSFETs enable the development of advanced microprocessors, memory devices, and communication systems, while their versatility extends to telecommunications, automotive electronics, renewable energy, and medical devices. Ongoing technological advancements ensure MOSFETs continue to drive innovation, shaping the future of computing and connectivity.

Despite widespread adoption, MOSFETs confront critical challenges, including power density constraints, heat dissipation limitations, and reliability concerns. As device dimensions scale down and operating frequencies increase, heightened power density exacerbates thermal management issues. Performance degradation at elevated temperatures further impacts longevity and stability. Research addresses these challenges by exploring novel semiconductor materials such as wide-bandgap compounds (GaN, SiC) and two-dimensional materials (MoS_2, graphene) to enhance electrical properties and efficiency. Furthermore, advancements in device architectures, including FinFETs, gate-all-around (GAA) transistors, and vertical-channel MOSFETs, offer promising solutions for improving power density and mitigating thermal effects. Persistent innovation in materials science, fabrication techniques, and transistor design remains essential for overcoming existing limitations, ensuring MOSFET evolution within next-generation electronic devices and integrated circuits.

4.3.2 Junction FETs

JFETs, also known as junction-gate field-effect transistors (JUGFETs), represent one of the simplest categories of FETs. Known for their inherent simplicity and high input impedance, JFETs are particularly suitable for low-noise amplification and switching applications.

These unipolar semiconductor devices, voltage-controlled and possessing three terminals, are fabricated in both *n*-channel and *p*-channel configurations. Structurally, JFETs comprise a semiconductor substrate (crystal pellet) with oppositely doped semiconductor regions on either side; this substrate can be either *p*-type or *n*-type. Unlike MOSFETs, JFETs regulate current flow via a *p*–*n* junction formed between the gate and the channel region, contributing to their reliability in

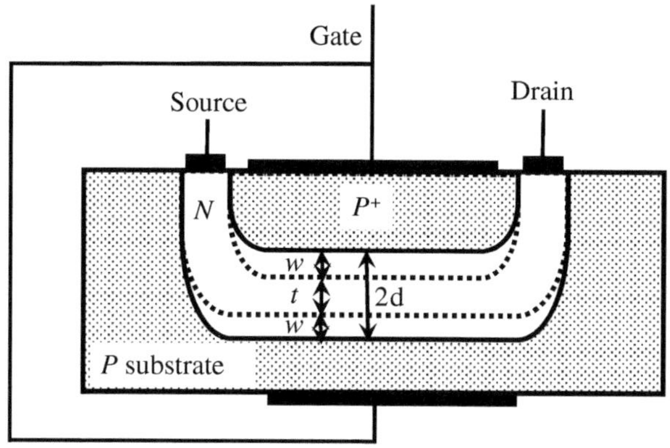

Figure 4.20 A schematic illustration of the cross section of a JFET.

analog and RF circuit applications. In BJTs, the current path between the emitter and collector terminals is defined by *p*–*n* junctions. Conversely, FETs do not utilize junctions for current conduction. Instead, they employ a narrow conductive "channel," composed of either *n*-type or *p*-type semiconductor, with electrical connections designated as the drain and source terminals at either end.

Figure 4.20 presents a schematic cross-sectional view of an *n*-channel JFET, highlighting its structural distinctions from BJTs. An *n*-type semiconductor layer is epitaxially grown on a p^+ semiconductor substrate. Subsequently, an additional p^+ region is formed atop the *n*-layer, creating two PN gate junctions. The two p^+ gate junctions, electrically connected, function as the gate control electrode. A reverse voltage (i.e., a negative bias for an *n*-channel device) is applied between the gate and the source electrode (S), while a positive drain voltage (V_D) is applied between the source (S) and drain (D). The reverse gate bias induces depletion regions extending into the channel. The depletion region width (w) is significant within the lightly doped *n*-channel and negligible within the heavily doped p^+ gate regions.

Modulation of the gate voltage alters the depletion region width, thereby controlling the current flow through the channel. As shown in Figure 4.20, the conductive channel thickness (t) is defined by the distance between the depletion edges and expressed as $t = 2(d–w)$, where $2d$ represents the initial channel thickness dimension, and $2w$ is the total depletion region thickness. The width of this critical channel, and consequently the drain current magnitude, is governed by the extent of the depletion regions. Application of a negative V_{GS} to an *n*-channel JFET widens the depletion regions, constricts the conductive channel, and increases the channel resistance, ultimately reducing the drain current. Conversely, decreasing the magnitude of the negative V_{GS} reduces the depletion region widths, expands the conductive channel, and lowers the resistance, thereby facilitating greater current flow from the source to the drain. Thus, adjusting the V_{GS} controls the drain current in a JFET. A *p*-channel JFET functions analogously to the *n*-channel device, albeit with reversed voltage polarities and holes serving as charge carriers instead of electrons.

Figure 4.21 schematically compares the simplified structures of *n*-channel and *p*-channel JFETs. Instead of using an *n*-type semiconductor as the channel, *p*-channel JFETs employ *p*-type material. For both JFET types, a terminal known

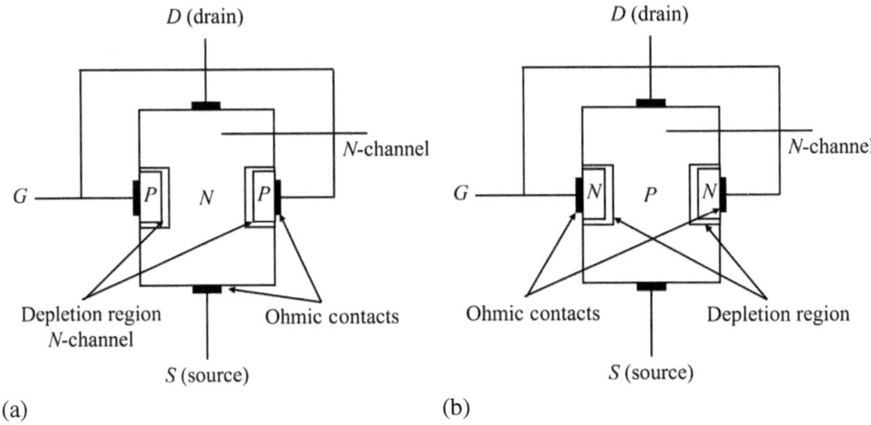

Figure 4.21 Comparison of (a) *n*-channel and (b) *p*-channel JFETs.

as the drain connects to the top of the channel via an Ohmic contact, while a terminal called the source connects to the lower end via another Ohmic contact. Materials with polarity opposite to the channel, adjacent to it, are connected together as the gate terminal. This structural difference fundamentally determines their operational characteristics: in *n*-channel JFETs, electrons function as the majority carriers. Current modulation is achieved by applying a reverse bias to the gate–channel PN junction; this controls the depletion region width and thus governs the conductive channel cross section. Conversely, *p*-channel JFETs utilize holes as majority carriers. Channel conductance is modulated by applying a reverse bias (with appropriate polarity) to the gate–channel junction, thereby expanding the depletion region into the channel. Therefore, although structurally analogous with source, drain, and gate terminals, *n*-channel and *p*-channel JFETs exhibit complementary behaviors regarding the polarities of operating voltages and the nature of the majority carriers responsible for drain current transport.

Like MOSFETs, besides categorized by *n*-type and *p*-type devices, JFETs can also be classified by their operational mode: enhancement mode or depletion mode. Enhancement-mode devices require a positive gate-to-source voltage bias to establish significant conduction and are normally nonconducting at zero bias. Conversely, depletion-mode devices exhibit significant conductivity at zero gate voltage and require a negative gate-to-source voltage to deplete the channel and reduce current flow. This operational classification highlights their fundamental similarity to MOSFETs, underscoring shared principles in semiconductor transistor design. This distinction enables tailored applications in analog circuits; depletion-mode JFETs are frequently employed as amplifiers due to their high input impedance, while enhancement-mode variants, though less common, are useful in specific low-power scenarios. Most JFETs operate in depletion mode, indicating they conduct current when no voltage is applied to the gate. Conversely, enhancement-type JFETs do not conduct current at zero gate bias, making them suitable for ultralow-power applications. The circuit symbols for *n*-channel

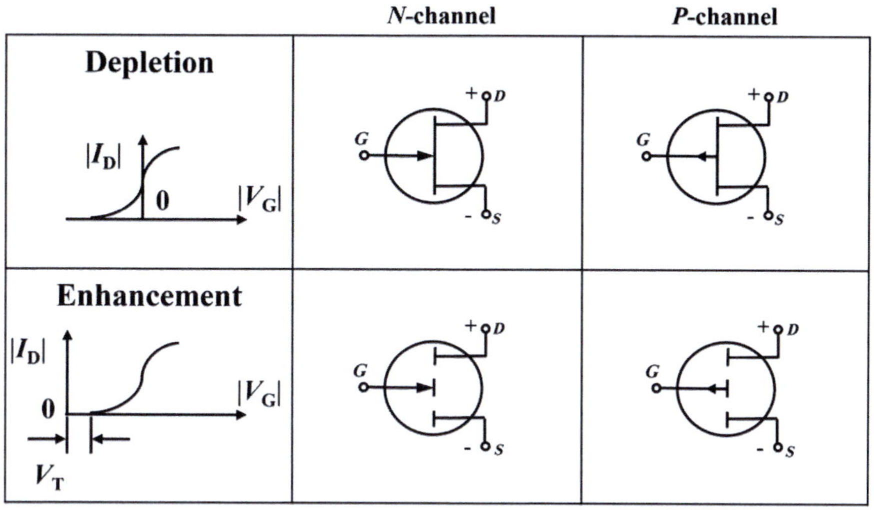

Figure 4.22 Symbols for the *n*-channel and *p*-channel JFETs in depletion and enhancement modes.

and *p*-channel JFETs in both enhancement and depletion modes are shown in Figure 4.22.

Resembling the output characteristics of MOSFETs, JFET output characteristics depict the I_D–V_{DS} relationship under various V_{GS} bias conditions. Consider an *n*-channel JFET: its output characteristics, illustrated in Figure 4.23, exhibit distinct operating regions: the cutoff region (also termed the pinch-off region), the Ohmic (linear) region, the saturation region, and the breakdown (avalanche) region. The cutoff region occurs when V_{GS} is sufficiently negative to fully deplete the conductive channel. In this state, channel resistance becomes extremely high, effectively halting current flow between drain and source; the JFET thus operates as an open circuit. In the Ohmic region, where V_{DS} is relatively small, the device functions as a voltage-controlled resistor, with drain–source resistance modulated

Figure 4.23 A typical output characteristic for *n*-channel JFETs.

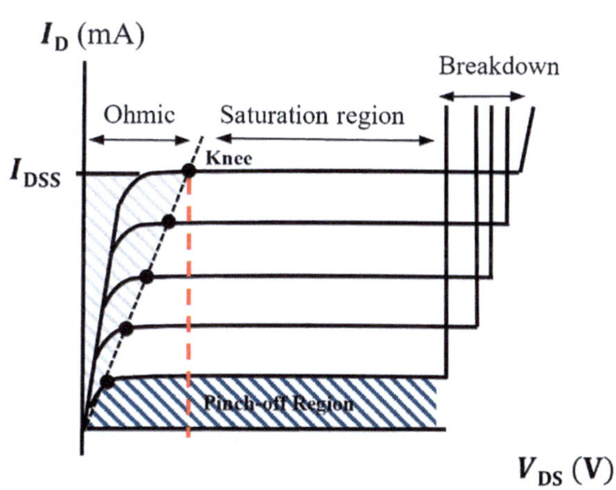

by V_{GS}. The saturation region exhibits drain current primarily controlled by V_{GS} and largely independent of V_{DS}; this is the optimal region for amplification. The breakdown region constitutes an undesirable mode of operation, where excessive V_{DS} triggers a sudden current surge, potentially causing device failure.

For depletion-mode n-channel JFET, when $V_{GS} = V_P$ (the pinch-off voltage, which is negative), I_D is zero. In typical operation, V_{GS} is biased within the range of V_P and 0. For any given bias point in the active or saturation region, I_D is given by

$$I_D = I_{DSS}\left(1 - \frac{V_{GS}}{V_P}\right)^2 \tag{4.11}$$

where I_{DSS} is the maximum drain current when $V_{GS} = 0$. The drain current varies from zero (at pinch-off) to I_{DSS}.

The high resistance presented by the reverse-biased gate–channel PN junction defines the input impedance (R_{in}) of a JFET, which is typically greater than 10^8 MΩ.

Given the drain current and the drain–source voltage, the channel resistance (R_{DS}) can be determined using

$$R_{DS} = \frac{\Delta V_{DS}}{\Delta I_D} \tag{4.12}$$

which is the inverse slope of the drain characteristic curve, typically falling within the range from 40 kΩ to 1 MΩ.

Transconductance, g_m, also referred to as mutual conductance, defined as the ratio of the change in the drain current (ΔI_D) to the corresponding change in the gate–source voltage (ΔV_{GS}) while keeping the drain–source voltage (V_{DS}) constant, is expressed by the following equation:

$$g_m = \left.\frac{\Delta I_D}{\Delta V_{GS}}\right|_{V_{DS} = \text{Constant}} \tag{4.13}$$

Similar to BJTs, when a JFET is configured as an amplifier, its transconductance directly influences the device's gain. Typically, g_m ranges from 1 to 10 mA/V, which is significantly lower than that of a BJT.

BJTs, MOSFETs, and JFETs exhibit distinct operating principles and characteristics. BJTs are current-controlled devices that utilize both electrons and holes as carriers; they offer high gain but exhibit relatively lower input impedance and require more complex biasing. In contrast, MOSFETs are voltage-controlled, unipolar devices characterized by high input impedance, low power consumption, and excellent scalability, making them ideal for digital integrated circuits despite challenges such as oxide trap effects. JFETs are also unipolar and voltage-controlled but are simpler in construction and typically operate in depletion mode; they offer very high input impedance and lower transconductance than MOSFETs. Each transistor type possesses specific advantages and limitations: BJTs excel in amplification applications, MOSFETs dominate in integration and low-power applications, and JFETs find preference in low-noise and high-impedance scenarios. Table 4.1 summarizes the key differences between BJTs, JFETs, and MOSFETs.

In summary, JFETs have a significant impact on modern integrated circuit technology due to their compatibility with a single power supply, high gain–bandwidth

Table 4.1 A comparison between BJTs, JFETs, and MOSFETs.

Characteristics	BJTs	JFETs	MOSFETs
Control type	Current (and voltage)-controlled	Voltage-controlled	Voltage-controlled
Controlling terminal	Base	Gate	Gate
Current flow	Bipolar (majority and minority carriers)	Unipolar (majority carriers)	Unipolar (majority carriers)
Junction	One forward- and one reverse-biased junction	One reverse-biased junction	One reverse-biased junction
Input resistance	Moderate	High	Very high
Gain	High	Low	Low
Noise	Moderate (low at low currents)	Low	Low
Mutual characteristic	Nonlinear	Nonlinear	Nonlinear

product, low power consumption, and suitability for direct coupled logic. This versatility has led to widespread application in various fields, including amplifiers, oscillators, switches, voltage-controlled resistors, mixers, modulators, and sensor circuits. These characteristics render the JFET an essential component in analog and mixed-signal circuit design.

4.3.3 Metal–Semiconductor FETs

MESFETs constitute a distinct class of FETs employing metal gate electrodes fabricated on semiconductor materials. Unlike JFETs, which utilize PN junctions, MESFETs regulate conduction via metal–semiconductor Schottky junctions, with the gate terminal controlling the depletion region width. These transistors are extensively deployed in high-frequency and high-speed applications, particularly wireless communication systems and microwave devices. In MESFET construction, a metal gate electrode directly interfaces with the semiconductor substrate, forming a Schottky barrier; this barrier modulates the conducting-channel cross section to control current flow between the source and drain terminals. Typically fabricated from compound semiconductors such as GaAs or InP, MESFETs exhibit superlative high-frequency performance and superior electron mobility characteristics, rendering them ideal for applications demanding rapid switching speeds and efficient high-frequency operation. Renowned for their elevated gain, diminished noise figure, and substantial power output, MESFETs are predominantly employed in microwave amplifiers, mixers, oscillators, and other high-frequency circuitry. Dependent upon their specific design, MESFETs operate in either depletion mode (conventionally "ON") or enhancement mode (conventionally "OFF").

The basic schematic structure of a MESFET is illustrated in Figure 4.24. The device is fabricated upon a semi-insulating GaAs substrate, which functions

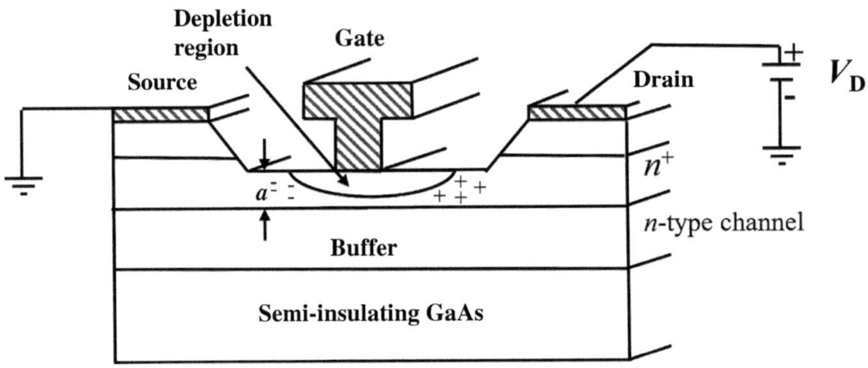

Figure 4.24 A schematic illustration of a MESFET.

as the structural foundation. To mitigate the impact of substrate defects on device performance, an epitaxial buffer layer is deposited initially atop this substrate. The active conducting channel comprises a lightly doped n-type (N-)semiconductor layer, epitaxially grown over the buffer layer. Given that GaAs possesses electron mobility substantially exceeding (typically by a factor of 5–10×) its hole mobility, microwave-optimized MESFETs typically incorporate an n-type conducting channel. To ensure low-resistance Ohmic contacts at the source and drain terminals, a heavily doped n^+-type (n^+) layer is integrated at the surface adjacent to these contacts. This surface layer optimizes carrier injection efficiency and minimizes extrinsic contact resistance.

In the channel region, the highly doped layer is selectively removed by an etching process. As an alternative, the n-channel and highly doped Ohmic contact regions can be formed directly within the semi-insulating substrate using ion implantation. A rectifying Schottky contact is established between the two Ohmic contacts. Typically, Au–Ge-based alloys are employed for the Ohmic contacts, while the Schottky contacts comprise Ti–Pt–Au. Figure 4.25 depicts the schematic symbols for n-channel and p-channel MESFETs. During depletion-mode operation, as illustrated in Figure 4.25, the expansion of the depletion layer necessitates a thicker representation for the gate terminal in the symbol. The arrows in the symbol indicate the current direction, oriented inward for n-channel MESFETs and outward for p-channel devices.

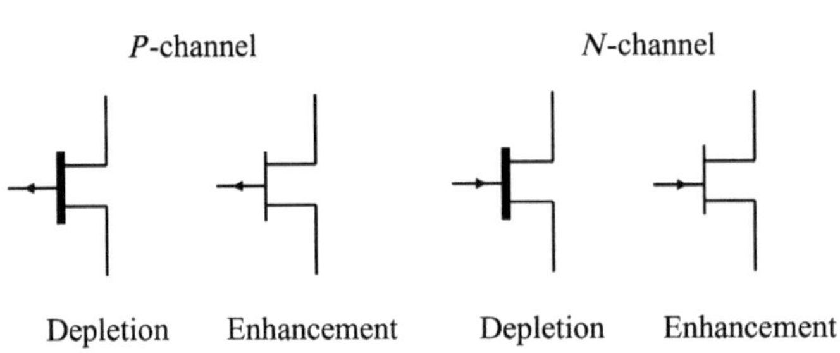

Figure 4.25 A schematic illustration of the symbols of MESFET.

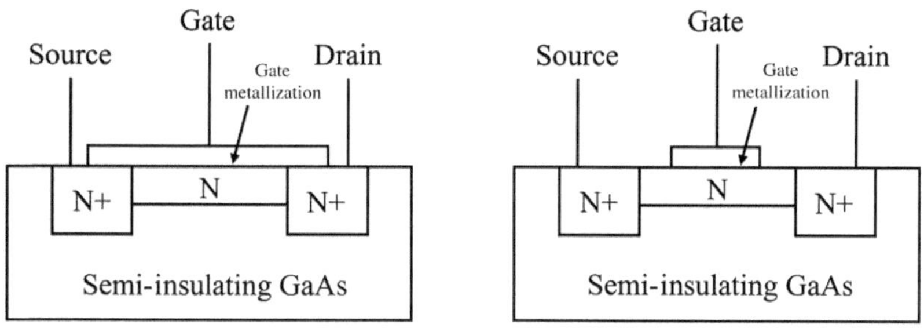

(a) Self-aligned structure of MESFET (b) Non self-aligned structure of MESFET

Figure 4.26 A schematic illustration showing (a) self- and (b) non-self-aligned structures of MESFETs.

Additionally, MESFETs can be categorized into two structural types based on their source and drain configurations: self-aligned and non-self-aligned, as illustrated in Figure 4.26. In the self-aligned structure (Figure 4.26a), the channel length corresponds precisely to the gate length. After gate formation, ion implantation is employed to create the source and drain regions, thereby ensuring that thermal effects during this process do not degrade the gate material. Consequently, the selection of materials suitable for the gate terminal is restricted. Furthermore, the source and drain terminals typically consist of heavily doped *n*-type semiconductor layers, with a Schottky gate contact positioned between them. Conversely, in the non-self-aligned structure (Figure 4.26b), the source and drain regions are fabricated prior to the gate. Consequently, the gate electrode partially overlaps the channel region rather than extending its full length.

To understand the fundamental principles and operation of MESFETs, we can first analyze its I–V characteristics without considering the gate contact, as illustrated in Figure 4.27. When a small voltage is applied between the source and drain, current flows through the device, initially following a linear relationship. In this region, the total resistance is determined by the sum of the source and drain Ohmic resistances (R_S and R_D) along with the channel resistance (R_{DS}). As the applied voltage increases, drain current also increases proportionally. The mathematical expression for the I_D is given by

$$I_D = \frac{V_D}{R_D + R_S + R_{DS}} \tag{4.14}$$

When the applied voltage increases, the resulting electric field can exceed the threshold for electron velocity saturation. Under strong bias conditions, the electron velocity reaches its maximum limit, as shown in Figure 4.28 for Si and GaAs, leading to a deviation from the linear relationship between current and voltage.

In this high-field regime, an alternative expression for the drain current becomes useful, as it directly relates the current to the channel parameters, providing a more accurate representation of device behavior.

$$I_D = Q(x)v(x) = Zb(x)qn(x)v(x) \tag{4.15}$$

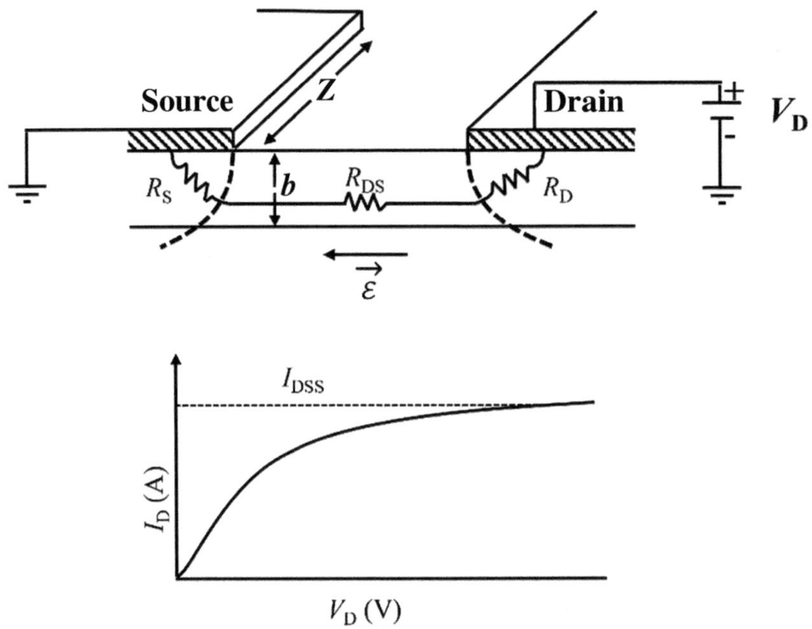

Figure 4.27 A typical I–V characteristic for an ungated MESFET.

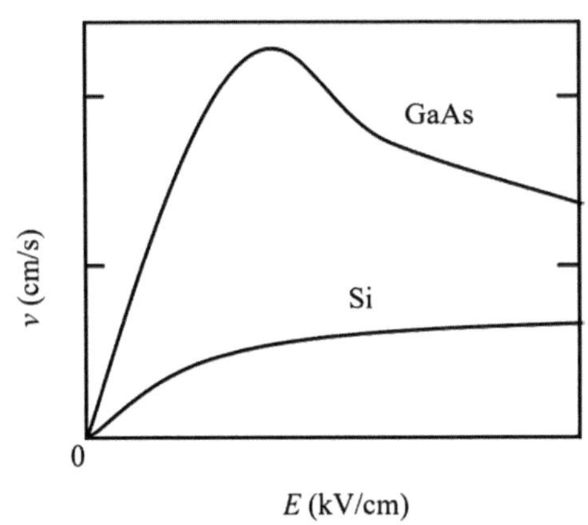

Figure 4.28 The drift velocity of electrons in GaAs and Si as a function of the electric field.

where Z represents the channel width, q stands for the elementary charge of an electron, $n(x)$ denotes the electron density, $b(x)$ corresponds to the effective channel depth, and $v(x)$ represents the electron velocity. The electron velocity is influenced by the electric field across the channel. Notably, when $v(x)$ reaches its saturation value, the drain current also saturates, a condition known as I_{DSS}.

In MESFET operation, the presence of a Schottky gate electrode over the channel, even without applied gate bias, creates a depletion region beneath the gate due to the built-in potential of the junction. This depletion reduces the effective channel depth, consequently increasing the channel resistance to drain current flow. Crucially, the extent of this depletion is governed by the local voltage drop across the Schottky junction. Because the drain–source current flows through the resistive channel, analogous to a distributed resistor, the voltage drop is nonuniform: it is smaller near the source and larger near the drain. Consequently, the depletion region becomes significantly deeper at the drain end of the channel compared to the source end.

This asymmetric channel depth profile impacts device characteristics in two key ways. First, the charge distribution becomes uneven, with a higher electron concentration near the source and a lower concentration near the drain. This charge imbalance (or dipole-like distribution) introduces a capacitance coupling between the drain terminal and the gate-controlled channel region. Second, the electric field resulting from this charge distribution augments the primary lateral electric field applied between drain and source. This augmented field facilitates channel pinch-off and thus induces drain current saturation at a lower applied drain-to-source voltage than would occur in a uniform channel.

By applying a bias to the gate junction, the depletion depth, the resistance to current flow between the source and drain, and the saturation current can all be controlled. When a sufficiently large negative gate bias is applied, the depletion region extends across the entire channel, effectively "pinching off" the current flow. The voltage at which this occurs, known as the pinch-off voltage, can be mathematically determined using the following expression:

$$V_p = \frac{qN_d}{2\varepsilon_0\varepsilon_r} b^2 \tag{4.16}$$

where b is the channel depth, N_d is the channel doping concentration, and ε_r is the relative dielectric constant of channel.

Under pinch-off conditions, the drain current reduces to a minimal level, enabling the transistor to function as either a voltage-controlled resistor or a switch. One primary advantage of MESFETs is their ability to amplify microwave signals and provide gain. Small variations in gate voltage induce significant changes in drain current, allowing MESFETs to be modeled as voltage-controlled current sources. The transconductance (g_m) of a MESFET is given by

$$g_m = \frac{\partial I_\mathrm{D}}{\partial V_\mathrm{GS}} \tag{4.17}$$

Using short-channel approximations, the transconductance can be expressed as

$$g_m = \frac{I_S}{2V_P} \left(\frac{I_S}{I_S - I_D} \right) \tag{4.18}$$

where I_S represents the maximum current that would flow if the channel were completely undepleted under saturated velocity conditions. This corresponds to the saturation current for a device without a gate electrode. Since I_S is proportional to channel depth while V_p is proportional to the square of channel depth, g_m is inversely related to channel depth. Additionally, minimizing parasitic resistances is crucial for maximizing I_S and g_m.

The most commonly used performance metrics for microwave transistors include the gain-bandwidth product, the maximum oscillation frequency (f_max), and the transition frequency (f_T), where the device's unilateral power gain becomes unity. The transition frequency is related to the electron transit time (τ) across the channel and is given by:

$$f_T = \frac{1}{2\pi\tau} = \frac{v_{sat}}{2\pi L} \tag{4.19}$$

For GaAs with typical doping levels, the saturation velocity (v_{sat}) is around 6×10^{10} µm/s. To achieve $f_T > 10$ GHz, the gate length (L) must be below 1 µm. The maximum oscillation frequency (f_{max}) can be estimated as

$$f_{max} = \frac{f_T}{2}\sqrt{\frac{R_{DS}}{R_G}} \tag{4.20}$$

where R_G is the gate resistance. These equations emphasize the importance of minimizing the gate length. However, due to fabrication constraints and the need to prevent avalanche breakdown, the lower limit for L is typically around 0.1 µm. Additionally, for effective gate control over the channel current, the ratio L/b (gate length to channel depth) must be greater than one, typically requiring a channel depth of 0.05 to 0.3 µm in GaAs MESFETs. Given the shallow channel depth, high carrier concentration is necessary to sustain a strong current.

MESFETs and MOSFETs are both FETs but exhibit fundamental structural and operational differences, as summarized in Table 4.2. The key distinction lies in

Table 4.2 Comparison between MESFETs and MOSFETs.

Feature	MESFETs	MOSFETs
Gate structure	Schottky metal–semiconductor junction	Metal–oxide–semiconductor (insulating oxide)
Semiconductor material	Compound semiconductors (e.g., GaAs, InP)	Silicon or silicon-based alloys (e.g., SOI)
Operating principle	Uses Schottky barrier for gate control	Uses oxide layer for capacitive gate control
Electron mobility	Higher electron mobility, especially in GaAs	Lower electron mobility (compared to MESFET)
Frequency range	High-frequency, microwave, RF applications	Low- to moderate-frequency applications
Noise performance	Low noise, suitable for high-fidelity circuits	Higher noise compared to MESFETs
Applications	Microwave amplifiers, satellite systems, RF systems	Digital circuits, logic gates, memory chips
Power Consumption	Generally higher power consumption	Lower power consumption in digital circuits
Fabrication	More complex, typically compound semiconductors	Easier fabrication with silicon technology
Cost	More expensive due to compound semiconductor material	Generally cheaper due to widespread silicon use

the gate architecture. While MESFETs utilize a Schottky metal–semiconductor junction at the gate, MOSFETs employ an insulating oxide layer (typically SiO_2) positioned between the gate electrode and the semiconductor channel, establishing a capacitive coupling mechanism. This structural variation profoundly influences their performance and application domains. Fabricated predominantly from compound semiconductors like gallium arsenide (GaAs), MESFETs exhibit superior high-frequency performance and low-noise characteristics, making them particularly suitable for microwave and RF applications. This suitability derives largely from the high electron mobility inherent in GaAs. In contrast, MOSFETs, fabricated primarily from silicon and available in both enhancement-mode and depletion-mode configurations, dominate digital electronics and integrated circuits.

MOSFETs offer significant advantages in scalability, low power consumption, and large-scale integration potential. MESFETs, however, generally demonstrate superior high-speed switching capabilities and higher power-handling capacities at elevated frequencies. Moreover, MESFETs offer several distinct advantages for microwave circuit design: notably high input impedance, precise geometric control, absence of detrimental oxide interface traps associated with the MOS structure, and excellent high-frequency operation. Compared to silicon, GaAs possesses a significantly higher electron mobility, enabling MESFETs with shorter gate lengths to operate effectively within the 1–10 GHz frequency range. A key limitation, however, is associated with the forward turn-on voltage of the GaAs Schottky barrier gate diode, typically approximately 0.7 V. This necessitates a correspondingly low device threshold voltage, which complicates circuit design within this frequency spectrum. Despite this constraint, MESFETs are extensively utilized in satellite communications, radar systems, microwave circuits, RF amplifiers, and mobile communication devices.

4.3.4 Modulation-Doped FETs

Modulation-doped field-effect transistors (MODFETs) represent a significant advancement in semiconductor technology, particularly for high-frequency and high-speed applications. As a specialized subclass of FETs, MODFETs utilize a heterojunction structure combining distinct semiconductor materials to optimize performance. A key characteristic of MODFETs is the formation of a two-dimensional electron gas (2DEG) at the interface between the channel and barrier layer, facilitated by selective doping. This 2DEG exhibits exceptionally high electron mobility, enabling faster and more efficient charge transport compared to conventional FETs.

Due to their superior electron mobility, MODFETs are ideally suited for applications requiring high-speed operation, such as RF amplifiers and microwave devices. They serve as critical components in wireless communication systems, satellite technologies, and cellular networks, where rapid signal processing and transmission are essential. Additionally, MODFETs find extensive application in high-frequency integrated circuits and millimeter-wave systems, where their performance contributes to enhanced device efficiency and signal integrity.

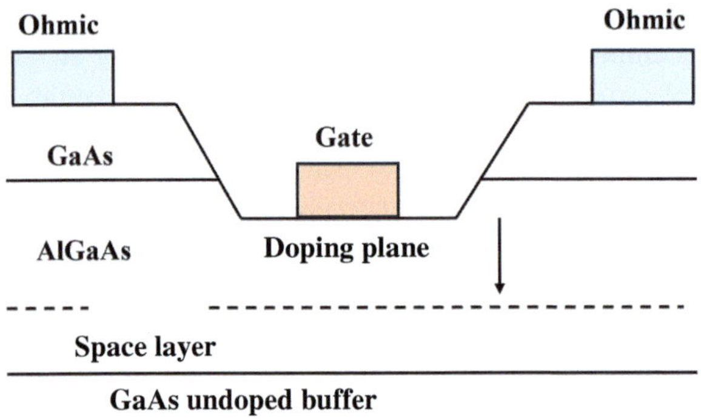

Figure 4.29 Simplified schematic diagram of a typical MODFET.

Another significant advantage of MODFETs is their low-noise operation, which is vital for applications demanding high signal fidelity and accuracy. By minimizing noise interference, MODFETs improve signal clarity, rendering them invaluable in radar systems and precision communication technologies. Furthermore, their inherent scalability and design flexibility enable engineers to tailor device performance for specific applications, reinforcing their versatility in advanced electronic systems.

The structure of a MODFET comprises several distinct layers contributing to its performance. As illustrated in Figure 4.29, the device consists of four primary layers: a cap layer, a doped AlGaAs donor layer, an undoped AlGaAs spacer layer, and an intrinsic GaAs buffer layer. The gate electrode and two Ohmic contacts provide the necessary electrical connections.

A more realistic structures of a GaAs/AlGaAs MODFET is schematically drawn in Figure 4.30. The active region forms an *n*-type channel, created by the interaction between the *n*-type AlGaAs donor material and the lower-bandgap intrinsic GaAs layer, resulting in 2DEG formation. Positioning the donors away from the interface facilitates improved electron mobility, particularly at low temperatures, by reducing Coulombic interactions between donor electrons in AlGaAs and the 2DEG.

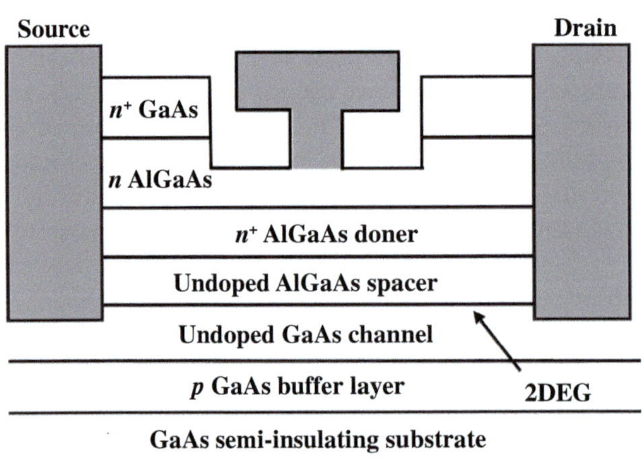

Figure 4.30 Detailed schematic illustration of a GaAs/AlGaAs MODFET.

The conduction band discontinuity between GaAs and AlGaAs, enhanced by higher AlAs mole fractions in AlGaAs, ensures superior electron confinement, thereby optimizing MODFET performance. The number of electrons within the 2DEG at the heterojunction interface is controlled by this conduction band discontinuity.

A 2DEG forms at the interface between the GaAs and AlGaAs layers. To maximize electron mobility by minimizing scattering from ionized impurities, donor atoms within the AlGaAs layer are intentionally situated at a distance from this interface. The adjacent undoped material eliminates impurity scattering, a factor that otherwise severely limits both electron mobility and saturation velocity. Consequently, conduction is governed almost exclusively by the intrinsic properties of the 2DEG, resulting in extremely high conduction mobility within GaAs, primarily limited only by lattice (phonon) scattering. An undoped spacer layer further enhances channel mobility by shielding the 2DEG from ionized impurities. A GaAs buffer layer is typically grown epitaxially on the substrate to provide a high-quality, defect-free surface, thereby isolating the transistor's active layers from the substrate and optimizing performance. Ohmic contacts are formed on the source and drain regions to facilitate electrical connection, while the device's performance critically relies on the heterojunction formed between the GaAs and AlGaAs layers.

A Schottky gate contact modulates the current flow between the source and drain contacts. This modulation constitutes a key operational characteristic of the MODFET. The fundamental operating principle involves modulating the concentration of the 2DEG at the heterojunction interface by applying a gate voltage. MODFET operation relies on controlling the electron population within the 2DEG channel through the gate voltage bias. This bias adjusts the channel conductivity by forming a Schottky barrier and depleting the channel carriers, thereby regulating the device's conductance. Application of the gate voltage therefore precisely governs the electron density within the channel, enabling MODFET operation analogous to other FETs while offering superior high-speed and high-frequency performance. Once the gate voltage exceeds a specific threshold, the source–drain channel ceases to be depleted, turning the device on.

In the modulation doping concept, impurities within the barrier layer are fully ionized, and the carriers are depleted from the region. This results in the formation of a potential variation across the depletion region, which can be derived from the energy-band diagrams (Figure 4.31). For a general doping profile $N_D(y)$, the potential variation in the depleted region is given by

$$\psi_p = \frac{q}{\varepsilon_s}\int_0^{y_0} N_D(y)y\,dy \tag{4.21}$$

For uniform doping, the built-in potential is

$$\psi_p = \frac{qN_D y_0^2}{2\varepsilon_s} \tag{4.22}$$

In the operation of a MODFET, the threshold voltage V_T is a critical parameter that determines the voltage at which the channel begins to conduct between the source and drain. The threshold voltage can be influenced by the doping profile and barrier height (Figure 4.31b). When the Fermi level (E_F) at the GaAs surface aligns

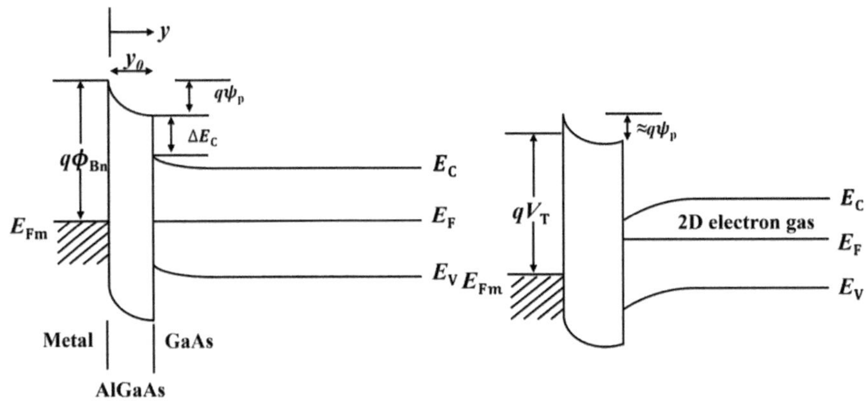

Figure 4.31 Energy-band diagrams for an enhancement-mode MODFET at (a) equilibrium and (b) onset of threshold.

with the conduction band edge (E_C), the channel forms, leading to the following bias condition of V_T:

$$V_T \approx \phi_{Bn} - \psi_p - \frac{\Delta E_C}{q} \tag{4.23}$$

where ϕ_{Bn} is the barrier height, ψ_p is the potential variation across the depletion region, and ΔE_C is the conduction band offset. The MOSFET-like behavior of the MODFET can be understood through this relationship and is useful for the subsequent analysis of the I–V characteristics.

When the gate voltage exceeds the threshold voltage, a charge sheet is induced in the channel, which capacitively connects to the gate and modifies the channel's electrical properties. The charge Q_n in the channel position x is given by

$$Q_n(x) = C_o(V_G - V_T - V_D(x)) \tag{4.24}$$

where C_o is the gate capacitance and V_G is the applied gate voltage. The current at any point x in the channel is given by

$$I_D(x) = WQ_n(x)v(x) \tag{4.25}$$

where W is the channel width, and $v(x)$ is the drift velocity. Integrating this expression over the length of the channel, the drain current is obtained as

$$I_D = \frac{W}{L}\int_0^L Q_n(x)v(x)\mathrm{d}x \tag{4.26}$$

The drift velocity $v(x)$ is related to the electric field $\xi(x)$ by

$$v(x) = \mu_n\xi(x) = \mu_n\frac{\mathrm{d}V_D(x)}{\mathrm{d}x} \tag{4.27}$$

Substituting this into the integral, the drain current equation becomes

$$I_D = \frac{W\mu_n C_o}{L}\left[(V_G - V_T)V_D - \frac{V_D^2}{2}\right] \tag{4.28}$$

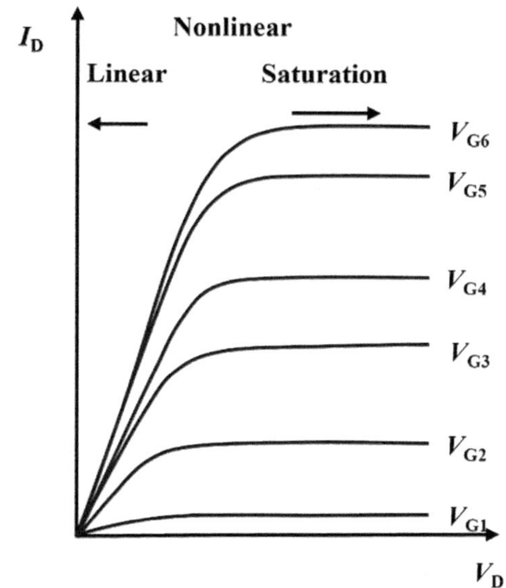

Figure 4.32 A typical output characteristics of an enhancement-mode MODFET.

As shown in Figure 4.32, at low drain voltages, the current behaves linearly with the drain voltage:

$$I_{\text{Dlin}} = \frac{W\mu_n C_o(V_{\text{G}} - V_{\text{T}})V_{\text{D}}}{L} \tag{4.29}$$

At high V_{D}, the current saturates, leading to a pinch-off condition where the current becomes independent of V_{D}:

$$V_{\text{Dsat}} = V_{\text{G}} - V_{\text{T}} \tag{4.30}$$

and the saturation drain current (I_{Dsat}) is given by

$$I_{\text{Dsat}} = \frac{W\mu_n C_o}{2L}(V_{\text{G}} - V_{\text{T}})^2 \tag{4.31}$$

The transconductance is computed as

$$g_m \equiv \frac{\partial I_{\text{Dsat}}}{\partial V_G} = \frac{W\mu_n C_o(V_{\text{G}} - V_{\text{T}})}{L} \tag{4.32}$$

MODFETs demonstrate superior performance in microwave and RF applications relative to conventional JFETs and MOSFETs. In MODFETs, electrons migrate from the n-type region, with a substantial fraction accumulating adjacent to the heterojunction to form a 2DEG. Electrons confined within this atomically thin interface region experience minimal scattering due to the absence of additional donor impurities within the conductive channel. This confinement confers extremely high electron mobility, essential for high-frequency operation. The MODFET, particularly within the $(AlGa)_2O_3/Ga_2O_3$ heterostructure, exhibits significant potential for high-power and high-frequency applications owing to its excellent transport properties and scalability. However, challenges such as poor Ohmic contacts and parasitic channel formation within the barrier layer require mitigation. Despite these issues, MODFETs have found application across diverse fields, including low-frequency noise characterization, optoelectronics, terahertz technology, and low-power electronics.

4.4 Thin-Film Transistors

4.4.1 Introduction

The history of thin-film transistors (TFTs) commenced in 1962 with the creation of the first transistor using polycrystalline cadmium sulfide (CdS) deposited on a glass substrate. This breakthrough catalyzed subsequent advancements in semiconductor technology. Initial research in semiconductor devices predominantly focused on BJTs and FETs. However, the successful development of TFTs redirected research trajectories, offering a compelling alternative to conventional transistor technologies. Over ensuing decades, the application scope of TFTs has broadened extensively beyond displays to encompass fields demanding cost-effective, large-area circuits, including robotics and automation.

TFTs have profoundly transformed the landscape of electronics, particularly within display technologies [7–9]. Their development marked a significant milestone by replacing traditional bulky cathode-ray tube (CRT) displays with more modern, energy-efficient, and versatile liquid-crystal displays (LCDs) and organic light-emitting diode (OLED) screens. TFTs have been instrumental in enabling flat-panel displays, which are significantly thinner, lighter, and more portable than conventional CRT models. Moreover, they have facilitated new design paradigms, such as flexible and bezel-less screens, enhancing user experience and adaptability to diverse applications. Beyond displays, TFTs have expanded into diverse sectors including healthcare, personal devices, security systems, environmental monitoring, and edge sensing. Their fundamental capability to control pixel units in display screens establishes TFTs as indispensable components within many ubiquitous technologies. The societal and economic impacts of TFTs are substantial, though challenging to quantify comprehensively as their adoption proliferates into areas previously dominated by other transistor types.

This subchapter explores the evolution of TFTs, elucidates their operational mechanisms, and examines their burgeoning potential for future applications. It emphasizes how TFTs have revolutionized electronics and continue to play a pivotal role in shaping contemporary technology.

4.4.2 Device Structures and Characterizations

Besides *n*-type and *p*-type according to the conduction channel polarities, or depletion mode and conduction mode according to the conductivity at zero gate bias, TFTs can be categorized according to the gate location, distinguishing top-gate from bottom-gate structures. Alternatively, they are classified based on the relative positions of the drain/source electrodes and the gate electrode with respect to the channel: coplanar configurations, where these electrodes reside on the same side, and staggered configurations, where they occupy opposite sides. Combining these classifications, TFTs can be constructed in four fundamental configurations, each possessing distinct characteristics as depicted in Figure 4.33.

For an *n*-channel TFT, applying a positive voltage to the drain electrode while connecting the source to ground attracts electrons from source to drain. Whether

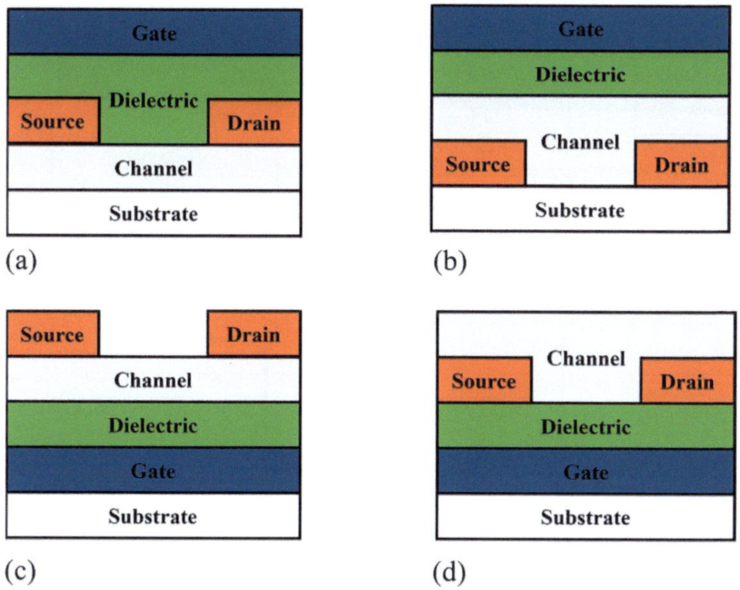

Figure 4.33 Schematic diagrams of TFTs with different configurations.

a current flow is allowed or not inside the channel is determined by the applied voltages, namely the operation regions, comprising the cutoff, linear, and saturation regions. The idealized operation of an *n*-type TFT (bottom-gate, top-contacts structure) is presented in Figure 4.34, displaying both the cross-sectional view and the corresponding output characteristics.

When the gate voltage (V_G) is below the threshold voltage (V_{TH}), the absence of electrons in the channel prevents the flow of drain current (I_D) between the drain and source electrodes, known as the cutoff region. This state is illustrated in Figure 4.34a,d.

Conversely, when V_G exceeds V_{TH}, carriers accumulate at the interface with the gate dielectric layer, forming a conductive channel that allows I_D flow. Under this condition, the transistor's operational state depends principally on the relationship between the drain voltage (V_D) and the gate overvoltage ($V_G - V_{TH}$). For a very small V_D ($V_D \ll V_G - V_{TH}$), I_D varies linearly with V_D according to Ohm's law ($I_D = V_D/R_{ch}$), where R_{ch} represents the channel resistance. This regime defines the linear region, shown in Figure 4.34b,e.

When V_D approaches $V_G - V_{TH}$, the channel potential reduces the overvoltage near the drain, thereby constricting the channel and saturating the current. This constitutes the saturation region, shown in Figure 4.34c,f.

The drain current before reaching saturation can be calculated by

$$I_D = \frac{W}{L} C\mu \left[(V_G - V_{TH})V_D - \frac{{V_D}^2}{2} \right] \tag{4.33}$$

where *W* is the channel width, *L* is the channel length, *C* is the gate capacitance per unit area, and μ is the mobility.

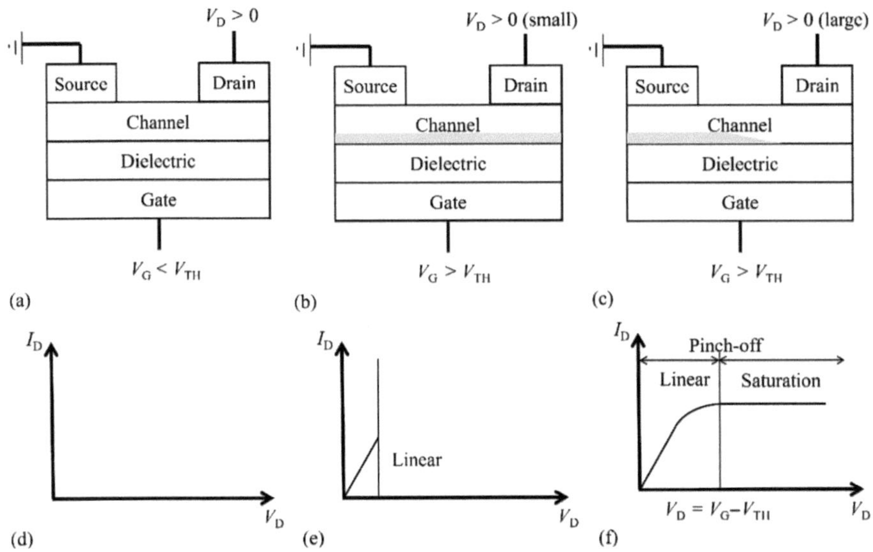

Figure 4.34 Cross-section view and the corresponding output curves of an ideal *n*-type TFT in (a, d) cutoff region, (b, e) linear region, and (c, f) saturation region.

When V_D becomes comparable with the gate overvoltage ($V_D = V_G - V_{TH}$), channel pinch-off occurs. Further increase in V_D will not push this pinch point backward significantly toward the source electrode. Therefore, the resistance of the channel from the source to the pinch point remains the same and I_D remains constant. The current equation for the saturation region is expressed as

$$I_D = \frac{W}{2L} C\mu (V_G - V_{TH})^2 \tag{4.34}$$

Here, an ideal case is considered, assuming that there is no contact resistance. However, in reality, the influence of contact resistance can be serious, especially for devices with a short channel. By taking contact resistance, R_C, into consideration, the equation for I_D should be extended as

$$I_D \approx \frac{W}{L} C\mu (V_G - V_{TH})(V_D - 2I_D R_C) = \frac{I_{D0}}{1 + 2\mu C \frac{W}{L}(V_G - V_{TH}) R_C} \tag{4.35}$$

where I_{D0} is the ideal drain current (no contact resistance). Contact resistance can also be extracted using the transmission line method, through

$$R_{total} = R_{sh} \frac{L}{W} + 2R_{SD} \tag{4.36}$$

where R_{sh} is the channel sheet resistance. By plotting the relationship between R_{total} and L at a fixed gate voltage, contact resistance can be extracted from the intercept at $L = 0$ and R_{sh}/W can be extracted from the slope.

The most important TFT characteristics can be extracted from the device's output and transfer characteristics, as depicted in Figure 4.35. These fundamental parameters include the current on/off ratio (I_{ON}/I_{OFF}), the turn-on voltage (V_{ON}), the threshold voltage (V_{TH}), the subthreshold swing (SS), and the mobility (μ).

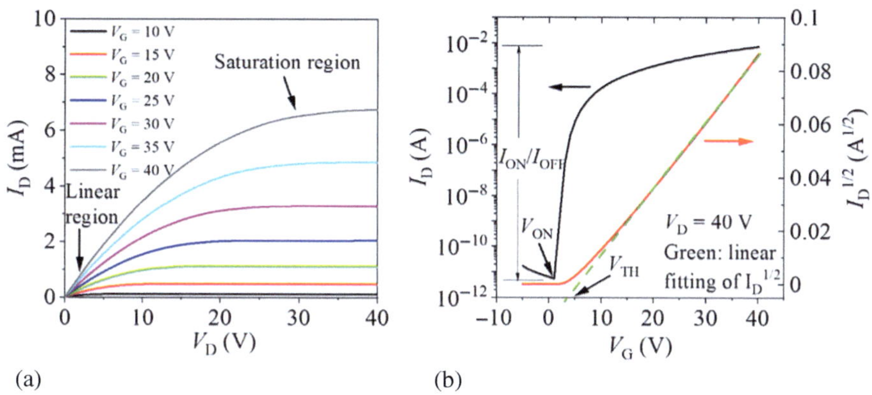

Figure 4.35 Typical (a) output and (b) transfer characteristics of *n*-type TFTs. *Source:* Cai et al. [10].

4.4.2.1 Current On/Off Ratio (I_{ON}/I_{OFF})

This denotes the ratio between the maximum and minimum drain current (I_D). Many parameters influence the maximum I_D, including the channel material, channel length and width, applied voltages, and gate capacitance. The minimum I_D is predominantly associated with the gate leakage current (I_G) and the channel layer resistivity. For applications such as electronic switches, a high I_{ON}/I_{OFF} ratio is essential.

4.4.2.2 Turn-On Voltage (V_{ON})

This represents the gate voltage (V_G) required to activate the transistor, specifically the value at which I_D begins to follow a logarithmic trajectory on the I_D-V_G plot.

4.4.2.3 Threshold Voltage (V_{TH})

This corresponds to the gate voltage (V_G) at which a conducting channel accumulates at the semiconductor/gate dielectric interface. Given TFT operation in distinct regions, V_{TH} can be determined via linear extrapolation to $I_D = 0$ using either an I_D–V_G plot under low drain voltage (V_D) (linear region) or an $I_D{}^{1/2}$–V_G plot under high V_D.

4.4.2.4 Subthreshold Swing (SS)

This parameter characterizes the magnitude of V_G necessary to increase I_D by one order of magnitude, thus indicating the ease of switching between on and off states. SS is calculated using

$$\mathrm{SS} = \left(\frac{\mathrm{d}\log(I_D)}{\mathrm{d}V_G} \bigg| \max \right)^{-1} \tag{4.37}$$

where the derivative represents the maximum gradient extracted from the $\log(I_D)$–V_G curve. SS also relates to trap density according to

$$\mathrm{SS} = \ln(10)\frac{\mathrm{d}V_G}{\mathrm{d}(\ln I_D)} = \ln(10)\frac{kT}{q}\frac{C + C_t}{C} \tag{4.38}$$

where C is the gate capacitance, k is the Boltzmann constant, q is the charge of an electron, T is the temperature, and C_t the trap-induced capacitance. $C_t = q^2 D_{\text{total}}$, where D_{total} represents the total trap density, can be estimated from the SS values of the TFTs using

$$\text{SS} = \ln(10)\frac{kT}{q}\left(1 + \frac{q^2 D_{\text{total}}}{C}\right) \tag{4.39}$$

Assuming a uniform bulk trap density, D_{total} can be expressed by

$$D_{\text{total}} = tN_b + D_{\text{it}} + D_{\text{top}} \tag{4.40}$$

where t is the channel layer thickness, N_b is the trap density in the bulk channel layer, D_{it} is the trap density at channel/dielectric interface, and D_{top} is the trap density at the other channel surface.

4.4.2.5 Mobility (μ)

Mobility governs carrier transport efficiency within the material, directly impacting the maximum I_{D} and the transistor's operating frequency. Under low electric fields, μ is proportional to carrier velocity, enhancing frequency response. Devices possessing higher mobility generate larger currents capable of charging capacitances more rapidly, further boosting frequency response. Additionally, for applications demanding fixed current, higher mobility permits device miniaturization. In display applications, smaller TFTs occupy less pixel area, thereby reducing power consumption. Scattering phenomena, such as interfacial charges, impurities, grain boundaries, and surface roughness, detrimentally affect mobility.

Mobility is typically evaluated in two operational regions: linear and saturation. Within the linear region ($V_{\text{D}} \ll V_{\text{G}} - V_{\text{TH}}$), the effective mobility, μ_{eff}, can be derived at low V_{D} by

$$\mu_{\text{eff}} = \frac{\partial I_{\text{D}}/\partial V_{\text{D}}}{C\frac{W}{L}(V_{\text{G}} - V_{\text{TH}})} \tag{4.41}$$

while the field-effect mobility, μ_{FE}, can be derived from the transconductance (g_m) at low V_{D} by

$$\mu_{\text{FE}} = \frac{g_m}{C\frac{W}{L}V_{\text{D}}} \tag{4.42}$$

Whereas, in the saturation region ($V_{\text{D}} > V_{\text{G}} - V_{\text{TH}}$), the saturation mobility, μ_{sat}, can be obtained by the transconductance at high V_{D} by

$$\mu_{\text{sat}} = \frac{\left(\frac{\text{d}\sqrt{I_{\text{D}}}}{\text{d}V_{\text{G}}}\right)^2}{\frac{1}{2}C\frac{W}{L}} \tag{4.43}$$

Each mobility extraction method presents advantages and limitations. Equation (4.41) is less commonly employed as it necessitates V_{TH} determination and exhibits sensitivity to contact resistance. While Eq. (4.42) also suffers from contact resistance sensitivity, it does not require V_{TH} and can be readily calculated from the derivative of transfer characteristics. In contrast, although Eq. (4.43) avoids V_{TH} requirements and demonstrates reduced sensitivity to contact resistance compared to Eqs. (4.41) and (4.42), it specifically describes device behavior solely within the saturation region where the channel is pinched off.

4.4.3 Representative Materials and Devices

TFTs can be manufactured using diverse materials and device architectures, demonstrating the broad spectrum of fabrication options. These components are integral to flexible electronic devices, finding extensive application in flexible displays, intelligent textiles, and conformal radio-frequency identification devices (RFIDs). The selection of device technology depends on factors including the semiconductor material utilized, fabrication constraints (such as substrate choice and equipment), electrical properties (i.e., carrier mobilities), and mechanical characteristics (i.e., flexibility). Representative materials encompass hydrogenated amorphous silicon (a-Si:H), polycrystalline silicon (poly-Si), oxide semiconductors (including amorphous indium gallium zinc oxide, a-IGZO), two-dimensional (2D) semiconductors (e.g., MoS_2, WS_2), organic semiconductors, and others.

As one of the earliest and most extensively adopted semiconductor materials for TFTs, a-Si:H is used widely within the flat-panel display industry. Introduced in the 1970s, a-Si TFTs became foundational to active-matrix liquid-crystal displays (AMLCDs) due to their compatibility with low-temperature deposition processes and glass substrates. Key advantages include low-cost fabrication, excellent large-area uniformity, and a well-established production infrastructure. However, a-Si suffers from relatively low carrier mobility (~0.5–1 $cm^2/(V{\cdot}s)$), limiting TFT circuit switching speed and integration density. Significant bias stress instability and light-induced degradation are further drawbacks. Although still employed in low-end displays, a-Si is increasingly being supplanted by more advanced materials in performance-critical applications.

The poly-Si emerged as a solution to the mobility limitations of a-Si. Introduced in the late 1980s, poly-Si TFTs exhibit substantially higher mobility (10–100 $cm^2/(V{\cdot}s)$), enabling high-speed operation and allowing the integration of peripheral circuits (such as gate and source drivers) directly onto display panels. This makes poly-Si ideal for high-resolution displays and OLEDs. However, poly-Si fabrication necessitates high-temperature processes or complex techniques like excimer laser annealing (ELA), increasing costs and imposing substrate limitations. Moreover, grain boundaries can cause device nonuniformity and elevated leakage current. Despite these challenges, advanced low-temperature poly-Si (LTPS) technologies have been developed to enhance process compatibility.

Oxide semiconductors, particularly amorphous indium gallium zinc oxide (a-IGZO), constitute a significant breakthrough in TFT development. First commercialized by Sharp in 2012 for high-resolution LCDs, IGZO offers an advantageous balance of moderate mobility (10–20 $cm^2/(V{\cdot}s)$), high transparency, low leakage current, and excellent large-area uniformity. Unlike a-Si and poly-Si, IGZO can be processed at relatively low temperatures, facilitating flexible and transparent electronics. Its amorphous nature ensures large-area uniformity, while its wide bandgap (~3.1–3.5 eV) provides low OFF-state current and improved stability under illumination. Nevertheless, oxide TFTs face challenges regarding stability under electrical stress and sensitivity to environmental factors (e.g., moisture, oxygen). Ongoing research focuses on enhancing intrinsic stability, optimizing dopant composition (including Ga, Sn, W), and developing solution-processable oxides for low-cost printing. Future applications are anticipated in flexible electronics, wearable sensors, and neuromorphic devices.

Two-dimensional (2D) semiconductors have garnered recent interest as promising candidates for next-generation TFTs, leveraging their atomically thin structure, exceptional gate control, and high intrinsic mobility. Transition metal dichalcogenides (TMDs), such as MoS_2 and WS_2, possess suitable bandgaps (~1–2 eV) and demonstrate moderate to high mobility (10–100 $cm^2/(V{\cdot}s)$), rendering them attractive for ultra-scaled, low-power transistors. While graphene lacks a bandgap, its exceptional mobility is being explored for electrode or interconnect applications within TFT architectures. Despite their potential, 2D semiconductors encounter barriers including synthesis scalability difficulties, challenges in uniform large-area growth, and the need for stable, low-resistance contacts. Future progress depends on developing scalable chemical vapor deposition (CVD) methods, deterministic layer transfer techniques, and heterostructure engineering to enable monolithic integration in advanced flexible and wearable electronics.

Organic semiconductors have attracted interest due to their inherent mechanical flexibility, low-temperature processability, and potential for large-area, printable electronics. Introduced in the early 2000s, organic TFTs (OTFTs) employ π-conjugated molecules or polymers as the active layer, deposited via solution-based techniques (e.g., inkjet printing, spin coating). Advantages include lightness, compatibility with flexible substrates, and relatively simple fabrication. However, OTFTs generally exhibit lower mobility (typically <1 $cm^2/(V{\cdot}s)$), limited operational stability, and sensitivity to environmental conditions. Performance variability arising from synthesis inconsistencies between batches also hinders mass production. Future development focuses on molecular design for higher mobility, encapsulation technologies for enhanced stability, and hybrid systems integrating organic and inorganic materials for multifunctional devices.

4.4.4 Emerging Applications

TFTs have become increasingly important in diverse applications due to their unique properties, including small size, lightweight, and low power consumption. These characteristics render them highly promising for emerging technologies.

4.4.4.1 Flexible and Wearable Electronics

TFTs are critical enablers due to their mechanical flexibility, low processing temperature, and lightweight nature. Fabricated on plastic, textile, or elastomeric substrates, they conform to curved or moving surfaces, unlike conventional rigid electronics, making them ideal for integration into wearable devices. Applications include smart clothing, flexible fitness trackers, electronic skin (e-skin), and patch-type biosensors for real-time physiological parameter monitoring. Low power consumption and compatibility with large-area fabrication techniques, such as roll-to-roll printing, enhance their suitability for next-generation wearable systems targeting continuous health monitoring, personalized medicine, and human–machine interaction.

4.4.4.2 Biosensors and Biomedical Applications

TFTs—particularly organic and oxide-based variants—are increasingly employed for detecting biomarkers, pH, glucose, or other biological signals. Advantages include label-free sensing, high sensitivity, and potential for miniaturization and integration with microfluidic platforms. TFT-based bioelectronics apply to point-of-care diagnostics, implantable devices, and remote patient-monitoring systems. Biocompatibility and conformability also suit them for skin-attachable or ingestible devices. With growing demand for low-cost, disposable diagnostics, TFT biosensors are anticipated to play a transformative role in personalized healthcare and digital medicine.

4.4.4.3 Active-Matrix Displays and Transparent Electronics

TFTs remain fundamental to the evolution of LCDs, OLEDs, micro-LED, and quantum-dot displays. Increasing resolutions and shrinking bezels require TFTs supporting faster switching speeds, higher currents, and compact integration. Innovations like oxide TFTs and LTPS TFTs enable high-refresh-rate and ultrahigh-resolution displays in smartphones, tablets, AR/VR (augmented reality/virtual reality) headsets, and foldable screens. Moreover, transparent TFTs using wide-bandgap semiconductors enable see-through displays and smart windows, expanding possibilities for user interfaces and smart environments.

4.4.4.4 Internet of Things (IoT)

The rapid expansion of the Internet of Things (IoT) demands low-cost, low-power, disposable electronics, an area where TFTs excel. Printed TFTs integrate into smart labels, RFID tags, environmental monitors, and wireless sensor nodes for data collection and communication. These systems benefit from TFTs' flexibility, lightweight nature, and compatibility with nontraditional substrates like paper and fabric. Integration with energy-harvesting components and low-power microcontrollers supports self-powered, autonomous IoT systems for agriculture, logistics, smart homes, and industrial automation.

4.4.4.5 Neuromorphic and Brain-Inspired Computing

Emerging research explores TFTs for neuromorphic and brain-inspired computing, mimicking neural structures to achieve efficient, low-power data processing. TFTs can be engineered to replicate synaptic behaviors through memory effects, hysteresis, and multilevel conductance states. Materials such as organic semiconductors, ferroelectric oxides, and electrolyte-gated systems construct artificial synapses and neurons. Such devices may revolutionize edge computing, enabling real-time analysis in smart sensors, robotics, and artificial intelligence (AI) systems within resource-constrained environments.

4.4.4.6 Stretchable and Implantable Electronics

Beyond flexibility, stretchable and implantable electronics exploit TFTs adapted to demanding mechanical deformation in soft robotics, prosthetics, and medical implants. Fabricated with serpentine interconnects or intrinsically stretchable materials, stretchable TFTs maintain electrical function under repeated strain. Biocompatible TFTs are investigated for neural interfaces, cardiac monitoring, and *in vivo* diagnostics. These applications require high integration, low power, and minimal invasiveness, aspects increasingly addressed through materials science advances.

In summary, TFTs drive significant technological innovation, enabling novel electronic devices with enhanced functionality, flexibility, and performance. However, underlying limitations hinder their full potential. A primary constraint is limited charge-carrier mobility, affecting device speed and efficiency. Devices fabricated on flexible substrates may also suffer performance and reliability degradation under mechanical strain over time. Mitigation strategies include developing novel materials like graphene and TMDs with higher carrier mobilities. Progress in fabrication techniques, such as atomic layer deposition, enhances precise control over material properties and structures, improving performance and durability. Furthermore, integrating TFTs with complementary technologies—sensors, energy-harvesting systems, and advanced signal-processing algorithms—can circumvent limitations and unlock new functionalities across emerging applications. Continued research remains essential to overcome present challenges and fully realize TFT potential across technological domains.

References

1 Bardeen, J. and Brattain, W.H. (1948). *Phys. Rev.* 74: 29–30.
2 Shockley, W., Sparks, M., and Teal, G.K. (1951). *Phys. Rev.* 83: 151.
3 May, G.S. and Sze, S.M. (2004). *Fundamentals of Semiconductor Fabrication*, vol. 19, 343–348. Wiley.
4 Ning, T.H. (2006). *IEEE Electron Devices Mag.* 1: 33–40.
5 Santos, L.A.P. (2022). *Sensors* 22: 1923.
6 Dastgeer, G., Shahzad, Z.M., Chae, H. et al. (2022). *Adv. Funct. Mater.* 32: 2204781.
7 Espera, A.H., Dizon, J.R.C., Valino, A.D. et al. (2022). *Jpn. J. Appl. Phys.* 61: SE0803.
8 Guo, X., Han, L., and Hou, X. (2021). *Mater. Chem. Front.* 5: 6760–6778.
9 Yan, A., Wang, C., Yan, J. et al. (2024). *Adv. Funct. Mater.* 34: 2304409.
10 Cai, W., Wilson, J., Song, A. et al. (2021). *Flex. Print. Electron.* 6: 043001.

5

Luminescent Materials and Devices

5.1 Introduction

Light is essential for human life and production activities. Sunlight is an ideal thermal emission source, but it is unavailable at night. Consequently, the exploration and development of artificial light sources are critically important. These light sources are applied across various fields, including display, lighting, and optical communication. Typical light sources in the history are illustrated in Figure 5.1. These sources are categorized into thermal (equilibrium) and non-thermal (nonequilibrium) emitters based on their emission mechanisms. Thermal emission involves converting the thermal energy of a source, such as an oil lamp or incandescent bulb, into photons, termed blackbody radiation. It is established that no electron transitions between distinct energy states occur in this process, resulting in low luminous efficiency and intensity. In contrast, nonequilibrium emission involves excited electrons with a nonequilibrium energy distribution. The relaxation of these high-energy electrons to lower-energy states enables photon emission (radiative recombination) or alternative energy dissipation paths (nonradiative recombination). Therefore, the direct nature of photon generation facilitates highly efficient emission, achieving high luminous power and luminous flux.

In this chapter, semiconductor-based light sources employing nonequilibrium emissive mechanisms—primarily light-emitting diodes (LEDs) and solid-state laser devices—are established as mainstream technologies. The optoelectronic characteristics of nonequilibrium light sources are introduced and discussed in detail.

5.2 Principle of Semiconductor Luminescence

5.2.1 Radiative Recombination of Nonequilibrium Carriers

The fundamental basis of luminescence in semiconductors is radiative recombination, a process wherein nonequilibrium charge carriers relax from high-energy states to lower-energy states, subsequently emitting photons with distinct energies. In semiconductors, five primary mechanisms of radiative recombination exist, outlined as follows.

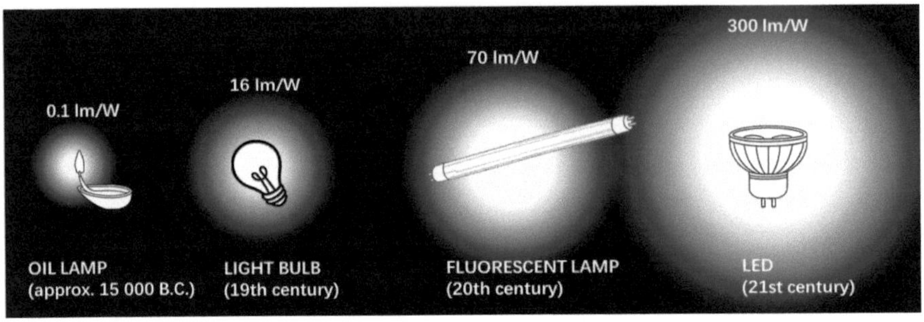

Figure 5.1 Historical development of lighting tools and their corresponding light flux.

5.2.1.1 Band-to-Band Recombination

During band-to-band recombination, excited electrons located at the conduction band minimum (CBM) and holes located at the valence band maximum (VBM) recombine to emit photons. The photon wavelength (λ_{em}) resulting from band-to-band recombination can be calculated as follows:

$$\lambda_{em} \approx \frac{1240}{E_g} \tag{5.1}$$

where E_g is the bandgap value in units of eV (electron volt), defined as the energy difference between the CBM and the VBM. Based on band structure analysis, band-to-band recombination mechanisms are categorized as direct or indirect transitions. As shown in Figure 5.2, when the CBM and VBM of a given semiconductor lie at the same k-point in $\boldsymbol{k}$ space, the radiative recombination process involving nonequilibrium carriers between these bands is referred to as a direct transition. Conversely, when the CBM and VBM are located at different k-points, the radiative recombination is termed an indirect transition of nonequilibrium carriers.

In terms of direct transition of nonequilibrium carriers, dynamics of the wave vector ($\vec{\boldsymbol{k}}$) of electrons complies with the law of conservation of momentum as follows:

$$\vec{\boldsymbol{k}}_f = \vec{\boldsymbol{k}}_i - \vec{\boldsymbol{k}}_\lambda \tag{5.2}$$

where $\vec{\boldsymbol{k}}_i$ and $\vec{\boldsymbol{k}}_f$ represent the wave vectors of electrons in the initial and final states, respectively, $\vec{\boldsymbol{k}}_\lambda$ is the wave vector of the photon, given by $|\vec{\boldsymbol{k}}_\lambda| = 2\pi/\lambda$. Since the lattice constant of the semiconductor is usually far smaller than the wavelength, the $\vec{\boldsymbol{k}}_\lambda$ value ($\sim 10^4$/cm) for photons in the visible range is significantly smaller than $\vec{\boldsymbol{k}}_i$ and $\vec{\boldsymbol{k}}_f$ for electrons (10^8/cm). As a result, for transition only involving electrons and holes, $\vec{\boldsymbol{k}}_i$ and $\vec{\boldsymbol{k}}_f$ are almost equal, which are called the direct optical transition. Consequently, direct-bandgap semiconductors including GaN, InN, InP, GaAs, InAs, GaSb, InSb, and all II–VI compounds have higher transition probability, which are advantageous for nonequilibrium carrier recombination and efficient photon emissions.

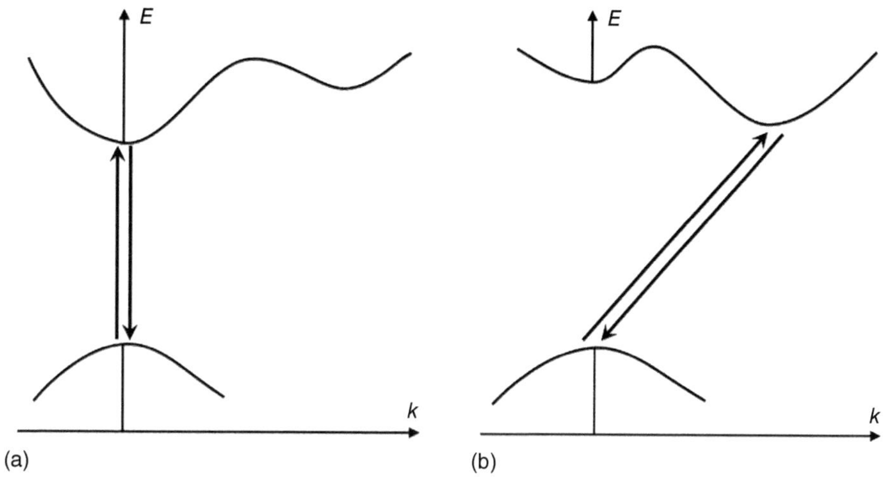

Figure 5.2 Diagram of (a) direct transition and (b) indirect transition of carriers in semiconductors.

Carriers can also recombine through indirect transitions. The wave vector of photons ($\vec{k}_\lambda$) during transitions is negligible. However, the wave vectors of electrons in the initial and final states differ because the VBM and CBM reside at distinct k points. Phonons, which describe the quantized vibrations of semiconductor lattices, facilitate the transition of electrons between the CBM and VBM to conserve momentum, as expressed in the following equation:

$$\vec{k}_f = \vec{k}_i \pm \vec{k}_p \tag{5.3}$$

where $\vec{k}_p$ is the wave vector of phonons. During the indirect transition, it is necessary for electron transition from the CBM to the VBM to absorb or release phonons. Therefore, the probability of indirect transition is relatively low, resulting in low-efficiency and broadband emission. Indirect-transition semiconductors include Si, Ge, BP, AlP, GaP, BAs, AlAs, AlSb, and others.

5.2.1.2 Band-to-Local-Energy-Level Recombination

In doped semiconductors, energy levels induced by donors and acceptors, as well as deep energy levels within the forbidden band, are considered local energy levels. Acceptors can recombine with electrons in CBM, while electrons in donor levels can recombine with holes in VBM. Additionally, recombination between electrons from the CBM and holes from the VBM may occur at deep energy levels. All the aforementioned local energy levels are referred to as recombination centers, as illustrated in Figure 5.3.

During band-to-local-energy-level recombination of carriers, energy of emissive photons is evaluated by the following equation:

$$h\nu = E_g - E_i \tag{5.4}$$

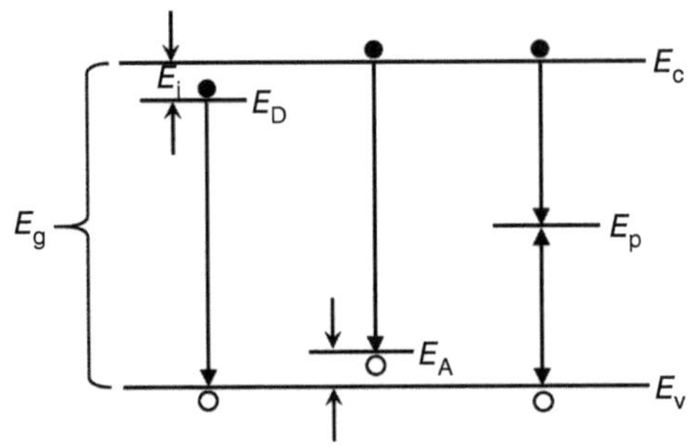

Figure 5.3 Diagram of band-to-local-energy-level recombination.

where h, ν, k, and T denote Planck's constant, the frequency of photons, Boltzmann's constant, and temperature, respectively. E_i is the ionization energy of dopants. When recombination centers in semiconductors are at shallow energy levels, the corresponding E_i value is determined using the hydrogen-like model as follows:

$$E_i = \frac{R}{\varepsilon^2}\frac{m^*}{m_0} \tag{5.5}$$

where R is Rydberg constant, $R = (m_0\, q^4)/(2\,\hbar^2) = 13.6$ eV, ε represents the dielectric constant of semiconductors, and m^* is the effective mass of electrons.

5.2.1.3 Donor–Acceptor Pair Recombination

For doped semiconductors, donors and acceptors serve as positive and negative charge centers, respectively. When both donors and acceptors are introduced into the crystal lattice, they may form bound donor–acceptor pairs at varying separation distances (r_n) due to the Coulomb interaction. As shown in Figure 5.4, the emission photon energy is determined by the following equation:

$$h\nu = E_g - (E_D + E_{A)} + \frac{q^2}{r_n\varepsilon} \tag{5.6}$$

where E_D and E_A represent ionized energy of donors and acceptors, respectively. q is elementary charge.

During donor–acceptor pair recombination, discontinuous spectral lines are exhibited in emissive spectra because the distance of electron–hole pairs is discrete. Additionally, due to dominant low r_n in emission energy, blueshifts of luminescence peaks are found with enhancement of excitation. After excitation in semiconductors, donor–acceptor pairs with large r_n values demonstrate long luminescent decay time because of their small recombination probability, resulting in redshifts of afterglow emission.

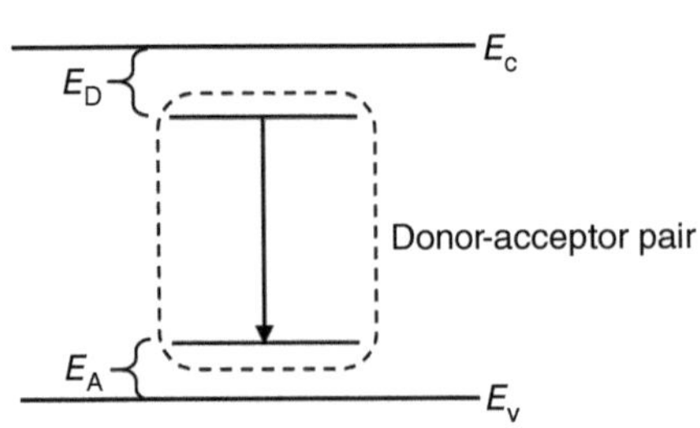

Figure 5.4 Diagram of donor–acceptor pair recombination.

5.2.1.4 Exciton Recombination

Excitons refer to bound electron–hole pairs that are electrically neutral in semiconductors. They are classified as free excitons and bound excitons. Free excitons are defined as those that can move freely through the crystal lattice. Conversely, bound excitons are formed when excitons are trapped by donors, acceptors, or other localized energy levels. To facilitate the energy analysis of excitons, free excitons are often modeled as hydrogen-like systems; accordingly, their energy levels can be calculated using the Schrödinger equation:

$$\left(-\frac{\hbar^2}{2\mu}\nabla^2 - \frac{q^2}{r\varepsilon}\right)\Psi = E\Psi \tag{5.7}$$

where μ is reduced mass calculated by $\mu = (m_c^* m_v^*)/(m_c^* + m_v^*)$. The eigenvalue of the aforementioned equation is the following:

$$E_n = -\frac{\mu R}{n^2\varepsilon^2 m_0} \tag{5.8}$$

where R is Rydberg constant, ε represents the dielectric constant of semiconductors, and m_0 is the free electron mass, and n is integer ranging from 1 to infinity. Thus, the exciton-binding energy (E_{exc}) can be evaluated as follows:

$$E_{\text{exc}} = |E_1| = \frac{\mu R}{m_0\varepsilon^2} \tag{5.9}$$

Usually, an exciton is confined in the range of several lattices, and the phonon assistances can be induced during exciton recombination. Thus, the photon energy is evaluated by the following equation:

$$h\nu = E_g - E_{\text{exc}} \pm mE_a \tag{5.10}$$

where E_a and m represent energy and number of phonons, respectively.

For bound excitons, photon energy of exciton recombination is calculated as follows:

$$h\nu = E_g - E_{\text{exc}} - E_b \pm mE_a \tag{5.11}$$

where E_b represents bound energy of excitons induced by dopants. It has been found that the relation between the bound energy of excitons and the ionization energy of dopants in doped semiconductors is evaluated as follows:

$$E_b = (0.1 \sim 0.4)E_i \tag{5.12}$$

Compared to the bandgap energy E_g, the other terms in Eqs. (5.10) and (5.11) are relatively small, meaning a low ionization energy for exciton. Therefore, the emission energy from exciton recombination is close to the corresponding bandgap energy. Moreover, in highly doped semiconductors, the high carrier concentrations weaken the exciton-binding Coulomb force via electrical screening effects, leading to exciton ionization and dissociation. As a result, excitons are generally observed exclusively in semiconductors with low carrier concentrations ($<10^{16}/\text{cm}^3$), approximating those characteristic of intrinsic semiconductors. Additionally, while the probability of exciton recombination is higher than that of band-to-band recombination in intrinsic semiconductors, such excitonic recombination is observable only at ultralow temperatures.

5.2.1.5 Isoelectronic Center Recombination

The incorporation of isoelectronic dopants, which share chemical valence states with host semiconductor atoms, creates isoelectronic centers within the lattice. Differences in size and electronegativity between dopants and host atoms enable these centers to function as carrier- and exciton-trapping sites, acting as donors or acceptors. Critically, isoelectronic centers modulate electronic wavefunctions, enhancing the spatial overlap of electron distributions in the conduction band and hole distributions in the valence band at equivalent k-points. Based on quantum mechanical uncertainty principles, the momentum of trapped carriers exhibits significant delocalization. This allows momentum conservation in radiative recombination through isoelectronic centers regardless of whether the host semiconductor possesses a direct or indirect bandgap. Consequently, this mechanism substantially increases radiative recombination probabilities in indirect-gap semiconductors, which inherently exhibit low band-to-band radiative transition rates. The resultant enhancement in radiative transition efficiency is schematically demonstrated in Figure 5.5.

Isoelectronic center recombination occurs when electrons or holes are captured by isoelectronic impurity atoms. Because the interaction between an isoelectronic dopant and an electron constitutes a short-range potential, a square-well potential with depth V_0 and radius a is employed to investigate the existence of bound states within these centers, as follows:

$$V_0 a^2 > \frac{\pi^2 \hbar^2}{8m^*} \tag{5.13}$$

Compared with narrow-bandgap semiconductors, wide-bandgap semiconductors possess a larger effective mass of carriers, facilitating the formation of isoelectronic centers. Furthermore, due to the higher effective mass of holes relative to electrons, bound holes are typically observed in semiconductors. Regarding isoelectronic centers, the depths of their potential wells depend critically on the distortion of the semiconductor lattice. Therefore, introducing dopants with atomic sizes larger or smaller than the host atoms causes lattice distortion, leading to the formation of stable, deep isoelectronic centers.

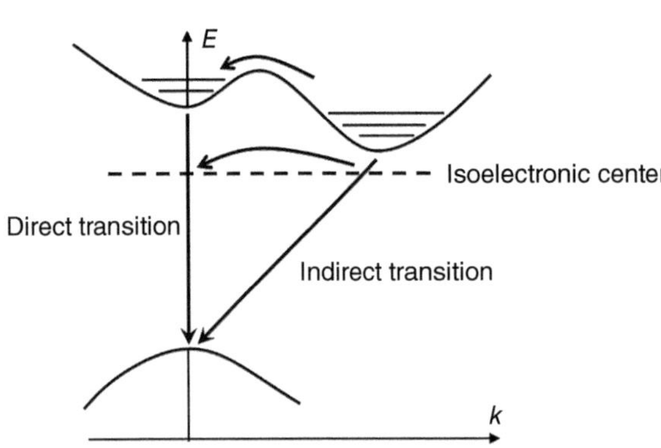

Figure 5.5 Diagram of change of electron distribution via isoelectronic centers.

5.2.2 Nonradiative Recombination Process

5.2.2.1 Multi-phonon Process

With increasing temperature, the increasing phonon population interacts with photons, facilitating energy transfer from photons to phonons, which reduces luminescence efficiency. This phenomenon is known as thermal quenching.

Recombination of carriers within deep energy levels in semiconductors may result in the emission of multiple phonons. Furthermore, for doped semiconductors, ionization of excited dopants induces charge alterations, leading to shifts in the equilibrium positions. Atomic vibrations facilitate the occurrence of vibrational states, with their characteristic frequency following a parabolic dependence. As illustrated in Figure 5.6, the ground state and excited state are represented by two distinct parabolic curves (curve I and curve II), each with unique vibrational equilibrium positions. Electrons transition from point A in the ground state to point B in the excited state. Subsequent energy release via phonons occurs as the electron relaxes to the corresponding vibrational equilibrium point B' of the excited state. The resultant energy dissipation is $(E(B)–E(B'))$. Finally, a portion of the excited electrons undergoes radiative relaxation from B' to point A' in the ground state, emitting light of energy $E(B')–E(A')$. The remaining electrons settle at point C, which is the intersection point of the excited-state and ground-state curves. From this point, electrons may undergo nonradiative recombination to point A in the ground state. The energy $(h\nu)$ associated with the electrons is then expressed as follows. The probability of process is proportional to

$$P \propto \exp\left(-\frac{\Delta E}{kT}\right) \tag{5.14}$$

where ΔE is energy difference between the B and C points. The probability increases with temperature, thus increased temperature may cause enhanced luminescent quenching. It has been found that the majority of electrons in deep-energy dopants undergo multi-phonon nonradiative recombination.

5.2.2.2 Auger Recombination

During Auger recombination, the energy released by recombining carriers is transferred to other carriers, exciting them into hot-carrier states. Subsequently, this

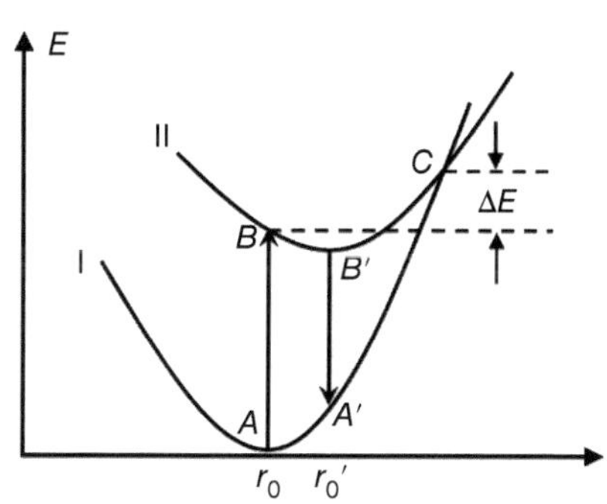

Figure 5.6 Diagram of configuration coordinate of multi-phonon process.

energy is dissipated non-radiatively through phonon emission. These fundamental Auger recombination processes are illustrated in the schematic diagrams of Figure 5.7.

During band-to-band recombination, the Auger recombination lifetime of electrons and holes depends on the concentrations of free carriers in semiconductors. For n-type and p-type semiconductors, the Auger lifetime (τ_{Au}) of minority carriers is determined by the following expressions, respectively.

$$\frac{1}{\tau_{Auger}} \propto n^2 p \tag{5.15}$$

$$\frac{1}{\tau_{Auger}} \propto p^2 n \tag{5.16}$$

where n and p represent the concentration of free electrons and holes. It is evident that the carrier lifetime for Auger recombination decreases with increasing carrier concentration. Consequently, the probability of Auger recombination is enhanced by the high concentration of majority carriers. This process can surpass radiative recombination in probability, leading to significant emission quenching.

5.2.3 Luminescent Semiconductors

The development of semiconductor luminescent devices is based on performance improvements in luminescent semiconductors. Compared with indirect-bandgap semiconductors, direct-bandgap semiconductors exhibit higher luminescent efficiencies. Consequently, most classic high-efficiency luminescent materials consist of direct-bandgap III–V and II–VI semiconductor composites as well as their alloys. In addition to these conventional semiconductors, novel materials, including organic polymers and metal halides, have emerged.

5.2.3.1 III–V Group Semiconductors

Of all the family of luminescent semiconductors, III–V Group alternatives received the most interest because of tunable emission ranging from UV to near-infrared

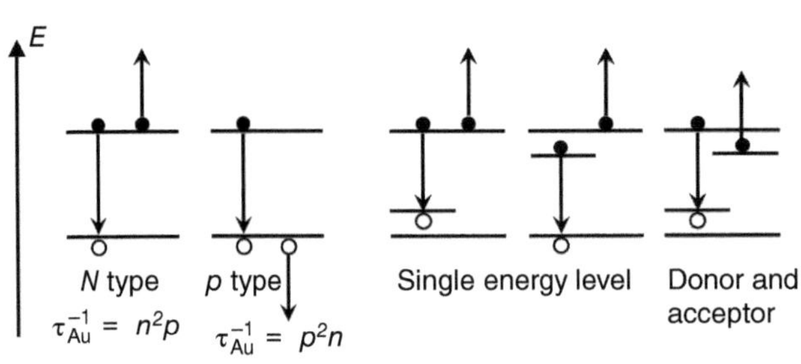

Figure 5.7 Diagram of Auger recombination of electrons and holes in semiconductors.

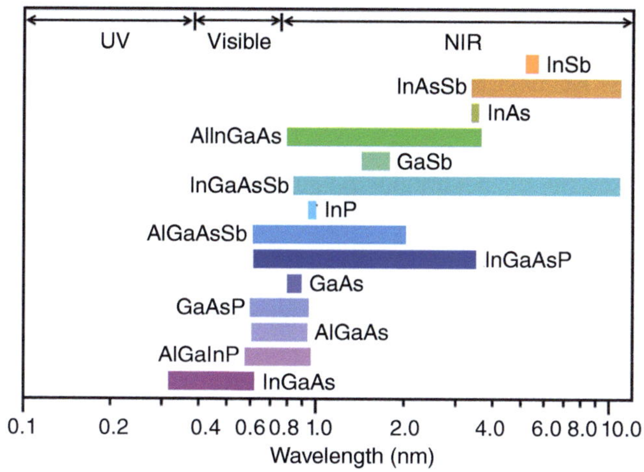

Figure 5.8 Emission wavelength ranges of III–V component semiconductors and their alloy in room temperature.

(NIR). The emission wavelengths of III–V Group semiconductors are summarized in Figure 5.8.

GaAs (Gallium Arsenide) As a typical direct-type III–V compound semiconductor, GaAs possesses a photon energy of ~1.42 eV, corresponding to an NIR wavelength of ~873 nm. GaAs finds application in microwave devices, semiconductor lasers, NIR LEDs, and solar cells. Additionally, it serves as fundamental material and epitaxy substrates for various light-emitting devices.

With respect to intrinsic GaAs, carrier mobility decreases with increasing carrier concentration. At room temperature, at a carrier concentration of $10^{18}/cm^3$, the electron and hole mobilities are 2000 and 150 $cm^2/(s\,V)$, respectively. The electron lifetime and diffusion length in n-type GaAs are 10^{-9} s and 2 μm, respectively. Silicon atoms can be incorporated into GaAs via liquid-phase epitaxy (LPE), occupying gallium and arsenic sites to act as donors and acceptors, respectively. Owing to higher electron mobility compared to holes, GaAs-based LEDs demonstrate p-region emission.

GaP (Gallium Phosphide) As a typical indirect-bandgap semiconductor, GaP exhibits a sphalerite structure with a wide bandgap of 2.26 eV. Consequently, dopants are introduced as isoelectronic centers to trap excitons, enabling efficient red, yellow, and green excitonic emission. LPE enables PN junction growth in GaP substrates, where Zn–O pairs function as isoelectronic centers in highly efficient (15%) GaP:ZnO red LEDs. Nitrogen atoms are incorporated into GaP through LPE and gas-phase diffusion methods, facilitating the fabrication of green LEDs emitting at 565 nm.

GaN (Gallium Nitride) As a prominent direct-bandgap semiconductor, GaN exhibits excellent characteristics including a wide bandgap, high thermal conductivity, and good chemical stability. At room temperature, GaN possesses a bandgap of 3.39 eV and a wurtzite crystal structure. Single crystals are typically grown on SiC or sapphire substrates using epitaxial methods. Intrinsic GaN is characterized as an *n*-type material with carrier concentrations ranging from 10^{16} to $10^{18}/cm^3$. *P*-type conductivity is achieved through zinc or magnesium doping, enabling its application in efficient blue LEDs. Furthermore, due to the similar properties of GaN, InN, and AlN, the ternary solid solution $Ga_{1-x-y}In_xAl_yN$ exhibits a tunable bandgap from 1.9 to 6.2 eV, meeting the requirements for diverse optoelectronic applications.

$GaAs_{1-x}P_x$ With a sphalerite structure, the $GaAs_{1-x}P_x$ solid solution is derived from GaAs and GaP components. At room temperature, $GaAs_{1-x}P_x$ exhibits a direct-bandgap transition for $x < 0.45$ and an indirect-bandgap transition for $x > 0.45$. Increasing the phosphorus composition ratio (x) reduces the external quantum efficiency (EQE) and emission wavelength of the material. The corresponding blueshift in the emission spectra follows the relationship with the composition parameter x as described by

$$\frac{1240}{\lambda} = 1.43 + 1.23x \tag{5.17}$$

Despite the reduction of EQEs of $GaAs_{1-x}P_x$ with increased x ratios, the sensitivity of human beings to the corresponding emission may enhance because of the reduction of luminescent wavelengths. As a result, emission brightness reaches the maximum value at $x = 0.40$, corresponding to the emission wavelength of 650–660 nm.

5.2.3.2 II–VI Group Semiconductors

II–VI Group semiconductors consist of atoms from Group II (Zn, Cd, Hg) and Group VI (S, Se).

ZnS (Zinc Sulphur) Among II–VI semiconductors, ZnS possesses a wide bandgap of 3.6 eV and has been applied in electroluminescence LEDs, where it is also doped to function as *n*-/*p*-type layers during LED fabrication.

ZnTe (Zinc Telluride) ZnTe is considered a direct-bandgap *n*-type semiconductor with a bandgap value of 2.26 eV at room temperature. This makes it suitable as an active material in visible and NIR optoelectronic devices; it is therefore employed in green LEDs and NIR modulation devices.

ZnO (Zinc Oxide) Intrinsic ZnO exhibits efficient band-to-band emission at a wavelength of 368 nm due to its large bandgap of 3.37 eV. This emission is attributed to the formation of free excitons within the ZnO lattice, stemming from its high exciton-binding energy of 60 meV. However, broadband green and yellow luminescence is observed in bulk and nanocrystalline ZnO. These emissions originate from intrinsic defects, such as oxygen vacancies, oxygen traps, and zinc vacancies, formed during the ZnO preparation process. Additionally, the incorporation of Li^+ and H^+ dopants enables green and yellow emission in ZnO, indicating its potential for tunable emission. Furthermore, ZnO's stable structure and chemical properties render it a promising emitter for UV and visible LEDs.

Cadmium Chalcogenide Nanocrystals Cadmium chalcogenides, such as CdS (~2.5 eV), CdSe (~1.7 eV), and CdTe (1.51 eV), have lower bandgap values. Combined with high defect densities in their bulk forms, these properties limit their suitability as efficient emitters in LEDs. Consequently, significant research interest has focused on their nanocrystalline forms. Benefiting from quantum confinement effects and large surface energies, the luminescent properties of cadmium sulfide nanocrystals can be finely modulated. A reduction in nanocrystal size increases their bandgap, resulting in a blueshift of their emission wavelength. Furthermore, coating strategies can suppress defects and facilitate the formation of quantum-well structures within these nanocrystals, significantly enhancing their luminescence efficiency. For example, increasing the size of CdSe nanocrystals causes a redshift in emission wavelength from 503 nm (radius = 1.35 nm) to 613 nm (radius = 2.7 nm). Similarly, increasing the reaction time to 180 minutes for CdS shells growth on CdSe cores results in a shell thickness of 2.3 nm, corresponding to approximately 7 molecular layers. This leads to a redshift in both photoluminescence (PL) and absorption spectra, along with variations in emission full width at half maximum (FWHM) and PL quantum yield (PLQY), as shown in Figure 5.9. It is noteworthy that the maximum PLQY of CdSe–CdS core–shell nanocrystals can reach 97%, substantially higher than that of intrinsic CdSe nanocrystals (~35%). This highlights the beneficial role of the core–shell structure in enhancing PLQY [1].

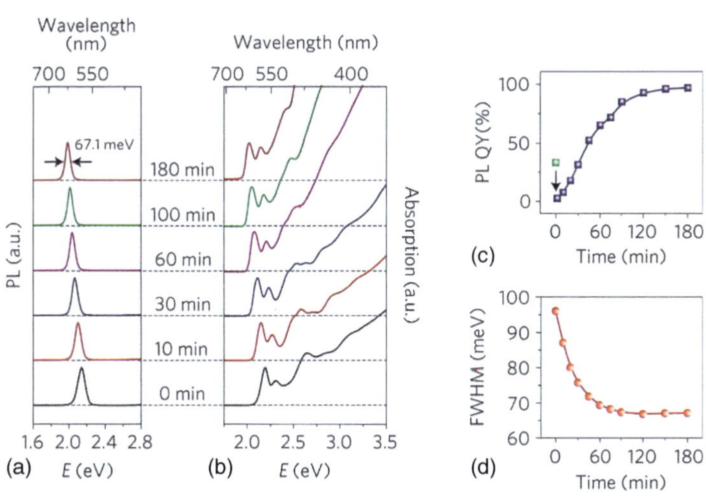

Figure 5.9 Temporal evolution of (a) photoluminescence, (b) absorption, (c) PLQY (green square shows the original PLQY of CdSe nanocrystals), and (d) FWHM of the PL peak of the CdSe–CdS core–shell nanocrystals during the shell growth reaction. *Source:* Chen et al. [1].

5.2.3.3 Silicon Carbide (SiC)

Silicon carbide (SiC) is a Group IV compound semiconductor with a stoichiometric ratio of silicon to carbon atoms of 1:1. Approximately 200 distinct crystal structures have been identified for SiC, exhibiting an indirect bandgap tunable within the 2–3 eV range. As a significant wide-bandgap semiconductor, SiC can incorporate various dopants to achieve emission spanning from red to near-ultraviolet (UV) wavelengths.

5.2.3.4 Organic Materials

Under optical excitation, singlet and triplet excitons generated within organic molecules undergo radiative relaxation, resulting in photon emission. For small-molecule organic emitters, conjugated heterocyclic rings and chromophores critically determine their photoluminescent properties. Thus, introducing unsaturated moieties such as double bonds, benzene rings, and chromophores modulates the optical performance. These materials find applications in organic light-emitting diodes (OLEDs), DNA diagnostics, optical sensors, and fluorescent dyes. However, emission quenching and aggregation phenomena in small-molecule organic materials may impede their practical utilization.

Additionally, polymeric emitters possessing conjugated systems or chromophores demonstrate excellent optical characteristics. Compared to small-molecule counterparts, polymers often exhibit higher luminous intensity, yet their complex processing due to low solubility and fusibility presents a significant challenge to commercialization.

5.2.3.5 Phosphors

Phosphors absorb high-energy radiation and subsequently release this energy as light (luminescence) or heat. Doping with lanthanide ions or other heterovalent ions introduces impurity energy levels within the phosphor host lattice. Energy transfer from the host matrix to these dopant ions facilitates excitonic emission from the excited dopant states. Consequently, by tailoring the chemical composition and dopants, phosphors can generate narrow-band emission across various wavelengths, enabling their application as efficient emitters in lighting and displays. Four primary host materials are used for phosphors in LEDs: yttrium aluminum garnet (YAG), silicates, nitrides, and sulfides. These phosphors are typically synthesized via high-temperature solid-state sintering or solution precipitation methods.

5.2.3.6 Metal Halides

Metal halides crystallize with the general chemical formula ABX_3, where A represents alkali metal cations (Cs^+, K^+, Rb^+, and Na^+), B denotes divalent metal cations (Pb^{2+}, Sn^{2+}, Zn^{2+}, Eu^{2+}, Mn^{2+}, etc.), and X signifies halide anions (Cl^-, Br^-, I^-) [2]. $CsPbBr_3$ exemplifies a classic ABX_3-type inorganic metal-halide perovskite (IMHP), adopting a cubic phase and a three-dimensional (3D) framework. In its structure, each Pb^{2+} cation is octahedrally coordinated by six Br^- anions, while Cs^+ cations occupy spaces formed by eight corner-sharing $[PbBr_6]$ octahedra. Beyond lead-based compounds, the synthesis of lead-free metal halides involves substituting Pb^{2+} with ions of various valences, including monovalent (Cu^{+}), divalent (Sn^{2+}, Zn^{2+}, Mn^{2+}), trivalent (Bi^{3+}, Sb^{3+}, In^{3+}), and tetravalent (Ti^{4+}, Zr^{4+}, Hf^{4+}) elements. Introducing monovalent, divalent, trivalent, or tetravalent B-site cations yields diverse crystal structures typically derived from the 3D ABX_3 perovskite motif. Metal halides with trivalent (B^{3+}) or tetravalent (B^{4+}) B-site cations feature the respective formulas $A_3B^{3+}{}_2X_9$ and $A_2B^{4+}X_6$, while a combination of monovalent and trivalent cations results in $A_2B^{+}B^{3+}X_6$-type halides [3].

In hybrid metal halides, organic molecules (possessing short/long chains or functional groups) occupy the A-site, often disrupting and spatially separating the metal-halide inorganic frameworks, leading to softer crystal structures. Consequently, most hybrid metal halides crystallize in lower-dimensional structures: two-dimensional (2D), one-dimensional (1D), or zero-dimensional (0D). Furthermore, these materials often undergo structural distortions under external stimuli (e.g., light, pressure, temperature), enabling tunable optoelectronic properties.

Metal halides exhibit defining characteristics of tunable emission wavelengths and variable PL decay times, distinguishing them from conventional semiconductors [4–6]. As noted previously, altering the halide composition modifies electronic band structures and bandgaps, enabling IMHP emission energy and wavelength shifts that span the entire visible spectrum (shown in Figure 5.10a). However, Figure 5.10b demonstrates that the PL decay time of metal-halide nanocrystals drastically decreases from approximately 34.5 to 2.5 ns as the halide anions change from I^- to Br^- to Cl^-. These decay times were derived by fitting PL decay transients using single- or double-exponential functions. Accordingly, the PLQYs of $CsPbX_3$ nanocrystals also decrease, falling from ~93% ($CsPbI_3$) and ~92% ($CsPbBr_3$) to ~10% ($CsPbCl_3$). The reductions in both PL decay time and PLQY correlate with the increasing concentration of halide vacancies, which act as nonradiative recombination centers, in $CsPbX_3$ crystals containing Cl^- anions. This increased vacancy concentration in Cl^--based nanocrystals arises due to the comparatively lower formation energy for vacancies in $CsPbCl_3$ than in $CsPbBr_3$ or $CsPbI_3$ [7].

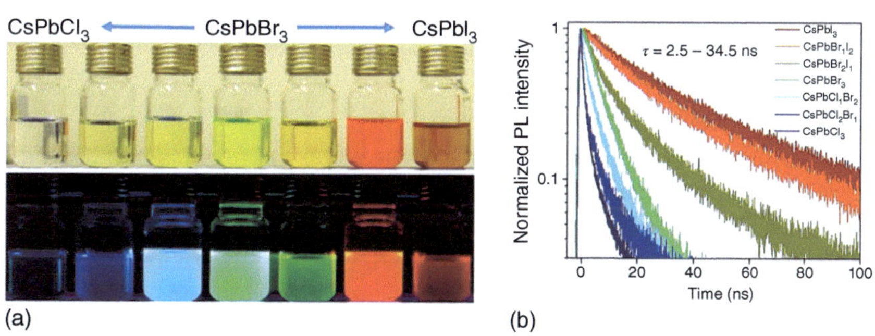

Figure 5.10 (a) Photograph of colloidal dispersions of $CsPbX_3$ nanocrystals with different halide (X = Cl, Br, and I) components in hexane under room light (top) and UV light (bottom, $\lambda = 367$ nm). (b) PL decay dynamics of the $CsPbX_3$ nanocrystals in (a). *Source:* Tong et al. [7].

Apart from lead halides, most lead-free metal halides exhibit novel luminescent properties governed by a self-trapped exciton (STE) emission mechanism. Upon photoexcitation, electrons transition from ground state to high-energy states above the CBM. These hot carriers subsequently relax to states proximate to the CBM via an ultrafast "cooling" process occurring within 1–100 ps. Owing to lattice flexibility and enabled electronic orbital hybridization in low-dimensional lead-free metal halides, the excited carriers interact with lattice phonons, reducing the system energy and inducing elastic structural distortion. This energetically favorable process, termed "Jahn-Teller distortion", establishes the energy disparity between the bandgap and emission energy, known as the Stokes shift. Beyond substantial Stokes shifts, STE emission in these materials demonstrates broadband luminescence spectra with large FWHM values and prolonged decay lifetimes. These characteristics collectively arise from strong electron–phonon interactions within their pliant crystal lattices.

5.3 Structure and Feature of LEDs

5.3.1 Principle and Structure of LEDs

The core structure of LEDs is the PN junction. The *p*- and *n*-regions are formed through heavy doping of the semiconductor, creating degenerate hole distributions in the *p*-region and degenerate electron distributions in the *n*-region. Without an external bias, a depletion region forms near the PN junction interface, devoid of free holes or electrons. Consequently, radiative recombination does not occur. In contrast, when a forward bias ($U_0 \sim E_g/e$) is applied across the LED, the applied voltage counteracts the depletion region potential, reducing its width. This enables holes from the *p*-region and electrons from the *n*-region to diffuse across the junction into the opposite regions. Consequently, holes and electrons coexist within the same region near the junction, facilitating radiative recombination. This process emits light with photon energy approximating the bandgap energy (E_g), as illustrated in Figure 5.11. It is important to note that the spectra of this radiative emission are characterized by narrow peaks rather than single-valued line spectra.

5.3.1.1 Homojunction LEDs

Classic semiconductors are fabricated through the epitaxial growth of doped semiconductor layers on compatible substrates, such as GaAs and GaP. As illustrated in Figure 5.12a, a *p*-type layer and an *n*-type layer are grown on the substrate. The relatively thin *p*-layer is crucial for minimizing light reabsorption and facilitating carrier injection. Critically, the *n*-layer is heavily doped to enhance radiative recombination within the *p*-layer. Interdigital electrodes are incorporated to improve current spreading, enhancing light-extraction efficiency. Additionally, a *p*-layer may be formed within an n^+-type layer via a diffusion doping process, enabling the fabrication of diffused-junction planar LEDs as depicted in Figure 5.12b.

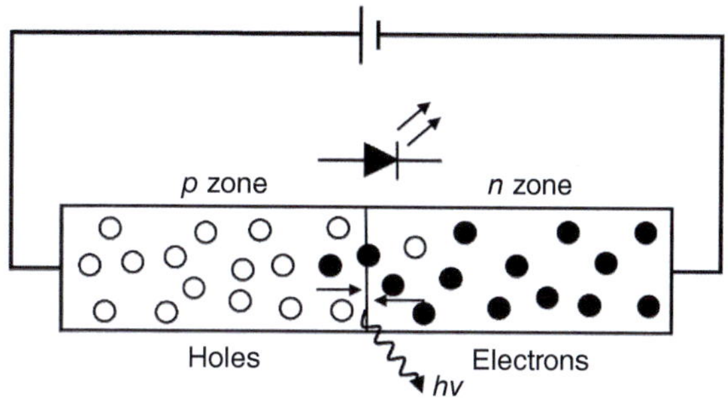

Figure 5.11 Schematic diagram of emission in PN junction.

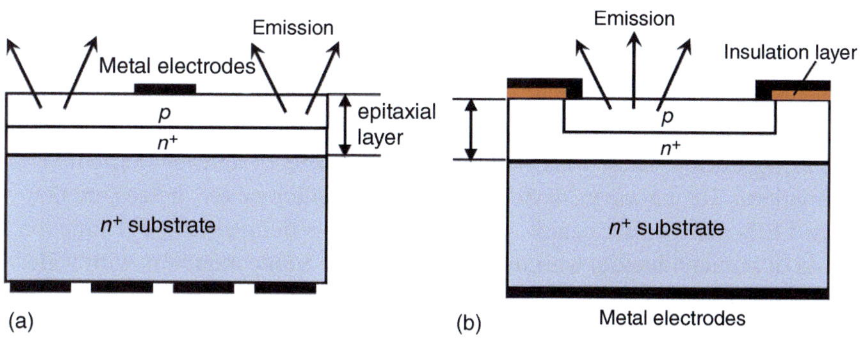

Figure 5.12 Schematic of structure of homojunction LEDs fabricated from (a) simultaneous growth of a n^+-layer and a p-layer and (b) a n^+-layer followed by a p-type doping strategy.

Furthermore, since the refractive index of semiconductors is higher than that of air, a portion of light generated within solid-state LEDs cannot escape the device. Light rays with emission angles exceeding the associated critical angle undergo total internal reflection at the solid/air interface, returning back into the LED material. This critical angle, defined as the angle between the light propagation vector and the interface normal, severely limits light extraction. For instance, the critical angle for a GaAs/air interface is only approximately 16°, meaning most generated light fails to escape. Consequently, GaAs-based LEDs exhibit inherently low luminous efficiencies, as illustrated in Figure 5.13a.

To improve light-extraction efficiency, shaping the emitting-layer surface into hemispherical geometries has been explored in Figure 5.13b; however, this approach significantly increases fabrication costs. Consequently, attaching separately fabricated transparent hemispherical plastic covers over planar LEDs has been employed to enhance light extraction while reducing manufacturing complexity, as demonstrated in Figure 5.13c.

5.3.1.2 Double-Heterojunction LEDs

In terms of homojunction LEDs, for example, top layer consisting of n-type layers, thin n-type layers are manufactured to decrease light reabsorption and enhance photon extraction. However, some injected electrons may recombine with holes on the surface of the n-type layers or become trapped by subsurface defects, potentially

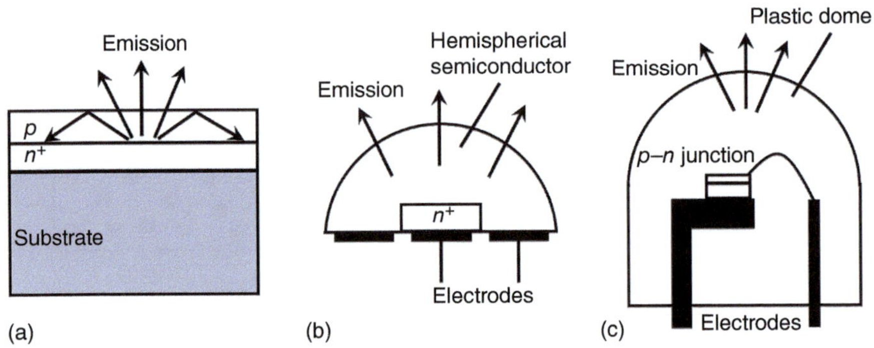

Figure 5.13 Schematic of LEDs with (a) planar structure exhibiting total internal reflection of part of photons, (b) hemispherical semiconductor enabling light injection, and (c) transparent plastic dome facilitating efficiencies of light injection and coupling.

leading to significant reductions in luminescent efficiency due to nonradiative recombination. To improve device luminescent efficiencies, heterojunction-structure LEDs can be fabricated. A heterojunction is defined as a PN junction composed of semiconductors with different bandgaps. Semiconductors with wider bandgaps have been found to exhibit low refractive indices. Consequently, the heterojunction acts as a dielectric waveguide, confining light emission from the emissive region.

To further enhance luminescent intensity, double-heterojunction structures are utilized in LED fabrication. Figure 5.14 depicts a typical double-heterojunction LED configuration, where a thin GaAs layer (bandgap 1.4 eV) is sandwiched between two AlGaAs layers (bandgap 2 eV). This structure forms two distinct heterojunctions: one between the n^+-type AlGaAs and p-type GaAs and another between the p-type GaAs and the n-type AlGaAs. In the unbiased state, a potential barrier with energy eV_0 exists between the n^+-type AlGaAs and the p-type GaAs layer, hindering electron diffusion from the n^+-type AlGaAs into the GaAs layer.

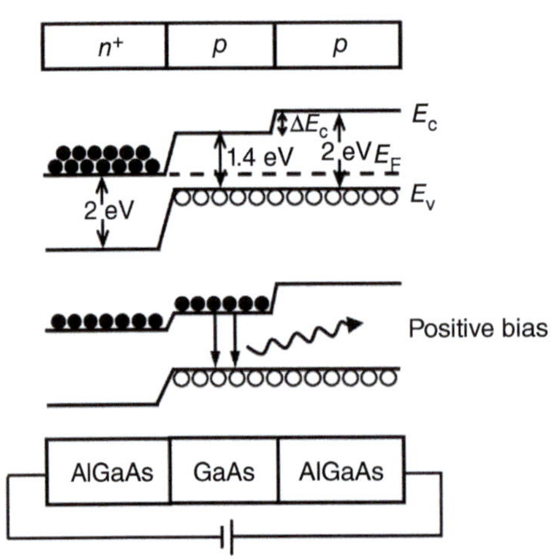

Figure 5.14 Device structure and energy band diagrams of double-heterojunction LEDs.

Additionally, a potential barrier with energy ΔE_c between the p-type GaAs and the p-type AlGaAs layer impedes electron diffusion from the GaAs to the AlGaAs layer. Applying a positive bias reduces the potential barrier between the n^+-type AlGaAs and p-type GaAs layer, facilitating electron diffusion from the n^+-type AlGaAs into the GaAs layer. Conversely, the potential barrier between the p-type GaAs and p-type AlGaAs layers remains, confining the electrons within the n-type GaAs layer. Here, they can recombine with holes in the valence band, generating radiative photons. Furthermore, the AlGaAs layer's larger bandgap compared to GaAs prevents reabsorption of photons emitted from the GaAs layer, thereby enhancing light-extraction efficiency. Negligible lattice mismatch between AlGaAs and GaAs also helps suppress interface defects. Consequently, the luminescence efficiency of double-heterojunction LEDs is significantly higher than that of conventional homojunction LEDs.

5.3.2 Optical Properties of LEDs

5.3.2.1 Efficiency of LEDs

Efficiency is identified as the most important parameter to evaluate the performance of LEDs, which can be divided into internal quantum efficiency (IQE), extraction efficiency, EQE, and luminous efficiency.

IQE of LEDs is defined as ratio of number of the emitting photons to the injected electron/hole pairs, which is calculated as the following equation:

$$\text{IQE} = \frac{P_{\text{int}}/h\upsilon}{I/e} \tag{5.18}$$

where P_{int} and I represent the light emission power in active region and the injected current, $h\upsilon$ is the photon energy, and e is the elementary charge of electrons.

Photons are subject to reflection at semiconductor–air interfaces and reabsorption within the semiconductor material. Consequently, the extraction efficiency is defined as the ratio of the number of photons emitted into the air to the number generated in the active region. It has been reported that achieving enhancement of extraction efficiency in LEDs remains challenging due to photon losses before emission, which can reach approximately 50%.

The EQE is defined as the ratio of the number of photons emitted into the air to the number of injected electron–hole pairs. The EQE equals the product of the IQE and the extraction efficiency, as expressed by the following equation:

$$\text{EQE} = \frac{P/h\upsilon}{I/e} \tag{5.19}$$

where P is emergent light power of LEDs.

Luminous efficiency is defined as the ratio of emergent light power to electric power of LEDs.

5.3.2.2 Luminescent Spectra

Energy distribution of electrons in conduction band and holes in valence bands may result in broadening of spectra, followed by a slight shift of the wavelength peak.

The wavelength peak (λ) in a given spectrum is calculated by

$$\lambda = \frac{hc}{E_g} \tag{5.20}$$

and the FWHM is used to evaluate broadening of a spectrum as follows:

$$\text{FWHM} = \frac{1.8\,kT\lambda^2}{hc} \tag{5.21}$$

In room temperature, the peak wavelength of LEDs based on GaAs locates at ~870 nm with FWHM of 28 nm.

5.3.2.3 Distribution of Emission Intensity

Due to the difference in refractive indices between semiconductors and air, as well as the effect of the LED surface on the propagation of photons, the emission intensity of LEDs is nonuniform. For planar LEDs, the emission intensity follows a Lambertian distribution, given by

$$I_{\text{air}} = \frac{P_{\text{source}}}{4\pi r^2}\frac{n_{\text{air}}^2}{n_s^2}\cos\phi \tag{5.22}$$

where n_{air} and n_s represent reflective indexes of air and semiconductors, respectively. P_{source} is total power of LEDs, and ϕ is an included angle. In terms of hemispherical LEDs, anisotropy of their emission intensity is obvious. Figure 5.15 demonstrates distribution of light intensity of LEDs with various types of surface structures.

5.3.2.4 Light-Extraction Efficiency

As shown in Figure 5.16a, because the refractive index of the emitting material (n_s) exceeds that of air (n_{air}), photons with higher incident angles cannot be extracted from the emitters, resulting in a reduction in the light-extraction efficiency

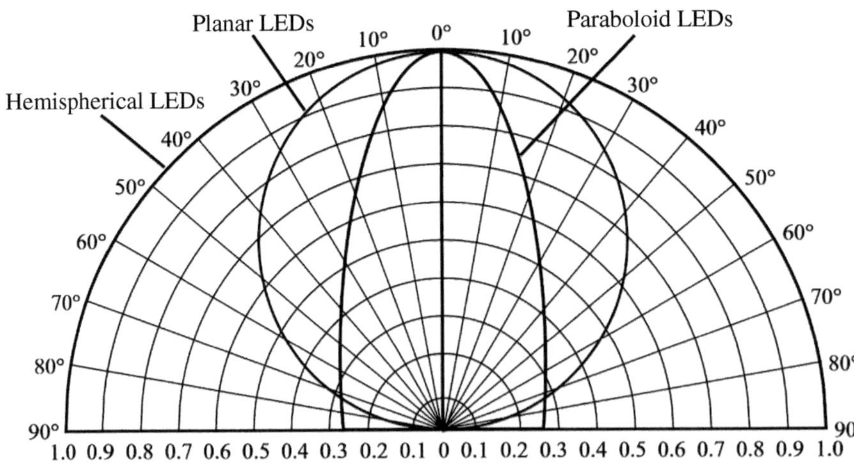

Figure 5.15 Distribution of light intensity of LEDs with various types of surface structures.

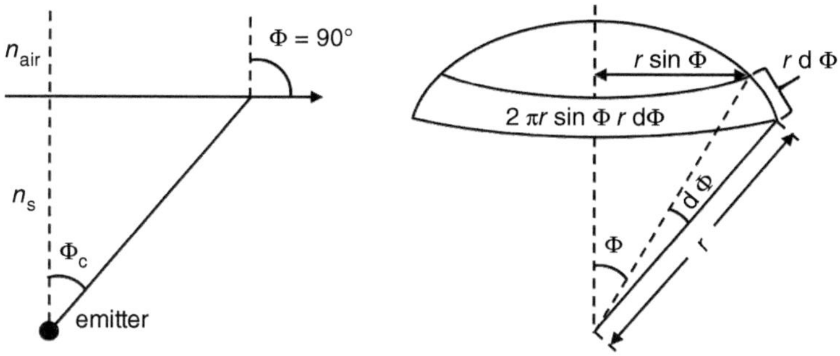

Figure 5.16 Schematic diagrams of (a) total reflection in the interface of a semiconductor and air and (b) surface area of the spherical crown.

of light sources. The incident angle (ϕ_i) and the refracted angle (ϕ_r) are determined by Snell's law at the interface between the semiconductor and air.

$$n_s \sin \phi_i = n_{air} \sin \phi_r \tag{5.23}$$

Thus, the critical angle of total reflection (ϕ_c) is evaluated by the following equation:

$$\sin \phi_c = \frac{n_{air}}{n_s} \sin 90° = \frac{n_{air}}{n_s} \tag{5.24}$$

As shown in Figure 5.16b, the emitting light cone of a given source, whose apex angle is $2\phi_c$, is used to explain the light-extraction efficiency. Firstly, surface area of the spherical crown (A) is calculated as follows:

$$A = \int dA = \int_{\phi=0°}^{\phi_c} (2\pi r \sin \phi) \cdot r d\phi = 2\pi r^2 (1 - \cos \phi_c) \tag{5.25}$$

Given uniform distribution of light emission, total power of light escaping from emitters to air (P_{escape}) is calculated by the following equation:

$$P_{escape} = P_{source} \frac{2\pi r^2 (1 - \cos \phi_c)}{4\pi r^2} \tag{5.26}$$

where P_{source} is radiant total power of light sources.

Thus, light-extraction efficiency of LEDs (η_{LEE}) is calculated as follows:

$$\eta_{LEE} = \frac{P_{escape}}{P_{source}} = \frac{1 - \cos \phi_c}{2} \tag{5.27}$$

Considering refractive index of semiconductor emitters is much higher than that of air, e.g., $n_{GaAs} = 3.4$ and $n_{GaN} = 2.5$, $\phi_c \approx 0°$, and the η_{LEE} is calculated as follows equation:

$$\eta_{LEE} = \frac{1 - \cos \phi_c}{2} \approx \frac{1}{2}\left[1 - \left(1 - \frac{\phi_c^2}{2}\right)\right] = \frac{1}{4}\phi_c^2 \approx \frac{1}{4}(\sin \phi_c)^2 = \frac{1}{4}\left(\frac{n_{air}}{n_s}\right)^2 \tag{5.28}$$

If light emission of both side and bottom of LED chips is negligible, light-extraction efficiency of planar GaN-based LEDs is about $\eta_{LEE} = \frac{1}{4}\left(\frac{1}{2.5}\right)^2 = 4\%$.

5.3.2.5 Luminous Flux and Radiant Flux

Luminous power describes the light energy emitted by a source over a specific duration. The standard unit of luminous flux is the lumen. Radiant flux (Φ_E), specific to LEDs, quantifies the total radiant energy emitted by an LED per unit time. When radiant flux varies with emission wavelength, the spectral radiant flux at a given wavelength, $\Phi_\mathrm{E}(\lambda)$, expressed in W/nm, can be determined using the following equation:

$$\Phi_\mathrm{E} = \int \Phi_\mathrm{E}(\lambda)\,\mathrm{d}\lambda \tag{5.29}$$

The wavelength-dependent sensitivity for human eyes is characterized by the spectral luminous efficiency function ($V(\lambda)$). Luminous flux (Φ_v) is derived from the following equation:

$$\Phi_v = 683 \int_{380}^{780} \Phi_\mathrm{E}(\lambda) V(\lambda)\mathrm{d}\lambda \tag{5.30}$$

5.3.2.6 Luminous Intensity, Luminance, and Illuminance

The ratio of the luminous flux (Φ) to a small solid angle (Ω) is defined as luminous intensity (I) of a given LED chip as the following equation:

$$I = \frac{\mathrm{d}\Phi}{\mathrm{d}\Omega} \tag{5.31}$$

The unit of luminous intensity is candela (Cd).

Luminance (L) is defined as the ratio of luminous intensity of a given small area to orthographic projection area as follows:

$$L = \frac{\mathrm{d}I}{\mathrm{d}S\cos\theta} = \frac{\mathrm{d}^2\Phi}{\mathrm{d}\Omega\,\mathrm{d}S\cos\theta} \tag{5.32}$$

where θ represents the included angle of obtaining direction and normal line of the area mentioned earlier.

In terms of a given accepted surface, its illuminance (E) is defined as the ratio of luminous flux and area as follows:

$$E = \frac{\mathrm{d}\Phi}{\mathrm{d}S} \tag{5.33}$$

The unit of illuminance is lux (lx). When light with luminous flux of 1 lm hits on a surface with an area of 1 m^2 uniformly, the illuminance is 1 lx. Thus, 1 lx = 1 lm/m^2.

5.3.2.7 Temperature Effect

It has been established that the light intensity of LEDs decreases with increasing temperature. This phenomenon primarily results from enhanced nonradiative recombination, increased surface recombination, and carrier loss within the heterojunction barrier. At room temperature, the relationship between LED light intensity and temperature is commonly described by the following empirical formula:

$$I = I_{300\,\mathrm{K}}\, e^{-\frac{T-300\,\mathrm{K}}{T_\mathrm{c}}} \tag{5.34}$$

where T_c is characteristic temperature in given LEDs.

5.4 Process of LEDs

5.4.1 Growth of Emitting Materials in LEDs

As described earlier, LEDs feature complex multilayer structures comprising PN junctions, transporting layers, and quantum wells. By altering their chemical composition and thickness, optoelectronic properties of LEDs—including emission wavelength, luminescence efficiency, and operating voltage—can be precisely regulated. Consequently, the fabrication of functional thin films is critical for LED performance. For inorganic-component thin films, high-quality layers are prepared using various epitaxial methods such as metal–organic chemical vapor deposition (MOCVD), molecular beam epitaxy (MBE), and LPE.

5.4.1.1 Growth Model of Emitting Materials in LEDs

When a significant lattice constant mismatch exists between the thin-film material and the substrate, adatoms adsorb and nucleate preferentially at low-surface-energy sites on the substrate surface. This nucleation is followed by the aggregation of atoms to form small atomic islands, initiating thin-film growth (Figure 5.17a). This process is governed by the three-dimensional (3D) growth mode (Volmer–Weber mode). As deposition continues, these islands develop into columnar structures that subsequently coalesce, ultimately forming a continuous film characterized by a rough surface morphology. This microstructural evolution adheres to an island-mediated growth mechanism, wherein the developing film incorporates a high density of misfit dislocations to accommodate lattice strain. In contrast, the two-dimensional (2D) layer-by-layer growth mode (also known as the nucleation–growth model or Frank–van der Merwe [FVM] growth mode) typically occurs in material systems exhibiting either lattice-matched configurations or minimal lattice mismatch, combined with strong interfacial bonding energy. This mechanism proceeds through sequential two-dimensional nucleation events followed by lateral expansion

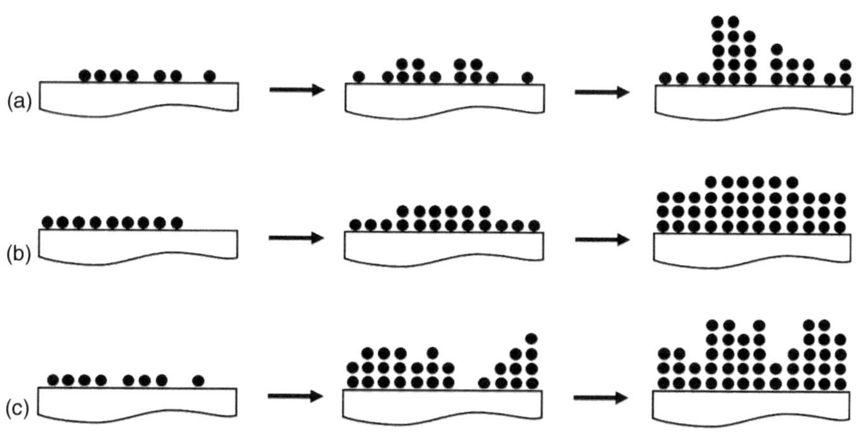

Figure 5.17 Schematic diagrams of epitaxy growth of films via (a) the island mode, (b) the layer–layer mode, and (c) the layer–island growth mode.

across the substrate surface; each atomic layer completes its formation before subsequent layer deposition commences (Figure 5.17b). This cyclical process enables the fabrication of epitaxial layers with exceptional structural perfection and atomically smooth surfaces. Furthermore, the layer-plus-island growth mode (Stranski–Krastanov growth mode) initially involves the formation of a thin, continuous two-dimensional wetting layer. Upon reaching a critical thickness for elastic strain energy relaxation, subsequently deposited material transitions into the nucleation and growth of three-dimensional islands, with preferential adatom accumulation on these islands (Figure 5.17c). In accordance with the principle of energy minimization, the facets formed on these islands typically correspond to low-index crystallographic planes characterized by lower surface energy.

5.4.1.2 Growth of Emitting Layers

Metal–Organic Chemical Vapor Deposition (MOCVD) MOCVD is a chemical vapor deposition technique that employs metal–organic compounds as precursors. The growth mechanism for III–V material epitaxy via MOCVD proceeds as follows: hydrogen or inert gases (nitrogen or argon) serve as carrier gases to transport both metal–organic (MO) sources containing Group III elements and nonmetal hydrides (NH_3, AsH_3, or PH_3) containing Group V elements to the heated substrate within the reaction chamber. Subsequently, a sequence of chemical and physical interactions occurs in both the gas phase and at the gas–solid interface, ultimately resulting in the formation of an epitaxial layer on the substrate.

MOCVD systems had been deployed globally for industrial LED production and technological research. Figure 5.18 illustrates a schematic of a typical MOCVD growth process. Within the MOCVD reaction chamber, precise control of key deposition parameters—such as temperature, pressure, total gas flow rate, and substrate rotation speed—is essential to establish a stable laminar flow near

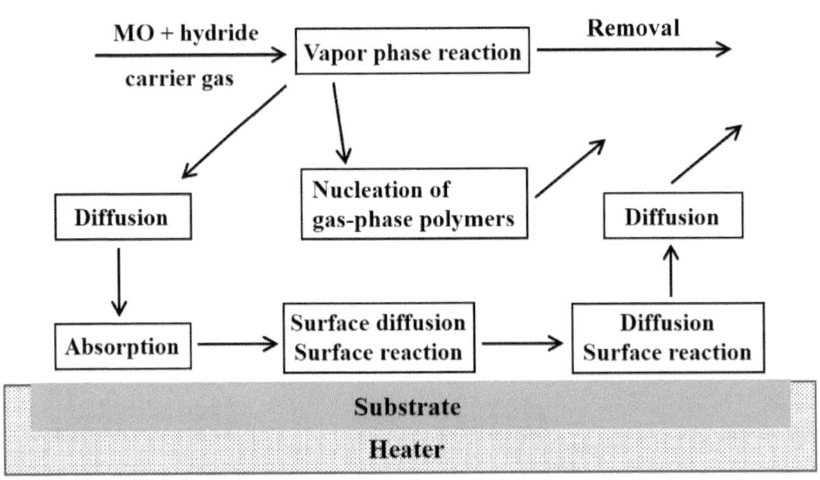

Figure 5.18 Schematic of a typical MOCVD growth process.

the substrate surface, referred to as the boundary layer, thereby ensuring stable, reproducible, and uniform epitaxial growth. The material growth process involves the following sequential stages:

(I) **Mass transport process:** The carrier gas delivers the MO precursors and hydride reactants into the reaction chamber. As the gas flows toward the heated substrate, its temperature increases gradually, and the reactants diffuse through the boundary layer toward the substrate surface.
(II) **Adsorption of reactants on the substrate surface:** MO compounds and nonmetal hydrides form intermediate adducts. Upon further temperature elevation, the MO precursors, hydrides, and adducts undergo progressive thermal decomposition, leading to gas-phase nucleation. The resulting decomposition products diffuse to and adsorb onto the substrate surface.
(III) **Surface reaction process:** The adsorbed reactants undergo surface migration and chemical reactions on the surface.
(IV) **Lattice incorporation:** The resulting III–V compounds form covalent bonds with the substrate surface through a "grafting" mechanism, subsequently becoming incorporated into the crystal lattice to form the epitaxial layer.
(V) **Byproduct removal:** Reaction byproducts desorb from the substrate surface, diffuse across the boundary layer, and are evacuated by the carrier gas. A fraction of unreacted gaseous species may also be swept directly out of the reaction chamber by the gas stream.

The MOCVD system and its associated techniques exhibit several notable advantages: (i) Versatility: The use of diverse MO precursors as source materials enables the growth of a wide range of compound semiconductors. (ii) Precise control: Optimization of growth parameters—including temperature, gas flow rates, V/III ratio, and chamber pressure—permits precise regulation of ultrathin layer thickness, composition, and interfacial properties. (iii) Reproducibility: The technology facilitates the repeated deposition of large-area uniform epitaxial layers, rendering it highly suitable for industrial-scale manufacturing. (iv) Cost-effectiveness: Compared to alternative epitaxial growth methods, MOCVD systems present distinct cost advantages.

Molecular Beam Epitaxy (MBE) MBE is an ultrahigh-vacuum (UHV) thin-film deposition technique enabling atomic-level control over crystalline material growth. Its fundamental principle relies on the reaction of molecular or atomic beams with a heated crystalline substrate under precisely controlled conditions. The MBE process involves several key stages:

(I) **Beam generation:** Thermal effusion cells containing ultrapure elemental sources (e.g., Ga, As, Al) undergo heating to generate directed beams of atoms or molecules. These beams traverse ballistically through the UHV environment (typically $<10^{-10}$ Torr) to reach the substrate surface.
(II) **Surface reactions:** Upon reaching the heated substrate (typically 400–600 °C for III–V compounds), the impinging atoms migrate across the surface until

they incorporate into the growing crystal lattice. Growth occurs layer by layer, governed by surface kinetics and thermodynamics.

(III) ***In-situ* monitoring:** Reflection high-energy electron diffraction (RHEED) provides real-time analysis of surface reconstruction and growth mode, enabling monolayer-level thickness control.

(IV) Intentional doping is achieved using dedicated effusion cells containing dopant materials (e.g., Si for *n*-type, Be for *p*-type doping in III–V semiconductors).

This precision renders MBE particularly valuable for fabricating sophisticated semiconductor heterostructures employed in quantum devices, high-electron-mobility transistors, and optoelectronic applications. The UHV environment minimizes impurity incorporation, while the comparatively low growth temperatures reduce interfacial interdiffusion.

Liquid-Phase Epitaxy (LPE) LPE is a solution-based epitaxial growth technique involving the controlled precipitation of crystalline materials from supersaturated metallic solutions onto single-crystal substrates. Its fundamental principle relies on the temperature-dependent solubility of solute atoms within a molten solvent, coupled with precise interfacial thermodynamics at the solidification front. The LPE process typically involves the following sequence:

(I) High-purity constituent elements (e.g., Ga, As for GaAs growth) are combined with a suitable solvent metal (typically Ga for III–V compounds) in precise stoichiometric ratios. The mixture undergoes heating beyond the saturation temperature (typically 500–900 °C for III–V materials) to form a homogeneous solution.

(II) **Substrate contact:** A carefully prepared substrate with an atomically clean and lattice-matched surface is brought into contact with the supersaturated solution.

(III) **Controlled cooling:** The system experiences precisely regulated cooling (typically 0.1–1.0 °C/min), inducing controlled supersaturation that drives epitaxial growth.

(IV) **Solution separation:** Upon achieving the desired thickness (typically 1–100 μm), the residual solution is removed from the substrate surface via mechanical wiping or rapid withdrawal from the growth zone.

The technique is particularly noted for producing materials with exceptional optoelectronic properties, establishing its historical significance for laser diodes and light-emitting devices. Although largely supplanted by vapor-phase methods for most advanced applications, LPE maintains relevance for specific material systems (e.g., III–V semiconductors, magnetic garnets) where its advantages in defect control and dopant incorporation are paramount. Critical process parameters—including cooling rate, solution composition, and substrate orientation—require careful optimization to govern layer thickness, surface morphology, and compositional uniformity.

5.4.2 Manufacture of LED Chips

Chip fabricationalso falls within the domain of semiconductor manufacturing, thus sharing fundamental processing principles with semiconductor chip production. However, it simultaneously exhibits distinctive characteristics. While specific fabrication sequences vary with different chip architectures, the underlying core processes remain consistent. In summary, LED chip manufacturing encompasses the following fundamental processes: evaporation, sputtering, photolithography, etching, and packaging.

Evaporation The fabrication of LED chip electrodes and interconnects employs metallic materials functioning as both electrical contacts and bonding interfaces. This process primarily utilizes evaporation techniques, a cornerstone of physical vapor deposition (PVD) technology. Through thermal evaporation under vacuum conditions, controlled deposition within the chamber produces uniform thin-film coatings on substrate surfaces, as shown in Figure 5.19. This methodology constitutes vacuum thermal evaporation, characterized by its high-purity deposition capabilities and precise thickness control at the micrometer scale. The evaporation process in thin-film metallization for optoelectronic devices typically comprises three sequential operational phases:

(I) **Source material thermal activation:** High-purity metals (e.g., aluminum, silver) undergo resistive heating in a crucible until reaching their sublimation points, initiating vapor-phase transition.

(II) **Establishment of vacuum environment:** The substrate assembly containing LED die arrays is positioned within a high-vacuum chamber, wherein internal pressure is reduced to the range of 10^{-5}–10^{-7} Torr via multistage pumping systems, effectively minimizing gas-phase molecular interference.

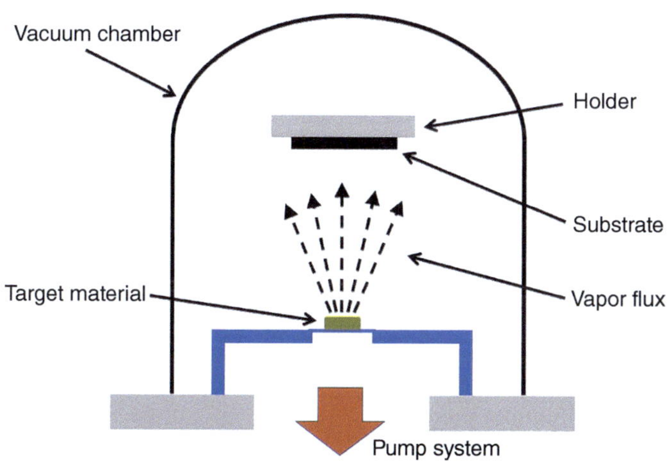

Figure 5.19 Schematic of thermal evaporation.

(III) **Directed vapor deposition:** Upon achieving thermal equilibrium, metallic atoms undergo ballistic transport through the mean-free-path–dominated vacuum environment. Subsequently, conformal polycrystalline coating formation occurs through surface diffusion-mediated nucleation on the semiconductor heterostructure interfaces.

This PVD methodology demonstrates superior film uniformity (coefficient of variation $\sigma < \pm 5\%$ across 200 mm wafers), exceptional thickness controllability (± 2 nm resolution via quartz crystal monitoring), and broad material compatibility. These attributes render it particularly suitable for fabricating advanced optoelectronic components, including ohmic contacts, current-spreading layers, and Bragg reflector stacks in III–V semiconductor devices.

Sputtering As a widely implemented PVD technique, sputtering is fundamentally employed for metallic thin-film formation on LED chip surfaces through momentum transfer mechanisms. The process initiates through the acceleration of plasma ions (ranging from 10^2 to 10^3 eV kinetic energy) within an electromagnetic confinement field. These energized particles undergo ballistic trajectories, colliding with the cathode target surface to induce atomic ejection via knock-on displacement cascades. The sputtered species exhibit a characteristic cosine angular distribution patterns in the horizontal plane, subsequently undergoing surface diffusion and nucleation processes that culminate in polycrystalline film growth on thermally regulated substrates. Figure 5.20 depicts the schematic representation of magnetron sputtering configuration, highlighting the balanced magnetic field confinement that enhances ionization efficiency and deposition rates.

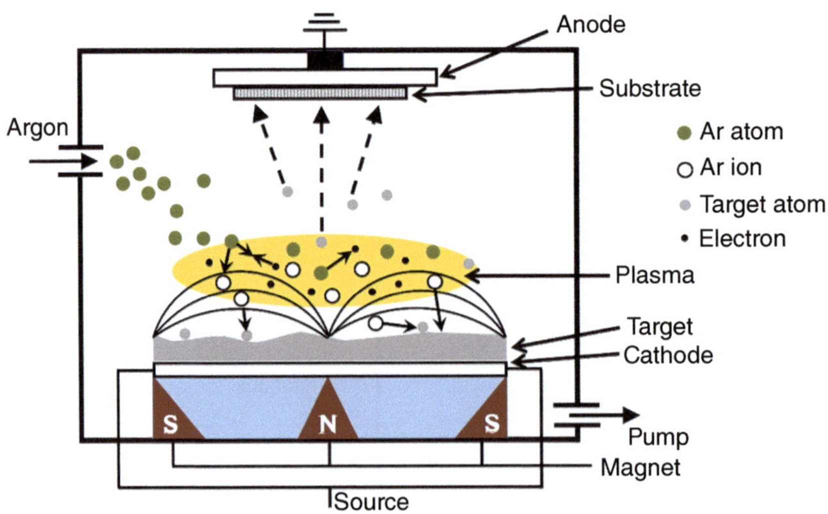

Figure 5.20 Schematic of magnetron sputtering process.

The magnetron sputtering deposition methodology comprises three primary stages:

(I) **Plasma generation and sustenance:** The substrate undergoes stringent surface pretreatment processes before being loaded into a high-vacuum deposition chamber (base pressure $\leq 10^{-6}$ Torr). Plasma initiation is achieved via radio-frequency (RF) or direct current (DC) magnetron discharge, facilitated by argon gas ionization under controlled partial pressure (typically 1–10 mTorr).

(II) **Thin-film deposition:** Energetic argon ions (50–1000 eV), generated in the plasma sheath, sputter target atoms through momentum transfer collisions. These ejected species subsequently condense on the LED substrate surface via PVD mechanisms, forming nanostructured metallic coatings with controlled stoichiometry.

Photolithography and Etching Photolithography constitutes a pivotal step in the fabrication of LEDs, enabling the precise patterning of semiconductor layers to define device geometries and electrical contacts. The process commences with the deposition of a photosensitive resist layer onto the epitaxially grown LED wafer, typically via spin-coating to ensure uniformity. Upon exposure to UV radiation through a photomask, the resist undergoes a photochemical reaction: regions exposed to UV light either polymerize (for negative resists) or degrade (for positive resists), depending on the resist chemistry.

Etching removes undesired material from the wafer surface through chemical or physical means, with the primary objective of accurately replicating the photolithographically defined mask pattern on the substrate. Etching techniques are broadly categorized into dry etching and wet etching.

Dry etching offers superior sidewall profiles, characterized by strong anisotropy, high selectivity, uniformity, stability, and precise controllability. Wet etching, conversely, exhibits lower cost, shorter process cycles, and suitability for industrial-scale production. In LED chip manufacturing, dry etching is predominantly employed for mesa structure formation and passivation layer etching, while wet etching is commonly used for indium tin oxide (ITO) removal, passivation layer stripping, and current-blocking layer (CBL) patterning.

Mesa dry etching is typically performed using inductively coupled plasma (ICP) etching, a hybrid process combining physical and chemical mechanisms. The physical component involves ion bombardment of the wafer surface, thereby enhancing chemical reactivity substantially. The chemical mechanism encompasses: (i) glow discharge of etching gases via inductive coupling, generating reactive radicals, metastable species, and atoms, along with their mutual reactions, and (ii) reactions between these active species and the solid substrate surface. Typically, a Cl_2/BCl_3 gas mixture is employed for GaN material etching. Under high-energy electromagnetic fields, Cl_2 dissociates into Cl, Cl^+, and Cl^- species, while BCl_3 decomposes into BCl, BCl_2, BCl^+, and B^+. These ionization products further undergo various chemical interactions.

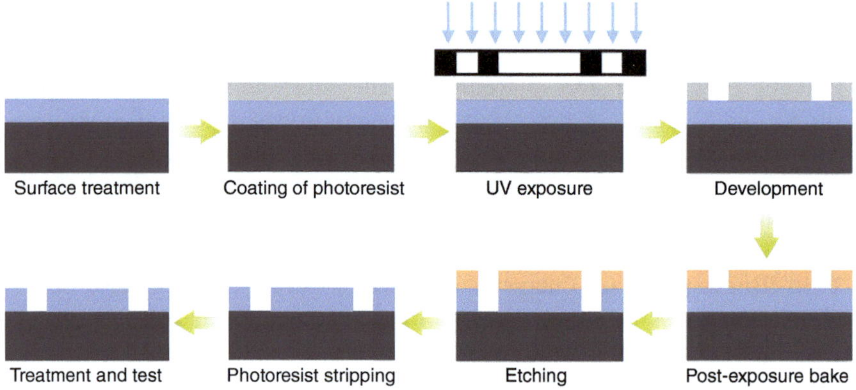

Figure 5.21 Schematic of photolithography process.

The lithography and etching processes in LED manufacturing are illustrated in Figure 5.21, both being critical for achieving high-resolution pattern transfer and material removal:

(I) **Surface treatment:** Standard procedures include solvent cleaning (acetone/IPA (isopropanol)) for organic contaminant removal, oxygen plasma treatment for surface activation, and dehydration baking (110–150 °C) for moisture elimination. For hydrophilic surfaces, hexamethyldisilazane (HMDS) priming enhances resist adhesion. These steps collectively improve surface energy (40–60 mN/m) and pattern fidelity while minimizing defects in subsequent processing steps.

(II) **Photoresist coating:** During photoresist coating, typical spin speeds range between 1000 and 5000 rpm, achieving film thicknesses of 0.5–2 μm with <5% variation. Parameters including viscosity, spin acceleration, and solvent content critically influence thickness uniformity and defect density. Post-application, a soft bake (90–120 °C) removes residual solvents prior to exposure.

(III) **UV exposure:** This step transfers the mask pattern onto the photoresist using precise UV radiation (typically 193–436 nm wavelength). Critical parameters include exposure dose (50–500 mJ/cm^2), numerical aperture (0.25–1.35), and focus depth (±0.2 μm). The photochemical reaction alters the resist's solubility, enabling pattern development. Proximity effects and diffraction limitations require compensation for high-fidelity pattern transfer.

(IV) **Development:** Development selectively removes exposed (positive resist) or unexposed (negative resist) regions using alkaline solutions (e.g., tetramethylammonium hydroxide [TMAH] 2.38%) or organic solvents. Critical parameters include development time (30–90 seconds), temperature (20–23 °C), and concentration, which collectively determine pattern fidelity. Overdevelopment may cause feature loss, whereas underdevelopment leads to residual resist. Post-development rinsing (deionized water) terminates the reaction and removes byproducts.

(V) **Post-exposure bake:** The post-exposure bake (PEB) thermally activates photochemical reactions initiated during exposure, typically at 90–130 °C for 60–90 seconds. This step critically enhances resist contrast by amplifying latent image formation through acid diffusion in chemically amplified resists. Precise temperature control (±1 °C) is essential for maintaining critical dimension uniformity while minimizing linewidth variation caused by excessive acid diffusion.

(VI) **Etching:** Etching transfers the lithographic patterns into the underlying substrate via wet chemical or dry plasma methods. Plasma etching, utilizing reactive gases (CF_4, Cl_2, or BCl_3), achieves anisotropic profiles with selectivity ratios exceeding 20:1 (resist:substrate). Critical parameters include RF power (100–1000 W), pressure (5–50 mTorr), and gas chemistry, collectively determining etch rates (50–500 nm/min) and sidewall angles (70–90°). Endpoint detection systems monitor process completion via optical emission spectroscopy.

(VII) **Photoresist stripping:** This process removes residual resist after pattern transfer using either oxidative plasma ashing (O_2/N_2 at 200–300 °C) or wet chemical dissolution (sulfuric acid–peroxide mixtures). Plasma methods achieve complete removal within 5–15 minutes while preserving substrate integrity, with selectivity controlled by RF power (300–1000 W) and temperature. Wet stripping employs piranha solution (H_2SO_4:H_2O_2, 3:1) for organic residue elimination, necessitating thorough rinsing to prevent surface contamination.

(VIII) **Final characterization:** Critical dimension measurement (scanning electron microscope [SEM]), defect inspection (bright-field/dark-field imaging), and surface characterization (atomic force microscopy [AFM], ellipsometry) are implemented. Automated metrology systems achieve sub-nanometer precision for linewidth control, while particle counters detect contaminants down to 50 nm. Electrical testing verifies pattern functionality, completing the process flow before wafer progression to subsequent fabrication stages.

Packaging Packaging constitutes a critical stage in LED device manufacturing, exhibiting comparable significance to LED chip fabrication. Identical LED chips may yield markedly divergent performance characteristics when processed using disparate packaging technologies. Through optimal package structure design, selection of high-performance materials, and implementation of advanced packaging processes, the luminous efficiency of devices may be enhanced while simultaneously reducing thermal resistance, improving system reliability, mitigating lumen depreciation, and extending operational life.

The evolutionary trajectory of LED packaging technology parallelizes that of other electronic device packaging. Originating to facilitate device interconnectivity via pin-attachment to common backplanes (typically circuit boards), the technology progressed to encompass surface-mount devices (SMD) for simplified board mounting. Recent advancements include integrated packaging and wire-bond-free technologies, enhancing board-level integration density and enabling broader design flexibility.

The current development trends in LED packaging technology are primarily characterized by the following:

(I) Highly integrated, miniaturized thin SMD optimized for portable applications.
(II) High-power devices featuring significantly increased power density and luminous flux per unit area, coupled with advanced thermal management employing enhanced structural designs and materials like aluminum nitride (AlN) ceramic substrates.
(III) Wire-bond-free chip-scale packaging technologies addressing miniaturization requirements for mass production and application. These devices obviate conventional lead frames or connection tabs, facilitating multi-device integration and reducing system costs.

5.5 GaN LEDs

5.5.1 Structure and Basic Properties of Nitrides

Group III nitrides primarily exhibit three crystal structures: wurtzite (hexagonal phase), zinc blende (cubic phase), and rock salt. Under ambient conditions, only the wurtzite and zinc-blende structures are typically observed, whereas the rock salt structure forms exclusively under extreme high-pressure conditions. The wurtzite structure represents the thermodynamically stable phase, while the zinc-blende structure is metastable and unstable at elevated temperatures. Consequently, research on the growth of polar surfaces in Group III nitrides predominantly focuses on the wurtzite phase.

Both wurtzite and zinc-blende structures share fundamental similarities: each Group III (V) atom bonds tetrahedrally with four nearest-neighbor Group V (III) atoms. The distinction lies in their stacking sequences of close-packed atomic planes. In the wurtzite structure, the {0001} planes follow an ABABAB stacking sequence along the <0001> direction, consisting of two hexagonal close-packed (HCP) sublattices offset by 3c/8 along the *c*-axis. In the zinc-blende structure, the {111} planes adopt an ABCABC stacking sequence along the <111> direction, formed by two face-centered cubic (FCC) sublattices shifted by 1/4 of the body diagonal length.

Nitride semiconductor materials, primarily comprising gallium nitride (GaN), aluminum nitride (AlN), indium nitride (InN), and their ternary alloys, constitute an important class of direct wide-bandgap semiconductors. Through precise compositional modulation of these elements, the bandgap energy can be continuously tuned across an exceptionally broad range from 0.7 to 6.2 eV, thereby covering the spectral region from infrared to UV. Table 5.1 summarizes fundamental properties of GaN, AlN, and InN at 300 K.

Table 5.1 Summary of properties of GaN, AlN, and InN with wurtzite structure at 300 K.

	GaN	AlN	InN
Crystal lattice/nm	$a = 0.3189$	$a = 0.3112$	$a = 0.3533$
	$c = 0.5189$	$c = 0.4982$	$c = 0.5693$
Atomic density/10^{22}/cm^3	8.9	9.58	6.4
Bandgap/eV	3.39	6.2	0.7
dielectric constant	8.9	8.5	15.3
Hole mobility/cm^2/(V s)	≤1000	300	≤3200
Electron mobility/cm^2/(V s)	≤200	14	
Thermal expansion coefficient/10^{-6}	5.59	4.2	3.8
	3.17	5.3	2.9
Refractive index	2.3	2.15	2.9

In terms of binary nitride alloys, their crystal lattices can be estimated via the Vegard equation. For example, the crystal lattices (a) of $In_xGa_{1-x}N$ alloys are calculated as the following equation:

$$a(\mathrm{In}_x\mathrm{Ga}_{1-x}\mathrm{N}) = xa(\mathrm{InN}) + (1 - x)a(\mathrm{GaN}) \tag{5.35}$$

5.5.2 Blue GaN LED

In 1994, Dr. Shuji Nakamura of Nichia Corporation achieved a groundbreaking advancement in the development of high-brightness gallium nitride (GaN)-based blue LEDs. This seminal work was later recognized with the 2014 Nobel Prize in Physics. The advent of blue LEDs enabled the synthesis of white light and catalyzed a transformative revolution in solid-state lighting technology.

However, mass production of GaN has been constrained by the challenges associated with manufacturing homogeneous substrates and prohibitively high costs. Consequently, GaN is typically heteroepitaxially grown on foreign substrates such as sapphire (Al_2O_3), silicon (Si), and silicon carbide (SiC). Presently, MOCVD remains the most widely adopted method industrially due to its balanced trade-off between growth efficiency and material quality. It should be noted, however, that MOCVD systems exhibit significantly greater complexity compared to MBE setups, posing additional engineering challenges. A schematic of the GaN LED structure is provided in Figure 5.22.

During GaN LED operation, application of an electric potential across the semiconductors drives charge-carrier dynamics where electrons move from the conduction band to the valence band, while holes move in the opposite direction. Radiative recombination of these electron–hole pairs subsequently emits photons, constituting the fundamental electroluminescence mechanism. Furthermore, quantum wells are sophisticated semiconductor heterostructures comprising repetitive layers of dissimilar semiconductor materials with varying

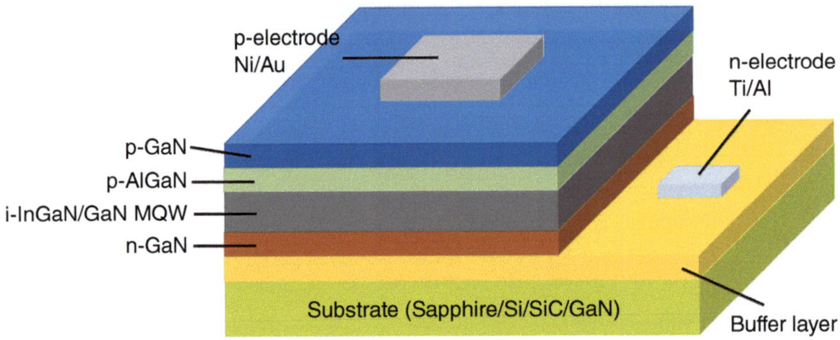

Figure 5.22 Schematic of representative device structure of GaN LED.

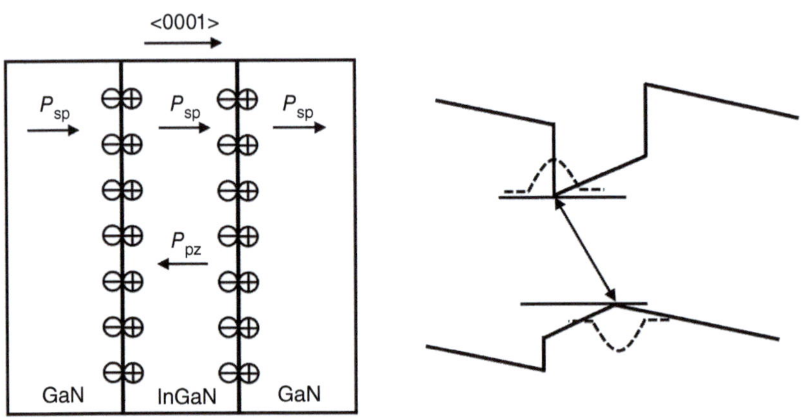

Figure 5.23 Schematic diagram of electron–hole distribution and polarization fields (P_{sp}: spontaneous polarization vector, P_{pz}: piezoelectric polarization vector) in a GaN/InGaN/GaN quantum well (left), as well as spatial separation of electron and hole wavefunctions in the quantum well induced by polarization electric fields (right).

bandgap energies. In GaN-based LEDs, these quantum confinement structures serve to confine the motion of charge carriers within nanometer-scale potential wells, as shown in Figure 5.23. This spatial confinement quantizes the carriers' energy states, modifying their density of states and creating discrete subbands.

The wurtzite structure of Group III nitrides exhibits strong spontaneous polarization fields due to the inherent lack of inversion symmetry and significant electronegativity difference between Group III metal atoms and nitrogen atoms, causing pronounced charge asymmetry within the crystal lattice. Additionally, piezoelectric polarization arises when the material experiences external strain, often induced by lattice mismatch.

In nitride-based quantum wells, substantial lattice mismatch generates particularly strong polarization fields. These fields induce band bending toward a triangular profile at the conduction and valence band edges, significantly modifying the quantum well's subband energy levels and bound-state wavefunctions. This phenomenon, known as the quantum-confined Stark effect (QCSE), primarily manifests through its impact on transition energies within InGaN quantum wells.

The polarization field induces significant band bending in quantum wells, spatially separating electrons and holes through asymmetric carrier localization. This spatial charge separation reduces the wavefunction overlap integral, thereby diminishing the radiative recombination rate, as illustrated in Figure 5.23. Furthermore, polarization-induced modifications to the quantum-well band structure represent a primary mechanism contributing to the efficiency droop phenomenon in LEDs.

To mitigate this droop effect, extensive research has focused on innovative quantum-well and barrier designs. Implementing polarization-matched quantum barriers constitutes an effective strategy for mitigating interfacial stress mismatch between wells and barriers, thereby reducing polarization-induced electric fields. This approach suppresses band bending in the quantum wells, consequently minimizing electron leakage. Furthermore, selective doping within specific barrier layers enables precise modulation of carrier distribution across the multiple quantum well (MQW) region. This engineered asymmetry in dopant profiles partially compensates for the inherent spatial separation of electrons and holes, leading to improved wavefunction overlap and significantly enhanced radiative recombination efficiency.

For InGaN/GaN MQW structures, piezoelectric polarization dominates due to the relatively small difference in spontaneous polarization coefficients between GaN and InN, compounded by the significant disparity in their lattice constants. The dependence of transition energies (E) on the polarization field in quantum wells can be approximated by the following equation:

$$E = E_g(\text{InGaN}) - eF_pL_w + 0.287(F_p)^{\frac{2}{3}} \tag{5.36}$$

where E_g(InGaN) represents the bandgap energy of InGaN, F_p denotes the total piezoelectric polarization field within the quantum well, and L_w is the well width. As the lattice mismatch between the well and barrier layers increases, the resultant polarization field strength is enhanced, leading to an elevation of the electron–hole transition energy and consequently a redshift in the radiative emission spectrum.

The modified electronic structure by MQW significantly enhances radiative recombination probability due to the increased spatial overlap between electron–hole wavefunctions and reduced non-radiative recombination pathways, as depicted in Figure 5.24. Crucially, these quantum mechanical effects collectively enhance IQE and improve the spectral purity of emitted photons. Consequently, implementing quantum-well architectures has become indispensable for achieving high-performance LED devices with superior luminous efficiency and chromatic characteristics.

5.5.3 Green and UV GaN LED

To achieve green light emission, it is necessary to increase the indium (In) composition within the InGaN/GaN MQWs. However, a higher In content exacerbates the lattice mismatch inherent to the InGaN/GaN system, leading to a high density of defects. These defects typically act as non-radiative recombination centers and carrier leakage pathways, thus degrading the luminescence efficiency of green

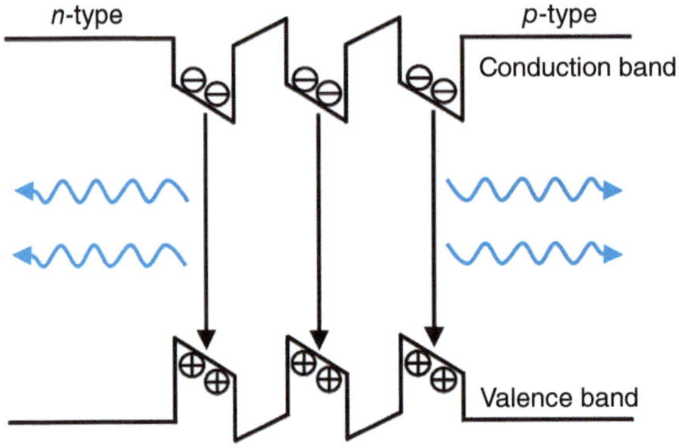

Figure 5.24 Energy-band structure of the blue-emission LED with InGaN/GaN multiple quantum wells.

LEDs. Furthermore, the lattice mismatch induces significant compressive strain, generating a strong built-in piezoelectric field within the InGaN/GaN MQWs. This results in a pronounced spatial separation of electron and hole wavefunctions, severely reducing the IQE of green emitters.

Considering emission across the 400–650 nm wavelength range, the EQE of GaN-based LEDs decreases with increasing wavelength, whereas the EQE of AlGaInP-based LEDs generally increases. Consequently, both GaN-based (~530 nm) and AlGaInP-based (~530 nm) green LEDs exhibit significantly low EQE—a phenomenon widely recognized as the "green gap". To address this limitation, researchers have proposed strategies aimed at enhancing the optoelectronic characteristics of green LEDs. One such approach involves inserting a superlattice into the electron transport layer (ETL). For instance, as shown in Figure 5.25, AlGaN, a 2 μm thick GaN, and a 2.5 μm thick n-GaN layers are sequentially sputtered onto substrates. Subsequently, a 24-pair $In_{0.04}Ga_{0.96}N$(3 nm)/GaN(3 nm) superlattice and a 12-pair $In_{0.25}Ga_{0.75}N$(3 nm)/GaN (12 nm) MQW structure are epitaxially grown. Finally, p-GaN, p-AlGaN/GaN superlattice layers, and a p-GaN contact layer are deposited on top of the superlattices and MQWs.

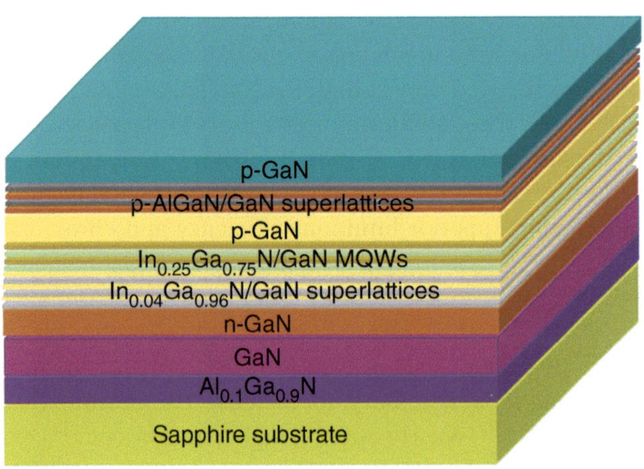

Figure 5.25 Schematic of epitaxy structure of green-emission GaN LEDs.

The underlying mechanisms behind green emission in GaN LEDs remain debated. Some researchers attribute the enhancement to improved electron capture efficiency within the InGaN/GaN MQWs due to the presence of the superlattice. Furthermore, the InGaN/GaN superlattice can function as a strain-relief layer, mitigating the piezoelectric field and reducing defect-related non-radiative recombination. Others propose that low-temperature-grown InGaN/GaN superlattices facilitate the introduction of V-pits, inverted pyramidal defects bounded by $(10\bar{1}1)$ facets. These V-pits propagate into the MQWs during epitaxial growth, influencing the LED's optoelectronic properties. High-In-content InGaN materials are metastable, exhibiting compositional fluctuations and quantum-well width variations that create localized states. Carriers trapped within these energy minima exhibit reduced susceptibility to non-radiative recombination at defect sites. Additionally, V-pits within InGaN/GaN MQWs nucleate primarily at threading dislocations, with their sidewalls composed of MQWs oriented along the $(10\bar{1}1)$ plane. These sidewall MQWs possess narrower well widths, lower In concentrations, and consequently wider bandgaps compared to their (0001)-oriented counterparts. V-pits thereby establish energy barriers encompassing dislocations, impeding lateral carrier diffusion towards dislocation cores. This suppression of carrier migration diminishes non-radiative recombination probabilities, leading to high-efficiency green emission.

Beyond green LEDs, GaN-based UV LEDs represent a critical semiconductor technology for applications ranging from sterilization to optical sensing, owing to their tunable emission wavelengths (200–400 nm) and compact solid-state design. Their operational principle relies on the radiative recombination of electron–hole pairs within the active region, where the emission wavelength is primarily governed by the bandgap energy of (AlGaN) $Al_xGa_{1-x}N$ alloys. Adjusting the aluminum composition (x) allows the bandgap to be engineered from 3.4 eV (GaN, 365 nm) to 6.2 eV (AlN, 200 nm), respectively, enabling precise spectral control across the UV range.

As illustrated in Figure 5.26, the epitaxial structure typically comprises an n-type AlGaN layer grown on an AlN/sapphire template, followed by an active region

Figure 5.26 Schematic of epitaxy structure of UV-emission GaN LEDs.

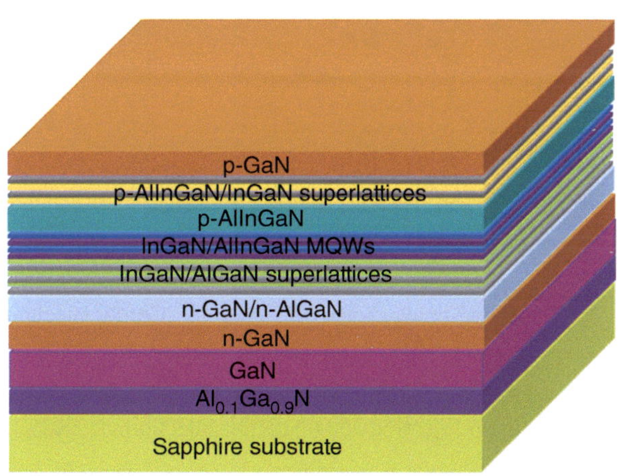

incorporating InGaN/AlGaN superlattices and InGaN/AlInGaN MQWs designed to enhance carrier confinement and radiative efficiency. The structure is completed with a magnesium-doped *p*-type AlInGaN layer. Key challenges in epitaxial growth include achieving high conductivity in both *n*- and *p*-type high-Al-content AlGaN, alongside minimizing threading dislocation densities employing techniques such as epitaxial lateral overgrowth (ELO) or pulsed atomic layer epitaxy (PALE).

5.6 Perovskite LEDs

While GaN- and GaAs-based LEDs currently dominate commercial markets, their widespread implementation is constrained by relatively high manufacturing costs and suboptimal optoelectronic performance. In contrast, solution-processed semiconductor LEDs have demonstrated remarkable progress in recent years, largely propelled by the exceptional cost-effectiveness and performance benchmarks established by OLEDs. Furthermore, emerging perovskite LEDs exhibit superior characteristics compared with both conventional inorganic semiconductor LEDs and OLEDs, particularly regarding minimal Stokes shifts ($\Delta E < 50\,\text{meV}$) and significantly enhanced photoluminescence quantum efficiencies (PLQEs >90% in optimized devices).

5.6.1 Structure and Progress of Perovskite LEDs

A typical perovskite LED incorporates a multilayer architecture consisting of (i) a perovskite emissive layer, (ii) an ETL, (iii) a hole transport layer (HTL), and (iv) top/bottom electrodes. Carrier recombination within perovskite LEDs follows two principal mechanisms [8, 9]. The first process involves direct carrier injection and recombination, illustrated in Figure 5.27a. Here, electrons and holes

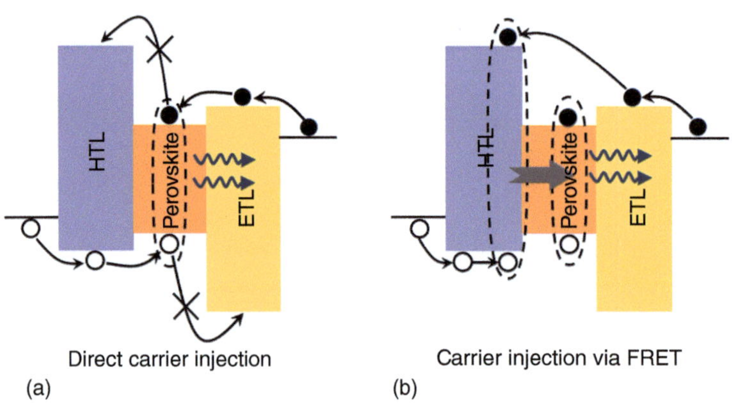

Figure 5.27 Schematic diagram of operation mechanisms in perovskite LEDs according to (a) direct carrier injection and (b) carrier injection via FRET.

are injected through the ETL and HTL, respectively, into the perovskite active layer under forward bias, subsequently forming excitons that undergo radiative recombination to emit photons. The second recombination mechanism operates *via* Förster resonance energy transfer (FRET). In this process, some charge carriers migrate beyond the perovskite layer into the opposing transport layer (electrons into the HTL or holes into the ETL). Energy is then transferred to the perovskite material through dipole–dipole coupling, generating excitons within the perovskite matrix that ultimately radiatively recombine and emit photons, as depicted in Figure 5.27b.

The evolution of perovskite LEDs has demonstrated remarkable progress since their inception. The first functioning PeLED device employing polycrystalline metal-halide perovskites (MHPs) with grain sizes exceeding 100 nm was reported in 2014, achieving a modest EQE of approximately 0.1%. A breakthrough was achieved in 2015 when grain size reduction strategies produced devices with EQE values approaching 8%. Subsequent advancements in 2018 enabled devices attaining ~20% EQE through three key innovations: (i) implementation of defect-passivated polycrystalline MHP layers, (ii) development of optimized nanocrystal synthesis protocols, and (iii) application of ligand engineering techniques. Among reported perovskite LEDs, blue devices present significant challenges due to the inherent difficulty in stabilizing wide-bandgap perovskites and mitigating non-radiative recombination processes. Recent progress has focused on compositional engineering, such as incorporating mixed halides (e.g., Cl^-/Br^-) or large organic cations (e.g., phenylethylammonium), to achieve blue emission (450–490 nm). Strategies involving dimensionality reduction (quasi-2D perovskites) and interfacial passivation have yielded efficiency improvements, with EQEs now exceeding 10% for sky-blue devices.

5.6.2 Efficiency Enhancement of Perovskite LEDs

5.6.2.1 Modulation of Carrier Dynamics

A critical challenge in perovskite LED operation originates from inherent charge injection imbalance, characterized by efficient electron injection coupled with inefficient hole injection. Under low current density operation, radiative recombination dominates due to constrained charge-carrier injection. However, at elevated current densities, increased carrier flux combined with energy-level misalignment between HTLs and emitter layers frequently results in exciton quenching. Under forward-bias conditions, charge injection in perovskite LEDs proceeds through defined energy pathways. Holes are injected from the highest occupied molecular orbital (HOMO) level of HTLs into the VBM of the emitter layers, while electrons transfer from the lowest unoccupied molecular orbital (LUMO) level of ETLs to the CBM of the emitter layers. Subsequent radiative recombination of these charge carriers occurs either through excitonic or bimolecular recombination processes within the emitter layers, producing photon emission with spectral characteristics dictated by the energy difference between the VBM and CBM. Consequently, balancing charge-carrier transport is essential, serving as a fundamental requirement for maximizing IQE. Energy levels of commonly used electrodes, HTLs, perovskite emitters, and ETLs are illustrated in Figure 5.28a [10]. Charge injection mismatches can lead to detrimental interfacial carrier accumulation and non-radiative leakage currents, both of which significantly degrade IQE and EQE metrics, as demonstrated in Figure 5.28b.

Reducing the hole-injection barrier for perovskite LEDs across various emission colors, particularly for blue perovskite LEDs, is critical for achieving high quantum efficiencies [12, 13]. Two principal strategies have emerged to enhance charge injection efficiency into HTLs: work-function modulation and graded energy-level engineering. The former approach involves the strategic incorporation of high-ionization-potential molecules into HTLs, inducing a vacuum-level shift and thereby modifying the layer's effective work function. This technique has proven effective in reducing hole-injection barriers, evidenced by EQEs exceeding 20% in both green- and red-emitting perovskite LEDs, as illustrated in Figure 5.28c [11].

5.6.2.2 Interfacial Engineering

The interfaces between the emissive layer and charge transport layers critically influence perovskite LED performance, as interfacial traps serve as dominant pathways for non-radiative recombination losses. Buried interfaces function as structural templates that govern perovskite nucleation and crystallization, thereby modulating key material properties including crystallinity, morphology, trap-state density, and work function. Recent studies have demonstrated that ultrathin polar interlayers, such as lithium fluoride (LiF), facilitate the growth of highly emissive and uniform perovskite films on hydrophobic polymeric charge transporters in OLED-inspired device architectures. This interfacial engineering approach has enabled the realization of efficient green quasi-2D perovskite LEDs exhibiting EQEs exceeding 19.1% at luminance levels surpassing 1500 cd/m^2. Alternative hydrophilic interfacial layers employing polymeric materials (e.g., polyethyleneimine, perfluorinated ionomer) or small-molecular modifiers (e.g., ethanolamine, aluminum oxide) have similarly demonstrated efficacy in promoting high-quality perovskite film formation. While these strategies have proven effective for performance enhancement, the fundamental mechanisms underlying these improvements require further systematic investigation to establish comprehensive structure–property relationships.

Beyond buried interface modifications, top-interface engineering enhances performance via surface-defect passivation and ion-migration suppression. Recent advances in perovskite LEDs demonstrate that bifacial passivation strategies targeting both top and bottom interfaces of the perovskite emissive layer are highly effective. This dual-interface engineering has yielded quantum-dot–based perovskite LEDs exhibiting peak EQEs of 18.7% alongside significantly improved operational stability. Furthermore, precise interfacial control enables modulation of charge-carrier balance, offering additional advantages beyond immediate performance metrics [14].

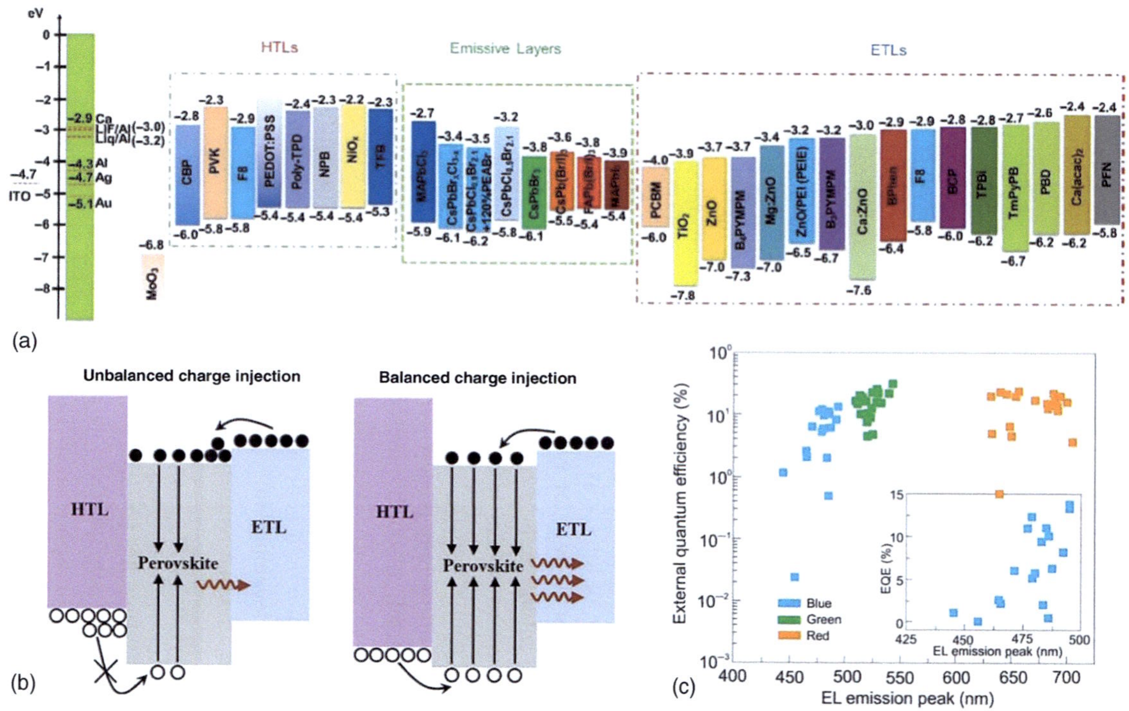

Figure 5.28 (a) Energy levels of commonly used electrodes, HTLs, perovskites emitters, and ETLs. *Source:* Lee et al. [10] / John Wiley & Sons / CC BY 4.0. (b) Schematic of charge injection and recombination in perovskite LEDs. (c) Representative achievements (EQE) of blue-, green-, and red-perovskite LEDs versus emission spectrum. *Source:* Jang et al. [11] / John Wiley & Sons / CC BY 4.0.

5.7 White LEDs

White light–emitting diodes (WLEDs), as highly efficient illumination sources, demonstrate a multitude of advantages, including low energy consumption, extended operational longevity, and environmental sustainability, thereby establishing themselves as the prevailing technology in lighting and display applications. Based on their excitation mechanisms, WLEDs can be categorized into photoluminescent and electroluminescent types. In both cases, the efficacy of white emission is assessed through key photometric parameters such as the color-rendering index (CRI), correlated color temperature (CCT), and luminous efficiency, all of which critically influence their suitability for illumination purposes.

The CRI quantifies a light source's ability to accurately reproduce the colors of objects relative to natural sunlight, while the CCT is defined as the temperature of an ideal blackbody radiator emitting light perceptually equivalent to that of the source. Depending on their CCT values, WLEDs are classified into warm-white (2700–3700 K), neutral-white (3700–6500 K), and cool-white (exceeding 6500 K) variants. However, cool-white light, which is dominated by blue and UV spectral components, has been found to pose potential risks to human ocular health. Consequently, warm- and neutral-white WLEDs are widely regarded as optimal candidates for lighting and display applications. To ensure high performance, an efficient WLED must exhibit a CRI approaching 100, an appropriate CCT (2700–6500 K), and superior luminous efficiency.

White-light generation in WLEDs typically originates from the spectral combination of multiple colors, such as the blending of yellow and blue light or the integration of tricolor emissions (red, green, and blue). In photoluminescent WLEDs, the yellow component or tricolor emissions are generally produced by phosphors or other emitting materials, which undergo optical excitation by blue or UV LED chips. Conversely, electroluminescent WLEDs rely on electrically driven device architectures incorporating specialized emissive and charge transport layers. Recent advancements have also introduced broad-spectrum white emitters capable of single-component white-light generation, representing a promising avenue for next-generation WLED development.

5.7.1 GaN-Based White LEDs

In phosphor-converted white LEDs, the fundamental structure comprises a high-efficiency blue-emitting InGaN/GaN quantum-well diode integrated with a wavelength-converting phosphor layer. The blue LED's active region consists of MQWs confined between *p*-type and *n*-type AlGaN cladding layers; this configuration is designed to achieve optimal carrier confinement and radiative recombination efficiency.

The phosphor coating, typically fabricated from cerium-doped yttrium aluminum garnet (YAG:Ce), absorbs a fraction of the blue emission and subsequently undergoes Stokes shift to produce broadband yellow emission, as illustrated in Figure 5.29. For phosphor-converted devices, this subsequent

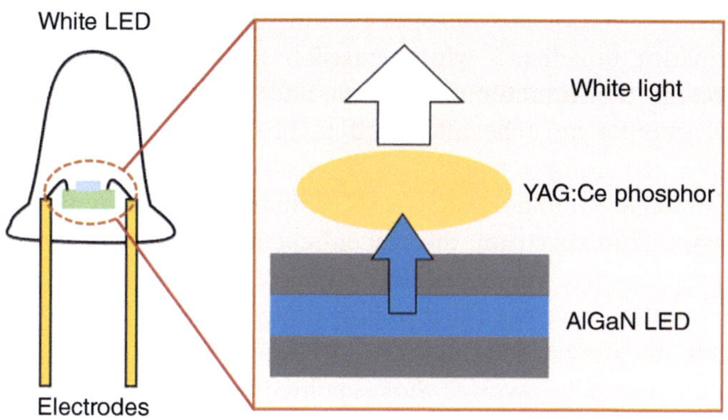

Figure 5.29 Schematic of the white LED based on a blue GaN-based LEDs and yellow phosphors.

photoluminescent-conversion process requires careful engineering to minimize Stokes-shift losses and thermal-quenching effects. The resultant additive mixture of residual blue emission and phosphor-generated yellow light produces the perception of white light. The CCT can be adjusted through precise control of the phosphor composition and concentration ratios.

Alternative approaches for white-light generation include RGB multichip configurations, where discrete red, green, and blue emitters are integrated within a single package. This architecture offers superior color-rendering capabilities, particularly addressing the deficiencies that phosphor-converted systems frequently exhibit in the red spectral region. The structural complexity increases significantly in such designs, necessitating careful thermal management and current balancing circuits to maintain chromatic stability. Recent advancements have introduced quantum-dot–based converters as an intermediate solution, combining the narrow emission lines of semiconductor nanocrystals with the operational simplicity of phosphor-conversion systems.

Device performance critically depends on several structural parameters, including quantum-well width and composition profile, doping concentration within carrier injection layers, and the design of current-spreading layers. Advanced structures may incorporate electron-blocking layers to prevent carrier overflow, patterned substrates to enhance light-extraction efficiency, and distributed Bragg reflectors to recycle emitted photons. The thermal management system, often involving ceramic substrates or metal-core printed circuit boards, forms an integral component of the complete device architecture due to the significant adverse influence of junction temperature on both LED efficiency and phosphor performance.

5.7.2 Perovskite-Based White LEDs

Owing to fascinating properties such as high PLQYs, tunable emission, and excellent absorption characteristics, perovskites have emerged as viable emitters coupled with blue or UV LED chips for constructing white LEDs. However, a primary limitation of perovskites is their narrow FWHM emission spectra, which restricts their ability to achieve high-performance broadband white emission in single-component configurations. To overcome this limitation, supplementary emitters possessing complementary spectral outputs must be integrated to broaden the composite emission profile. Notably, the composition-tunable emission characteristics of perovskites permit the strategic combination of different halides to cooperatively generate a broadband white-light spectrum, thereby enhancing their applicability for WLEDs. Furthermore, lead-free metal halides, characterized by broadband emission via STE recombination, have been successfully utilized as emitters. These materials offer compatible integration with blue/UV LED chips, facilitating white LED fabrication. Their PL typically involves Stokes-shifted emission exhibiting large FWHM (>100 nm), enabling high CRIs (CRI > 80) and tunable correlated color temperatures (CCT: 3000–6000 K). In contrast to lead-halide perovskites, which often necessitate multiple emitters for broadband coverage, lead-free metal halides inherently generate single-component white emission.

Although photoluminescent white LEDs utilizing inorganic halide perovskites (IHPs) have been investigated over extended periods, their current efficiency, defined as the ratio of luminance to drive current, remains relatively low, primarily due to thermal relaxation and energy losses within the chips. Compared to photoluminescent approaches, emission from electroluminescent white LEDs originates from the direct radiative recombination of carriers confined within the emissive layers, representing a substantially more efficient utilization of electrical current and energy [15, 16].

As depicted in Figure 5.30a, the fabricated WLEDs incorporate a composite thin film comprising both α-phase and δ-phase $CsPbI_3$ [17]. This dual-phase architecture exploits the distinct properties of each polymorph: the α-phase facilitates efficient charge transport due to its superior charge-carrier mobility, while the δ-phase serves as the dominant luminescent center. Operating at 5.2 V, these devices exhibit spatially uniform white emission (Figure 5.30b), with CIE chromaticity coordinates of (0.35, 0.43) indicative of high color quality. The spectral profiles presented in Figure 5.30c reveal a broadband white emission, where the distinct contributions of each phase are identifiable: the δ-phase produces the prevailing broad emission while the α-phase contributes predominantly to the red spectral component. Current–voltage analysis exhibits conventional diode characteristics, with luminance saturating at 12,000 cd/m^2 under a bias of 6.6 V. However, both the EQE and current efficiency demonstrate characteristic roll-off with increasing current density, reaching peak values of 6.5% and 12.23 cd/A, respectively.

Furthermore, lead-free metal halides have been utilized in electroluminescence white LEDs. $CsCu_2I_3/Cs_3Cu_2I_5$ composite thin films, which are prepared by a one-step method, demonstrate composition-dependent PL characteristics [18]. As illustrated in Figure 5.30d, systematic variation of the CsI/CuI precursor ratio enables precise control over the relative phase composition, with increased CuI content promoting the formation of the yellow-emitting $CsCu_2I_3$ phase. This compositional engineering yields optimized warm-white emission, achieving a CRI of 91.6 and a CCT of 4264 K, which are particularly suitable for indoor lighting applications.

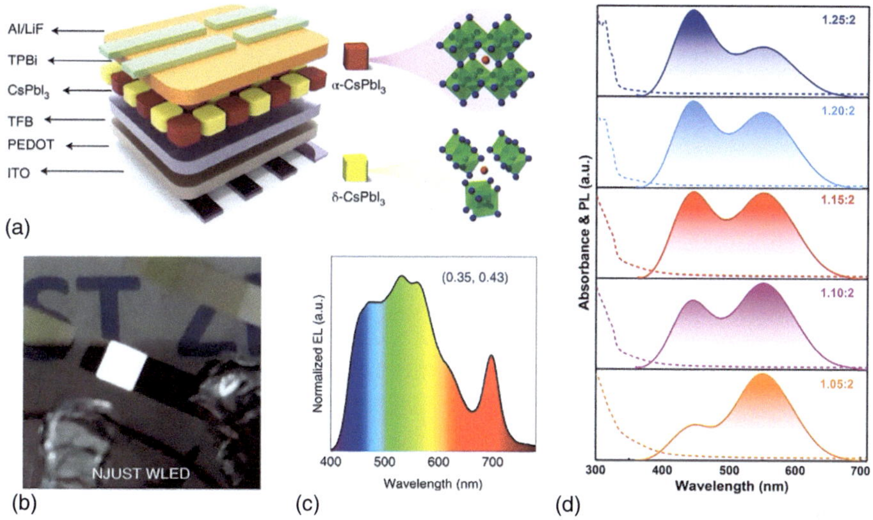

Figure 5.30 (a) Schematic of device structure, (b) photograph of an operating device, and (c) EL spectrum of the white LEDs with the emitting of blend of α- and δ-$CsPbI_3$. *Source:* Chen et al. [17]. (d) Absorption and PL spectra of $CsCu_2I_3/Cs_3Cu_2I_5$ thin films with different CsI/CuI ratios. *Source:* Ma et al. [18].

References

1 Chen, O., Zhao, J., Chauhan, V.P. et al. (2013). *Nat. Mater.* 12: 445–451.
2 He, X., Deng, Y., Ouyang, D. et al. (2023). *Chem. Rev.* 123: 1207–1261.
3 Liu, X.K., Xu, W., Bai, S. et al. (2021). *Nat. Mater.* 20: 10–21.
4 Akkerman, Q.A., Raino, G., Kovalenko, M.V. et al. (2018). *Nat. Mater.* 17: 394–405.
5 Shamsi, J., Urban, A.S., Imran, M. et al. (2019). *Chem. Rev.* 119: 3296–3348.
6 Akkerman, Q.A., D'Innocenzo, V., Accornero, S. et al. (2015). *J. Am. Chem. Soc.* 137: 10276–10281.
7 Tong, Y., Bladt, E., Aygüler, M.F. et al. (2016). *Angew. Chem. Int. Ed.* 55: 13887–13892.
8 Jeong, J.E., Park, J.H., Jang, C.H. et al. (2020). *Adv. Mater.* 32: e2002176.
9 Fu, Y., Zhu, H., Chen, J. et al. (2019). *Nat. Rev. Mater.* 4: 169–188.
10 Lee, J., Xie, Z., Wang, L. et al. (2024). *Chem. Eur. J.* 30: e202400372.
11 Jang, J.H., Li, S., Kim, D.H. et al. (2023). *Adv. Electron. Mater.* 9: 2201271.
12 Xiang, H., Wang, R., Chen, J. et al. (2021). *Light Sci. Appl.* 10: 206.
13 Chen, Z., Li, Z., Hopper, T.R. et al. (2021). *Rep. Prog. Phys.* 84.
14 Xu, L., Li, J., Cai, B. et al. (2020). *Nat. Commun.* 11: 3902.
15 Quan, L.N., Garcia de Arquer, F.P., Sabatini, R.P. et al. (2018). *Adv. Mater.* 30: e1801996.
16 Zhao, S., Mo, Q., Wang, B. et al. (2022). *Photonics Res.* 10: 1039.
17 Chen, J., Wang, J., Xu, X. et al. (2020). *Nat. Photonics* 15: 238–244.
18 Ma, Z., Shi, Z., Yang, D. et al. (2021). *Adv. Mater.* 33: e2001367.

6

Superluminescent Light-Emitting Diodes

6.1 Introduction

Superluminescent light-emitting diodes (SLEDs) combine high output power with a wide bandwidth, exhibiting properties intermediate between those of laser diodes (LDs) and conventional light-emitting diodes (LEDs). Consequently, SLEDs have attracted significant interest for optical fiber systems such as optical gyroscopes [1], optical coherence tomography (OCT) [2], and wavelength-division multiplexing (WDM) [3]. As essential light sources, SLEDs have experienced rapid development due to their extensive applications.

Scientifically, SLEDs operate via an electrically driven PN junction. When the injection current flows from the *p*-region through the active region—sandwiched between the *p*- and *n*-regions as depicted in Figure 6.1a—and into the *n*-region, amplified spontaneous emission (ASE) is generated across a broad spectral range (typically exceeding several tens of nanometers in spectral width) [4]. ASE originates from spontaneously emitted light that undergoes optical amplification via stimulated emission upon the establishment of a population inversion within the gain medium. Achieving ASE necessitates the suppression of lasing through minimized facet reflectivity. Common techniques to reduce facet reflectivity include utilizing a tilted waveguide [5], incorporating an unpumped absorbing region near the rear facet [6], or applying antireflection (AR) coatings on both facets [7]. Figure 6.1a illustrates a ridge-waveguide SLED structure employing AR coatings on both the front and rear facets. These AR coatings substantially suppress optical reflection or feedback within the device. Internally, two counter-propagating ASE beams travel along the active region, as demonstrated in Figure 6.1b.

Achieving high output power represents a critical objective for SLED development, prompting substantial research efforts to enhance this parameter. However, output power and spectral bandwidth represent inherently conflicting demands, making their simultaneous optimization challenging.

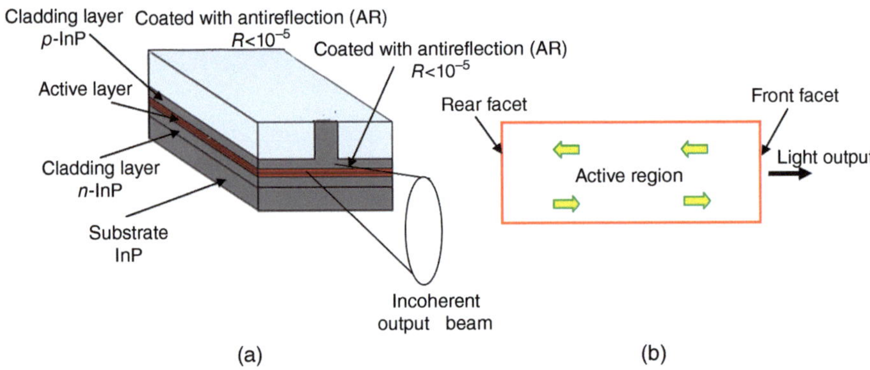

Figure 6.1 (a) Ridge-waveguide SLEDs and (b) top view illustrating waveguide without feedback.

6.2 History of SLED

Achieving high-power operation in SLEDs necessitates the suppression of lasing oscillation and intracavity feedback. To effectively suppress lasing oscillation and minimize optical feedback, facet reflectivities must be reduced to values less than the order of 10^{-5}. Below the power saturation level, SLED output power increases proportionally with applied injection current. However, elevated injection currents present challenges, potentially inducing significant wavelength shifts and heat generation [4]. Consequently, the development of methods to achieve high-power emission while mitigating these effects is essential.

Several techniques have been developed to realize the requisite low facet reflectivity for high-power SLEDs. These include tilting the active stripe relative to the facets [5], incorporating an unpumped absorbing region near the rear facet [6], applying AR coatings to one or both facets [7], implementing a bent stripe configuration near the output or rear facet [6], and fabricating the rear facet via dry etching at an oblique angle relative to the stripe [8].

Alphonse [5] first reported a tilted active-stripe waveguide SLED emitting at 0.83 μm wavelength, achieving 28 mW of output power. By tilting the stripe waveguide at an angle θ relative to the facets, the power reflected into the active region of a narrow-stripe diode was reduced to values below 6×10^{-5} in their design. As illustrated in Figure 6.2, this configuration ensures that light reflected from the back facet does not reenter the waveguide due to angular misalignment. The selection of θ is critical to simultaneously enable light emission and suppress optical feedback; θ must be less than the critical angle at the facet to avoid total internal reflection while maintaining sufficient tilt to decouple reflected light. For their device, an angle of 5° was implemented [5].

High-power continuous-wave (CW) operation at 1.55 μm was demonstrated in 2001 for SLEDs featuring a tapered active region [9]. A schematic diagram of the tapered SLED is presented in Figure 6.3. Tapering the active region alleviates carrier-density saturation at the output port. The taper angle is 1.4°. To eliminate

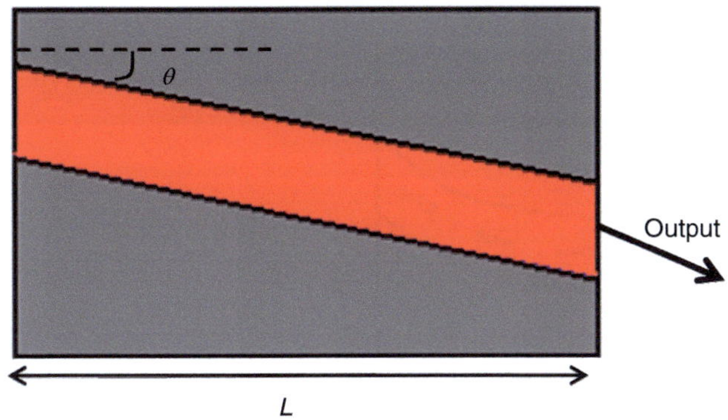

Figure 6.2 Schematic of SLEDs with a tilted waveguide. Adapted from Alphonse et al. [5].

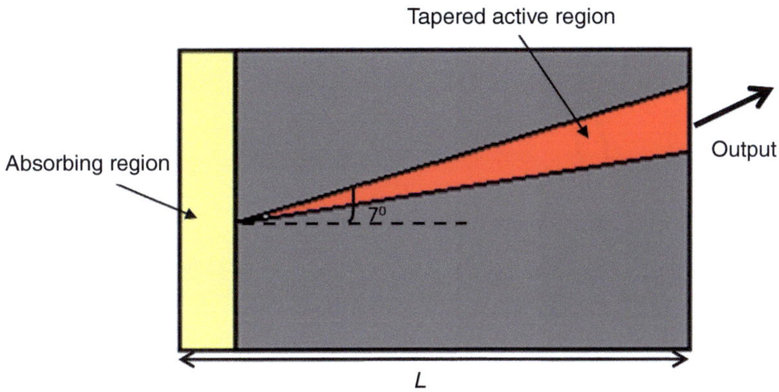

Figure 6.3 Schematic of tapered SLEDs. Adapted from Yamatoya et al. [9].

facet reflections, the device's center axis is tilted by 7.0°, and an unpumped absorbing region is positioned at the end of the tapered active region. The SLED length is typically 3.5 mm with an output width of 125 μm. An output power of 210 mW was achieved under CW operation at 200 °C. A disadvantage of this tapered SLED design is the difficulty in efficiently coupling the output beam into a single-mode fiber.

Nagai [6] reported a SLED operating at a wavelength of 1.3 μm. Reflectivity is effectively suppressed by incorporating an unpumped, buried, bent waveguide structure behind the active region, as depicted in Figure 6.4. The pumped region length L_p is 400 μm, and the absorbing region length L_a is 200 μm. The bent waveguide radius R is 500 μm. The absorbing region absorbs radiated light propagating toward the rear facet. In SLEDs with a straight absorbing region, laser oscillation readily occurs due to reduced absorption caused by photoexcited carriers, leading to increased radiation. However, the bent waveguide structure intersects the cleaved facet at an angle exceeding the critical angle for total reflection. Consequently, optical propagation toward the end facet does not recouple into the absorbing region, even under high-excitation conditions. This configuration enables effective

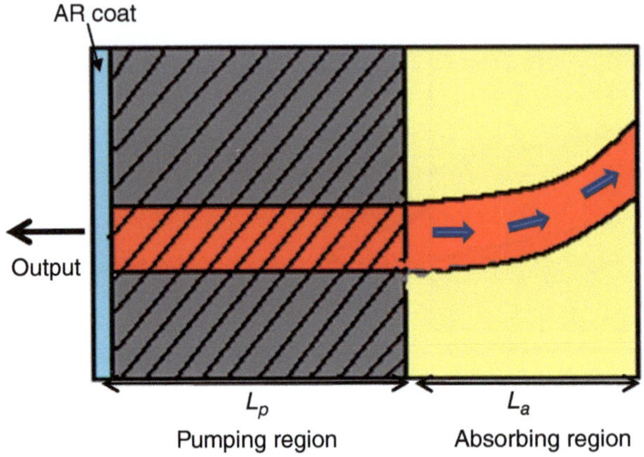

Figure 6.4 Schematic of buried bent absorbing waveguide SLEDs. Nagai et al. [6].

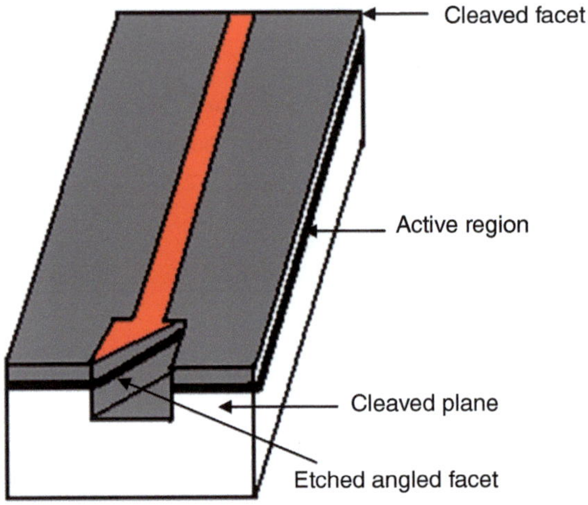

Figure 6.5 Schematic of SLEDs with angled facet etched. Lin [8].

suppression of laser oscillation with a compact absorbing region. An output power of 11.5 mW was obtained with a 3 dB bandwidth of 40 nm at an injection current of 200 mA and a temperature of 250 °C. A coupled power of 1 mW into a single-mode fiber was also achieved.

To maintain sufficiently low spectral ripple, Lin [8] demonstrated in 1991 a SLED featuring one facet etched at an angle using chemically assisted ion-beam etching (CAIBE), which significantly reduces facet reflectivity. Figure 6.5 illustrates the schematic of SLEDs with an angled, etched facet. The principle underlying facet reflectivity reduction in this structure resembles that of a tilted waveguide but demonstrates superior lasing suppression efficacy. Without AR coating, the reflectivity of light back into the waveguide from the etched, angled facet is less than 1×10^{-5}. A maximum output power of 10 mW was achieved, with 85% of the total output power coupled into a multimode fiber without any AR or high-reflection (HR) coatings on either facet.

Since an absorbing region alone proved insufficient for adequately reducing optical feedback, a V-groove etched structure [10] angled at 100° relative to the

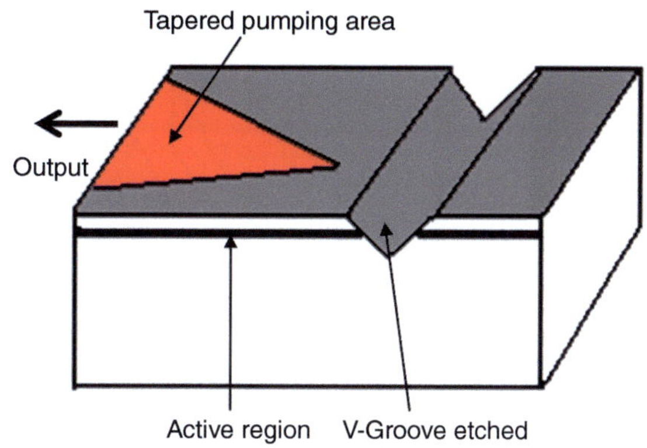

Figure 6.6 Schematic of SLEDs with V-groove etched. Adapted from Middlemast et al. [10].

facet was incorporated into the SLED design, as shown in Figure 6.6. The V-groove etched deflector was fabricated using a chemical wet etching technique that penetrates completely through both the active layer and the lower cladding layer. The V-shaped groove was formed by selective etching of the (001) crystal plane. The etched V-groove scatters backward-propagating light emitted from the active layer, thereby decreasing the light feedback into the active region. Furthermore, an unpumped absorbing region at the rear of the device further suppresses any undesired reflected light from returning to the active region. Under pulsed operation, an output power of 500 mW with a 3 dB bandwidth of 11 nm was observed at an injection current of 4 A.

The techniques described earlier are effective methods for reducing internal optical feedback. However, the high gain and residual facet reflectivity in such devices, even with AR coatings or tilted waveguides, can still produce spectral ripple due to Fabry–Perot longitudinal modes in the SLED cavity. Consequently, combining one or more of these techniques is often necessary to further enhance output power by minimizing feedback. For example, Paterson et al. [7] reported a 1.3 μm SLED employing a ridge-waveguide geometry with an integrated absorber. Both front and rear facets were coated with AR coatings. As shown in Figure 6.7, the device comprises an active region followed by a passive absorber section. Combining this passive absorber section with AR coatings achieves a very low spectral ripple, below 10^{-5}.

High coupling efficiency from SLEDs into single-mode fiber is essential for long-distance transmission in WDM systems and for achieving fiber optic gyroscopes (FOGs) with higher sensitivity. Although single-mode output can be attained by controlling the waveguide geometry, tapered waveguides suffer from lower coupling efficiency into single-mode fibers due to an astigmatic effect. This effect arises because the lateral component of the beam diffracts within the tapered region, making the light appear to originate from a virtual point inside the structure. Consequently, the output field profile deviates from circularity [9], reducing coupling efficiency.

Active multimode interferometer (MMI) SLEDs offer the dual advantages of high output power, characteristic of tapered SLEDs, and a regular single-mode output

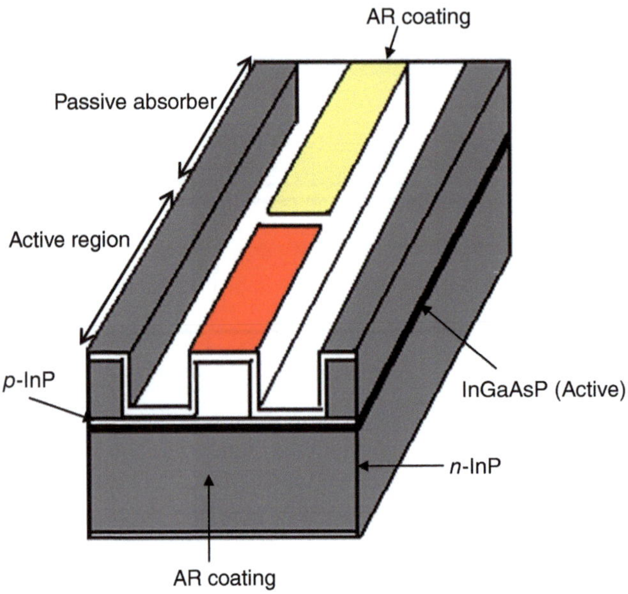

Figure 6.7 Schematic of SLED with a passive absorber section and AR coating. Adapted from Paterson et al. [7].

typical of conventional SLEDs. This configuration enables coupling into standard single-mode fibers without requiring an additional mode-control device [11]. The inherent circularity of the output field profile leads to significantly higher coupling efficiency compared to simple tapered SLEDs.

6.3 Principle of Active-MMI SLED

6.3.1 Self-Imaging Consideration

Although numerous active-MMI LDs have been demonstrated, applying active-multimode interference for superluminescent diodes remains unreported. Consequently, significant interest exists in developing active-MMI SLEDs, particularly for high-power applications where they represent promising candidates for WDM systems. Active-MMI lasers are typically achieved by meeting the requisite conditions for self-imaging formation [12]. The feasibility of utilizing active-MMI for SLEDs requires prior evaluation before configuration design.

The active-MMI SLED design employs two 1×1 MMI couplers interfaced with standard single-mode waveguides, maximizing the proportion of active MMI region within the cavity, as illustrated schematically in Figure 6.8a. Passive MMI relies on self-imaging formed by interference between distinct eigenmodes. However, since SLEDs exhibit minimal cavity oscillation and emit broad-spectrum light with short coherence lengths, questions arise regarding the achievability of self-imaging in an MMI waveguide. Through optimized design of MMI section length and width,

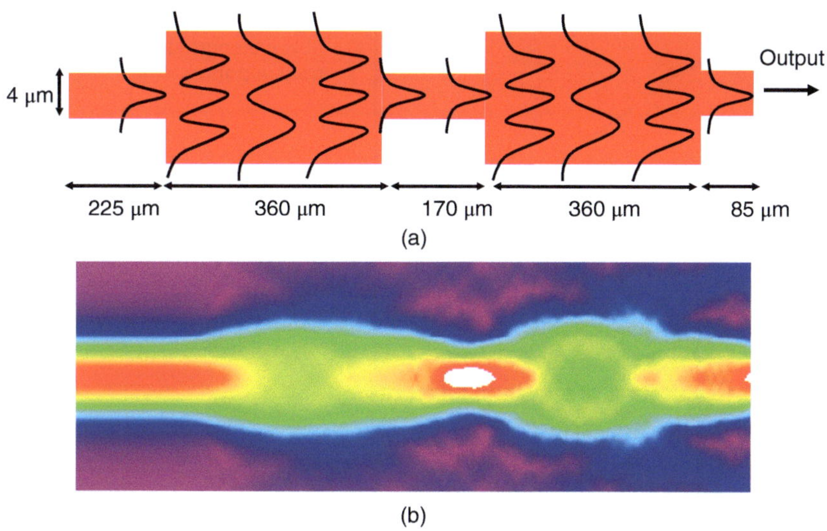

Figure 6.8 (a) Mode profile in active-MMI SLEDs and (b) optical field in active-MMI SLEDs.

ASE is expected to manifest active-MMI phenomena, yielding a stable fundamental transverse output mode. This is confirmed in the computational optical field profile shown in Figure 6.8b, simulated via beam propagation methods. Furthermore, ensuring self-imaging formation in active MMI devices necessitates satisfying specific phase and gain conditions across the supported modes.

6.3.2 Wide-Spectrum Consideration

The optical spectrum of SLEDs resembles that of LEDs but is typically narrower in bandwidth. Achieving a wide bandwidth, resulting in a short coherence length, is a fundamental requirement for SLEDs utilized in OCT systems demanding high depth resolution (typically several micrometers) and in WDM systems (typically requiring bandwidths on the order of 100 nm). The coherence length of a SLED is inversely proportional to its optical bandwidth; consequently, a larger bandwidth yields a shorter coherence length, enabling higher resolution in OCT systems. This principle drives the development of SLEDs with significantly broader emission spectra.

When designing high-power, wide-bandwidth SLEDs, suppressing lasing oscillations is essential to ensure that the output corresponds to the ASE spectrum. Since SLED emission is primarily based on amplified spontaneous radiation, the bandwidth and shape of its spectrum correlate directly with the material's gain spectrum. Based on this relationship, broadening the output spectrum using conventional bulk materials presents significant difficulties. One approach involves stacking or staggering multiple quantum-well (MQW) layers of differing thicknesses and compositions within the active layer structure [13]. Each layer emits a unique central wavelength, leading to a total SLED emission spectrum that results from the superposition of the individual emission spectra from each MQW layer.

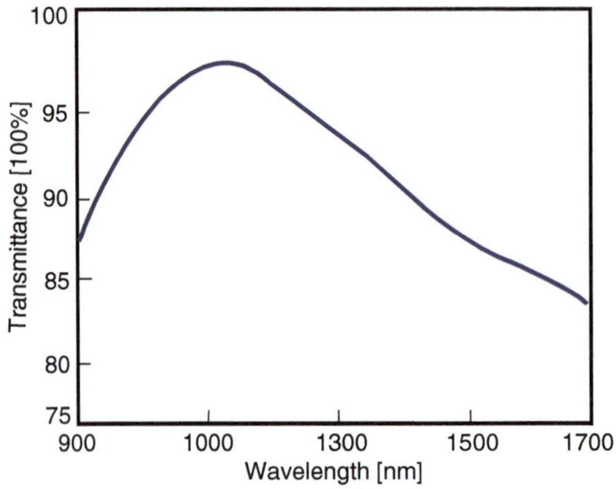

Figure 6.9 Simulation result of wavelength dependence of transmittance.

Consequently, SLEDs incorporating an MQW structure typically exhibit spectra two to three times wider than those utilizing a conventional bulk double-heterostructure (DH) design [14].

For active-MMI SLEDs, a critical issue concerns the feasibility of achieving large bandwidth operation, a key characteristic for these devices. This concern arises because interference or MMI filter effects within the MMI region could potentially limit the operational bandwidth. However, as demonstrated by the structure in Figure 6.8a, high transmittance can be achieved across a broad wavelength range. Figure 6.9 illustrates the simulated wavelength dependence of transmittance. These results confirm that a wide operational bandwidth can be maintained by optimizing the width and length of the MMI section. Therefore, active-MMI SLEDs can leverage large active areas to achieve high output power while simultaneously preserving a wide bandwidth with high transmittance.

6.4 Merit and Results of Active-MMI SLED

In this section, to confirm the feasibility of using the MMI configuration for SLEDs, we demonstrate the active MMI effect within SLEDs [15]. Figure 6.10a illustrates the schematic structure of active-MMI SLEDs. The 1×1 MMI couplers are connected to regular single-mode waveguides to achieve high-power output while maintaining a single-transverse mode. The single-mode waveguide width is 4 μm, while the MMI region is designed with a width $W_{\mathrm{MMI}} = 10\,\mu\mathrm{m}$ and a length $L_{\mathrm{MMI}} = 360\,\mu\mathrm{m}$. Figure 6.10a shows that two MMI sections are incorporated to maximize the coverage of the cavity length, thereby increasing the active area. Furthermore, the achievable MMI length is constrained by the MMI region width; consequently, two shorter MMI sections are utilized rather than a single section. For comparison, regular single-stripe SLEDs with a 4 μm stripe width were fabricated simultaneously on the same wafer, as depicted in Figure 6.10b. Both device structures possess a cavity length of 1200 μm.

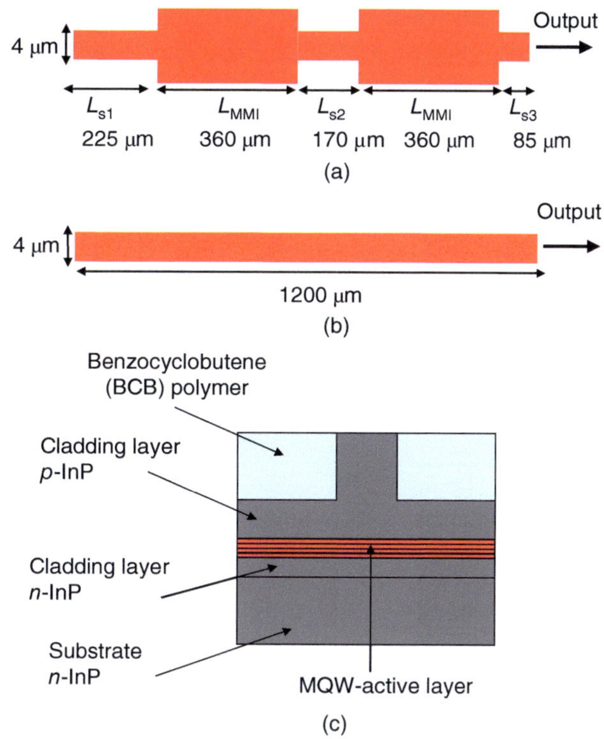

Figure 6.10 Schematics of the fabricated SLEDs: (a) active-MMI SLEDs and (b) regular SLEDs.

The fabricated SLEDs employ a ridge-waveguide structure. For actual device implementation, MQWs operating at $\lambda = 1.55\,\mu m$ served as the active layer, forming the waveguide core. The active layer is sandwiched between cladding layers of lower refractive index. Benzocyclobutene (BCB) polymer fills the region between ridges, acting as an electrical isolation layer. The upper cladding layer (p-InP) is positioned atop the active layer, while a lower cladding layer (n-InP) separates the active layer from the substrate. Both cladding layers and the ridge structure possess a larger energy bandgap than the active layer, effectively confining charge carriers within the quantum wells. The upper cladding layer thickness between the ridge and the active layer is intentionally thin, typically on the order of 0.2–0.3 μm or less. This thin configuration reduces current spreading laterally near the ridge edges. Optical mode confinement is achieved by the effective refractive index step induced by the ridge structure [16, 17]. The waveguide structure was intentionally tilted [18] at an angle with respect to the facets to achieve very low facet reflectivity ($R < 10^{-5}$) [19]. This design, combined with AR coatings applied to both facets [20], minimizes external optical feedback.

Facet reflectivity is significantly suppressed for SLEDs within the cavity, reducing oscillation and feedback. Consequently, SLED emission occurs via a single-pass ASE mechanism with short coherence length. The feasibility of self-imaging in active-MMI waveguides for SLEDs may be questioned, since self-imaging in passive MMI waveguides arises from interference mechanisms. Thus, it remains uncertain whether active-MMI phenomena occur in MMI-configured SLEDs. By evaluating the slope efficiency of active-MMI SLEDs alongside conventional SLEDs, the occurrence of active-MMI phenomena in SLEDs is demonstrated and analyzed.

The slope efficiency, dP/dI (where dP is the change in output power and dI the change in injection current), of fabricated SLEDs is defined as the ratio of optical power P to injection current I above the threshold current I_{th}. This parameter is critical for assessing SLED performance [21], representing the percentage of carriers injected above I_{th} that contribute photons to the output power [22]. A high slope efficiency reduces operating currents and enhances device efficiency. Typically, SLEDs exhibit diminished slope efficiency with increasing operating temperature. At constant current, a temperature rise reduces emitted optical power. The slope efficiency is determined by calculating the gradient of the measured power–current (P–I) curve above I_{th}, involving identification of a linear region and selection of two points for the calculation:

$$\eta_s = \frac{P_0}{I - I_{th}} \tag{6.1}$$

where P_0 is considered to be the point with higher output power, I is the injection currents above the threshold current of I_{th} ($I>I_{th}$), and the threshold current is defined as

$$I_{th} = \frac{N_{th}eV}{\tau_{sp}} \tag{6.2}$$

where N_{th} is the threshold carrier density, e is the electronic charge, V is the volume of the active region, and τ_{sp} is the carrier life time. Equation (6.1) indicates the optical power (in watts) generated per ampere of injected current in the linear regime prior to power saturation at high currents. As saturation occurs, the P–I curve rolls over to a plateau, marking peak output power and initiating a decrease in slope efficiency. The units of η_s are W/A or mW/mA, contingent on SLED optical power calibration.

The slope efficiency of SLEDs can be further expressed as

$$\eta_s = \frac{dP_{out}}{dI} = \eta_i \cdot \beta\Gamma \cdot \frac{hv}{e} \cdot \exp\left(\Gamma g_0 \eta_i \frac{J}{d} L\right) \cdot \exp(-\alpha L) \tag{6.3}$$

where η_i is the internal quantum efficiency, which is restricted by carrier losses caused by carrier leakage and recombination losses; Γ is the mode - confinement factor; $\exp(\Gamma g_0 \eta_i J/dL)$ is the overall gain of the optical field after propagating through the active region; α is the internal loss, which is related to the photon losses; hv is the photon energy; and L is the cavity length.

In SLEDs, rear and front facet reflectivities (R_1 and R_2) are below 10^{-5}, rendering facet losses negligible. Hence, for SLEDs of identical length, slope efficiency primarily depends on internal quantum efficiency, internal loss, and active region width. Maximizing η_s requires minimizing carrier and photon losses. Enhanced carrier confinement reduces leakage, while photon absorption is reduced through lower doping in confinement layers.

Figure 6.11 illustrates the measured output power of fabricated SLEDs under CW operation at 25 °C [23]. Slope efficiencies derived from P–I curves are 0.11 W/A (active-MMI SLEDs) and 0.12 W/A (conventional SLEDs) at 500 mA injection current (above I_{th}), signifying negligible difference between device types. As previously established, η_s depends critically on η_i and α. The absence of efficiency degradation confirms that no excess internal loss occurs within the cavity of active-MMI SLEDs.

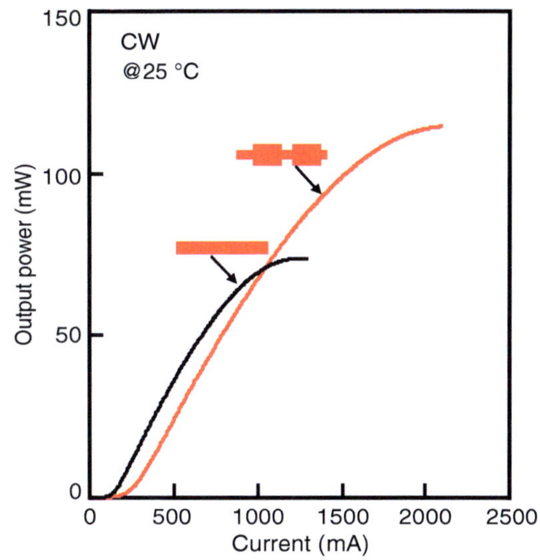

Figure 6.11 Light output power against current. The slope efficiency can be obtained from the power–current curve. Adapted from Zang et al. [23].

This indicates that active-MMI SLEDs operate without significant excess loss and function via active-MMI phenomena.

6.4.1 High-Power and Low Wall-Plug Consumption

In this section, the active-MMI phenomena in SLEDs are demonstrated. This confirms the applicability of the active-MMI configuration to these devices. Furthermore, it is established that the MMI configuration not only enhances the maximum output power but also reduces electrical power consumption.

For a detailed investigation, an active-MMI SLED structure (see Figure 6.10a) was employed. Conventional SLEDs (see Figure 6.10b) were also fabricated and examined for comparative assessment. Both device types were mounted p-side down onto a chemical vapor deposition (CVD) diamond heat sink. This mounting orientation significantly shortens the heat-flow path from the active region to the heat sink compared to the p-side-up configuration. Owing to the exceptional thermal conductivity of CVD diamond, the heat generated in the active region is rapidly dissipated across the entire heat-sink surface.

Subsequently, the devices were mounted onto copper submounts bonded to thermoelectric coolers (TECs) to maintain a constant heat-sink temperature of 250 °C for performance evaluation. This temperature represents realistic SLED operating conditions. The devices were driven by a current source to evaluate output power, while a voltmeter was used to monitor electrical power consumption.

Figure 6.12 presents the light output power versus current (P–I) and voltage versus current (V–I) characteristics for the fabricated SLEDs under CW operation. The P–I curves, measured using an optical power meter, exhibit no knee or discontinuity. Combined with the tilted waveguide structure and AR coating on the facets, this nonlinear characteristic confirms superluminescent operation, indicating effective ASE performance throughout the operational current range for both device structures.

The active-MMI SLED achieved a maximum output power of 115 mW at 2100 mA, representing the highest reported output power to date for a 1.55 μm SLED. In contrast, the conventional SLED attained a maximum output power of 75 mW. These results demonstrate a substantial enhancement of 40 mW, corresponding to a 54% improvement over the conventional SLED of identical cavity length. As anticipated, the optical power saturation level is higher for the active-MMI SLED. This increased saturation originates from the wider actively pumped area inherent to the active-MMI design, facilitating enhanced output power generation.

Examination of the P–I curves reveals a marginally higher threshold current for the active-MMI SLED compared to the conventional SLED, arising from the difference in active area size. However, this difference does not substantially affect device performance, as both types demonstrate comparable slope efficiencies above their respective threshold currents. Notably, SLED threshold currents generally exceed those of LDs with identical structures due to the significantly reduced facet reflectivity within the SLED cavity. This factor delays the onset of ASE, thereby increasing the threshold current [24].

As clearly observed in Figure 6.12, the maximum output power and the linearity of the P–I curve are constrained by saturation roll-off, occurring at injection currents of 2100 and 1300 mA for the active-MMI and conventional SLEDs, respectively. This roll-off is primarily attributable to self-heating within the active region under high injection currents. Key SLED operating mechanisms, including optical gain, power saturation, and nonradiative recombination, exhibit strong dependence on such self-heating effects. Despite the occurrence of saturation roll-off, the active-MMI SLED demonstrates significantly superior efficiency under CW operation compared to the conventional device.

Figure 6.12 also depicts the measured V–I characteristics. These characteristics indicate that active area size does not influence the turn-on voltage, which remains slightly below 0.8 V and is essentially identical for both SLED types. However, above the turn-on voltage, the driving voltage of the active-MMI SLED is substantially lower than that of the conventional SLED. At a common driving current of 1000 mA,

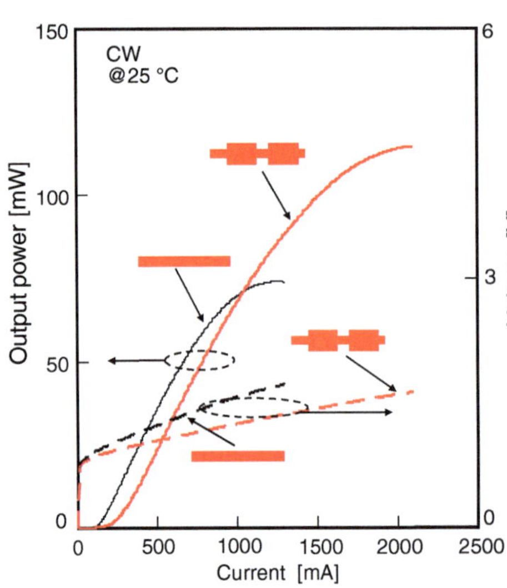

Figure 6.12 Light output power against current and voltage against current characteristics. Adapted from Zang et al. [15].

the driving voltage for the active-MMI SLED is reduced to 1.25 V, compared to 1.57 V for the conventional SLED. This voltage reduction stems from a comparatively lower differential resistance (R_d) in the active-MMI SLED. The differential resistance of SLEDs is defined as

$$R_d = \frac{dV}{dI} \propto \frac{1}{A} \tag{6.4}$$

where I and V represent the injection current and voltage, respectively, and A denotes the active area. Equation (6.4) demonstrates that a larger active area contributes to a reduced R_d, consequently leading to a lower driving voltage.

Active-MMI SLEDs offer a significant advantage over conventional SLEDs through reduced driving voltage, resulting in superior wall-plug efficiency. Wall-plug efficiency quantifies the proportion of electrical input power converted into useful optical output power. Given that SLEDs constitute primary power-consuming components in many systems, minimizing their electrical power consumption is imperative. Enhanced efficiency facilitates increased output power while reducing heat-sink requirements and associated cooling costs for end users. Moreover, lower electrical dissipation minimizes thermal generation within the device under high-current operation.

Figure 6.13 illustrates wall-plug efficiency as a function of output power. At 70 mW output, active-MMI SLEDs consumed 1.29 W electrical power, whereas conventional SLEDs consumed 1.58 W. This represents a 19% reduction in power consumption for the active-MMI design. Furthermore, active-MMI SLEDs achieved a maximum electrical-to-optical conversion efficiency of 6% at 1000 mA injection current.

6.4.2 Spectrum and Ripple Analysis

To achieve high-power SLED operation, suppression of lasing oscillation or optical feedback within the cavity is essential. For superluminescent emission, facet reflectivity should typically be reduced to less than 10^{-5} to prevent parasitic oscillations. In conventional SLEDs, this requirement is readily addressed, as documented in the literature [12]. However, for active-MMI SLEDs, one might suspect that internal reflection or excess loss occurs at the multimode-to-single-mode waveguide

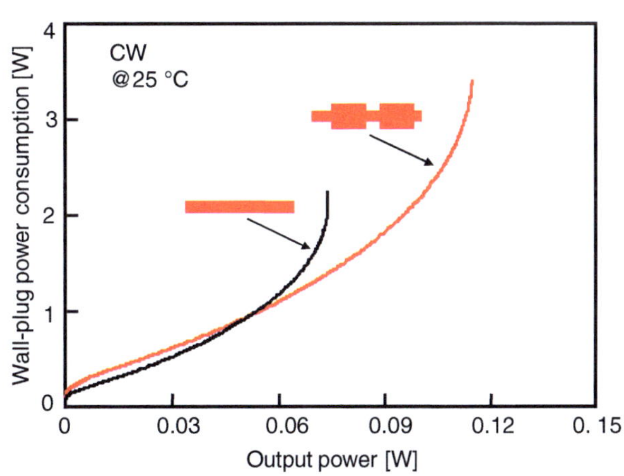

Figure 6.13 Measured wall-plug consumption. Adapted from Zang et al. [23].

interface, which could lead to failure in achieving superluminescent operation. By evaluating the optical spectrum in conjunction with ripple analysis, the absence of internal reflection in active-MMI SLEDs is demonstrated.

During spectral measurements, the optical emission from the SLEDs was coupled into an optical spectrum analyzer via a single-mode fiber and a lens system. The analyzer operated with a resolution of 0.05 nm to characterize the spectral ripple. Spectral properties of SLEDs are typically characterized by three key parameters: full width at half maximum (FWHM), center wavelength, and spectral ripple. FWHM quantifies the spectral width defined as the bandwidth at half the peak spectral intensity. The center wavelength is the midpoint of the FWHM range. Spectral ripple flatness is characterized by the maximum oscillation depth within the spectrum intensity.

Figure 6.14 presents the emission spectrum of the fabricated SLEDs at a current density of 9 kA/cm^2. Active-MMI SLEDs exhibit a spectrum with comparable width and flatness to conventional SLEDs. This indicates that despite the multimode-to-single-mode waveguide interface, superluminescent emission occurs effectively in active-MMI devices. The bandwidth (FWHM) and center wavelength for both device structures are 50 nm and λ_{center} = 1550 nm, respectively. Spectral bandwidth is primarily governed by the material gain bandwidth; consequently, the broad bandwidth arises from the overlap between the spontaneous emission spectrum and the MQW gain profile. Furthermore, the tilted waveguide structure with AR coating contributes to this characteristic by ensuring low facet reflectivity. Active-MMI SLEDs exhibit a spectral ripple depth of 0.03 dB, which is lower than that of conventional SLEDs, indicating smoother wideband output.

6.4.3 Field Patterns and Fiber-Coupling Efficiency

Since the emission mechanism of SLEDs is based on single-pass ASE with a short coherence length, it is essential to determine whether self-imaging occurs within the active region of MMI SLED and whether this device operates similarly to a passive

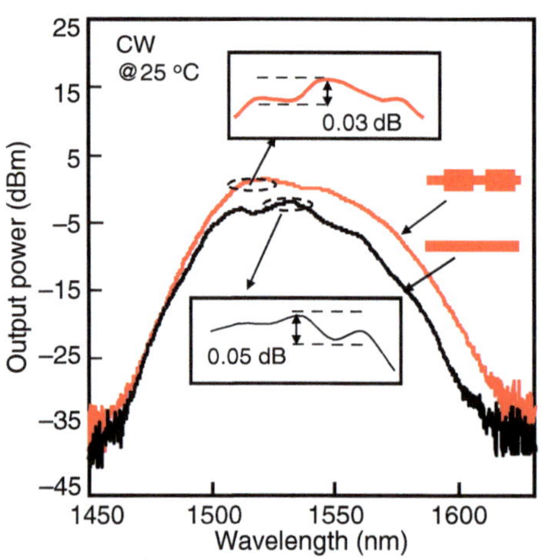

Figure 6.14 Spectra of the fabricated SLEDs. Adapted from Zang et al. [15].

MMI structure. Crucially, does the optical output of an active-MMI SLED generate a single-transverse-mode beam? To address these questions, the occurrence of both self-imaging and single-transverse-mode operation in active-MMI SLEDs should be demonstrated through measurements of their far-field patterns (FFP) and near-field patterns (NFP). Upon emerging from the waveguide facet, the optical wave diffracts into free space. This light is typically captured by an external optical system after propagating a defined distance. Measurements obtained at distances greater than w^2/λ from the facet, where w represents the transverse dimension of the emitting aperture for light of wavelength λ, constitute FFP. Conversely, measurements taken at distances less than w^2/λ define NFPs. The intensity distributions in the FFP and NFP directly reflect the beam divergence characteristics of the SLED. High spatial beam quality and stability of the output beam are critical requirements for most practical SLED applications. The quality of the FFP and NFP is governed by the refractive indices of the active and cladding layers, as well as the width of the active layer. Typically, the observed field patterns vary with increasing injection current due to carrier-density–induced changes in the refractive index [25].

To analyze the optical beam quality, the beam profiles of the SLEDs are investigated by measuring their NFPs and FFPs. The measurement system for NFPs and FFPs typically utilizes a dedicated optical setup coupled with a digital CCD camera to acquire two-dimensional images of the optical beam. Figure 6.15 presents the lateral FFPs of the fabricated SLEDs at varied output power with an operating temperature of 250 °C. The measured lateral far-field angles $\theta_{//}$ (where $\theta_{//}$ is defined as the full angle at $1/e^2$ of the maximum intensity in the FFP) were approximately 81° and 83° for the active-MMI SLEDs and the regular SLEDs, respectively. The measured far-field angle $\theta_{//}$ is largely invariant with respect to output power or driving current. Both device types operated in a stable single mode, exhibiting symmetric intensity distributions across the entire output power range. This characteristic ensures

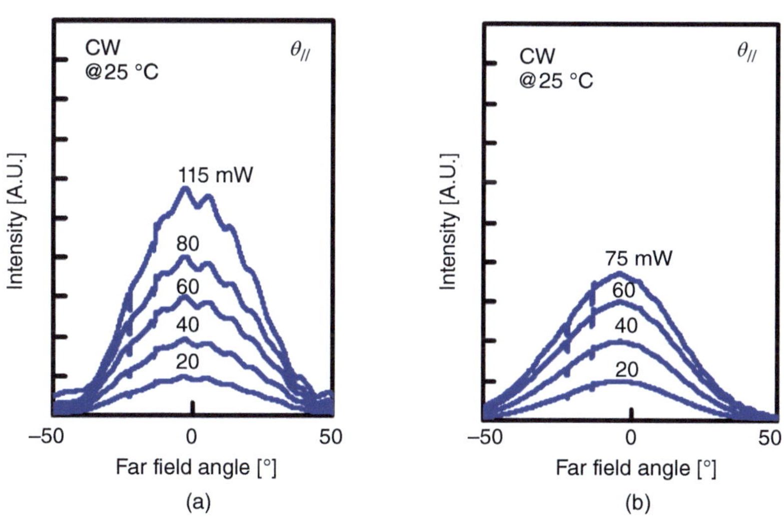

Figure 6.15 Far-field patterns of (a) active-MMI SLEDs and (b) regular SLEDs.

efficient coupling of the output light into a single-mode fiber. The measured intensity distributions for both structures align well with a Gaussian profile. Crucially, no side lobes were observed even under the maximum output power condition for each device.

These results indicate the successful establishment of self-imaging within the active-MMI structure, even during ASE mode operation, i.e., conditions of non-oscillation. Minor ripples observed in the FFPs of the active-MMI SLEDs are likely attributable to a small fraction of radiation emanating from the MMI region. Crucially, these ripples did not impair high fiber-coupling efficiency, as evidenced by a measured fiber-coupling efficiency of 66% attained using a conventional hemispherical lens with a single-mode fiber. Furthermore, it can be deduced that the optical output field profile of the active-MMI SLEDs, like that of the conventional devices, remains unaffected by carrier-induced refractive index fluctuations up to the maximum output power. This conclusion is supported by the negligible change observed in the FFPs with increasing injection current.

To gain further insight into the optical behavior of active-MMI SLEDs, we also investigated NFPs of the fabricated SLEDs. The near-field emission patterns of the fabricated SLEDs were obtained by focusing the SLEDs emission into a CCD camera with a lens imaging system. Figure 6.16 shows the measured NFPs of the fabricated SLEDs at an injection current density of 9 kA/cm^2 under CW operation. The measured output NFP images show that the optical light is guided along the cavity under the ASE. As can be seen in Figure 6.16, all of them clearly emit in a regular single-transverse-mode output with a highly symmetrical distribution, without any higher modes appearance.

Therefore, based on the previous analysis and discussion, we could successfully confirm that SLEDs with MMI configuration should also work as active-MMI phenomena, due to their wide bandwidth with extremely low ripple and stable single-transverse-mode output.

Coupling with hemispherical lens fiber is one of the ways to confirm the stability of the output optical field, as the coupling efficiency is sensitive to the far-field angle deviation [23]. Figure 6.17 shows the calculated coupling efficiency as a function of $\theta_{//}$ (here, $\theta_{//}$ is the far-field angle parallel to the junction plane and is defined as full angles at 1/e^2 of the maximum power of the FFP).

Figure 6.18 illustrates the calculated fiber-coupling efficiency as a function of the radius R for hemispherical lens fibers integrated with active MMI SLEDs. During this

(a) (b)

Figure 6.16 Near-field patterns of (a) active-MMI SLEDs and (b) regular SLEDs.

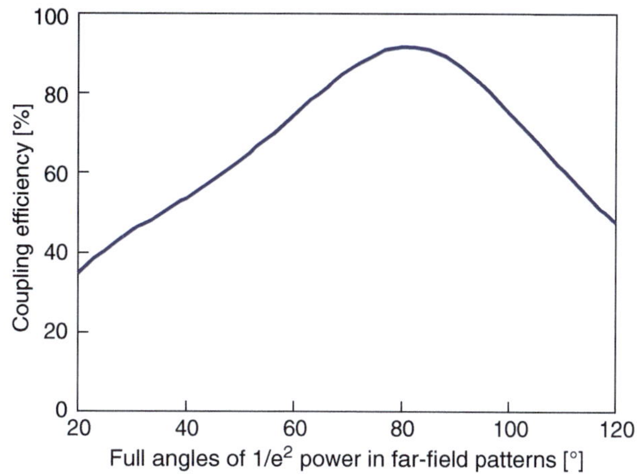

Figure 6.17 Calculated coupling efficiency as a function of far-field angles $\theta_{//}$. Zang et al. [23].

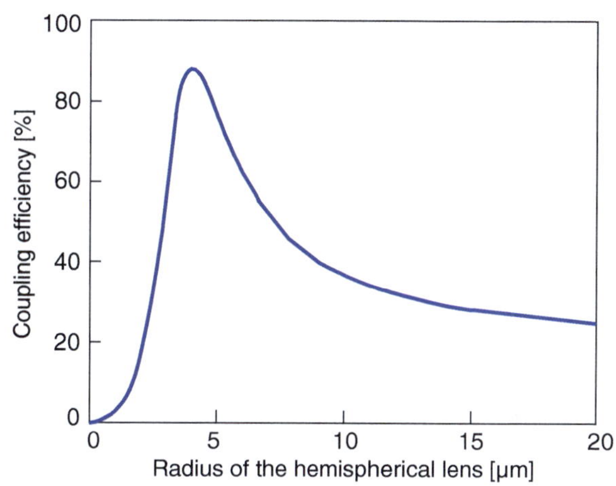

Figure 6.18 Calculated coupling efficiency as a function of lens radius. Zang et al. [23].

calculation, the far-field angles perpendicular ($\theta_{\perp}$) and parallel ($\theta_{\perp}$) to the junction planes were measured as 69° and 81°, respectively, based on the FFPs of fabricated devices. The coupling efficiency increases significantly with increasing lens radius R, reaches a maximum, and subsequently decreases. This calculated dependency indicates that a hemispherical lens radius $R = 4\,\mu\text{m}$ delivers the highest coupling efficiency.

Figure 6.19 displays the output power and fiber-coupled power versus drive current under CW operation at 25 °C for the devices. The active MMI SLEDs achieved a maximum output power of 115 mW at 25 °C. This represents a significant improvement of 40 mW (54%) compared to the maximum output power of 75 mW obtained from regular SLEDs. The enhanced power performance is attributed to the wider actively pumped region enabled by the active-MMI configuration, which elevates the power saturation level. Additionally, these active-MMI SLEDs demonstrated a high fiber-coupling efficiency of 66%, resulting in a maximum fiber-coupled power of 77 mW.

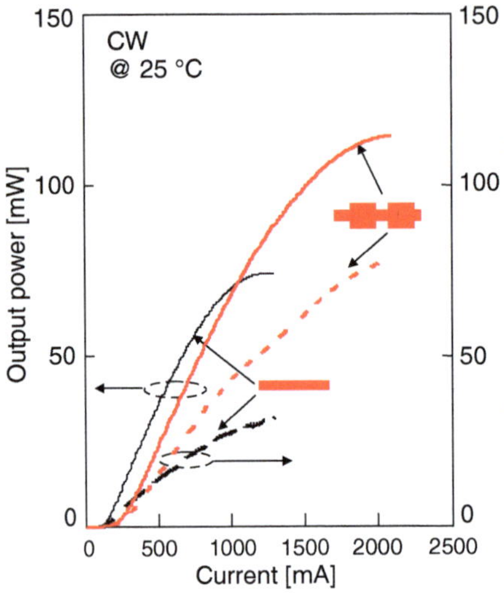

Figure 6.19 Light output power and fiber-coupled power. Zang et al. [23].

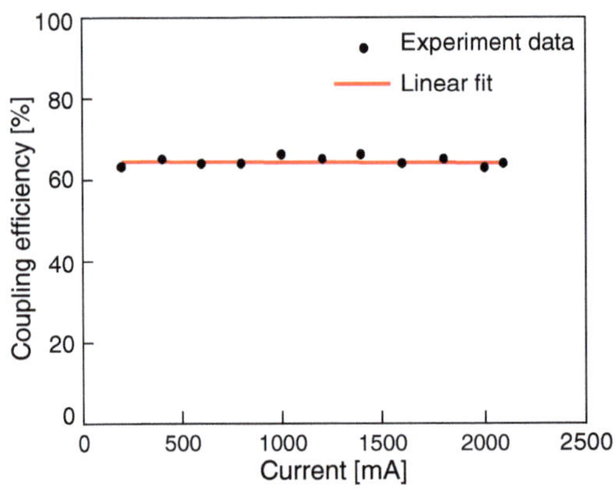

Figure 6.20 Coupling efficiency as a function of injection current. Zang et al. [23].

The measured coupling efficiency of the active-MMI SLEDs remained nearly constant up to the maximum output power, as demonstrated in Figure 6.20. It was confirmed that active-MMI SLEDs maintain a stable output profile at maximum power levels. Consequently, these devices achieve stable and high coupling efficiency when coupled to a hemispherical lens single-mode fiber.

In addition, the measured coupling efficiency was significantly higher for the active-MMI SLEDs (66%) compared to the regular SLEDs (43%). To investigate this disparity in fiber-coupling efficiency, we experimentally evaluated coupling efficiencies using hemispherical lens fibers with four different radii ($R = 4$, 6, 8, and 13 μm). Figure 6.21a,b presents the measured data alongside calculated coupling efficiencies for comparison. The theoretical fiber-coupling efficiencies assumed $\theta_{//} = 83°$ and $\theta_{\perp} = 57°$. Discrepancies between theoretical predictions and experimental results are likely attributable to reflection losses and spherical aberration from the lens surface, as the theoretical model assumed an ideal lens transformer. Furthermore, the tilted waveguide structure contributed to the observed difference. Importantly,

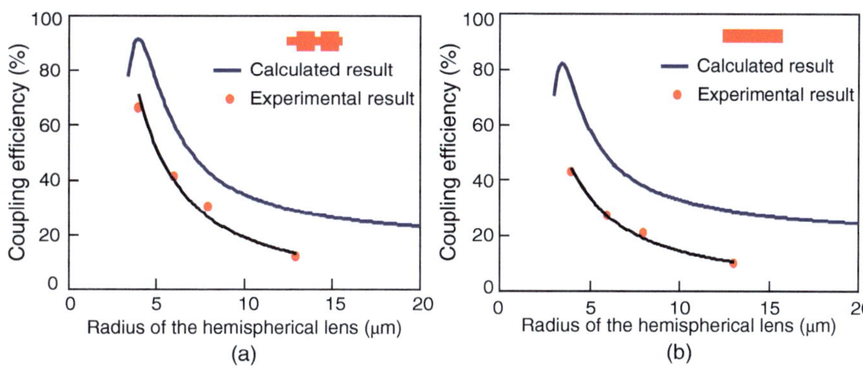

Figure 6.21 Calculated and experimental results about coupling efficiency as a function of lens radius. Zang et al. [23].

both experimental and theoretical results consistently demonstrate higher coupling efficiency for active-MMI SLEDs across all lens radii.

This enhanced coupling efficiency arises from a difference in the transverse far-field angle between active-MMI and regular SLEDs, potentially caused by the active-MMI phenomenon. The following origin for this perpendicular angle difference is proposed. While the layer structure of the active-MMI device is identical to the regular SLED, the light propagation differs significantly. Specifically, within the active-MMI region, light undergoes diffraction at the MMI input followed by refocusing at the MMI output. This process may generate a marginally defocused distribution across the entire spectral range at the MMI output, resulting in a more circular far-field profile. For active-MMI SLEDs, the measured $\theta_\perp$ (69°) was notably closer to $\theta_{//}$ (81°), whereas for regular SLEDs, $\theta_\perp$ was 57° and $\theta_{//}$ was 83° (where $\theta_\perp$ and $\theta_{//}$ denote the full angles at $1/e^2$ of the maximum power in the SLED FFP). Consequently, the more circular output far-field profile of active-MMI SLEDs facilitates higher coupling efficiency compared to regular SLEDs.

References

1 Bohm, K., Marten, P., Peterman, K. et al. (1981). *Electron. Lett.* 17: 352–353.

2 Fercher, A.F., Drexler, W., Hitzenberger, C.K. et al. (2003). *Rep. Prog. Phys.* 66: 239–303.

3 Wagner, S.S. and Chapuran, T.E. (1990). *Electron. Lett.* 26: 696–697.

4 Kashima, Y., Kobayashi, M., and Takano, H. (1988). *Electron. Lett.* 24: 1507–1508.

5 Alphonse, G.A., Gilbert, D.B., Harvey, M.G. et al. (1988). *IEEE J. Quantum Electron.* 24: 2454–2457.

6 Nagai, H., Noguchi, Y., and Sudo, S. (1989). *Appl. Phys. Lett.* 54: 1719–1721.

7 Paterson, B.D., Epler, J.E., Graf, B. et al. (1994). *IEEE J. Quantum Electron.* 30: 703–712.

8 Lin, C.F. (1991). *Electron. Lett.* 27: 968–970.

9 Yamatoya, T., Sekiguchi, S., Koyama, F. et al. (2001). *Jpn. J. Appl. Phys.* 40: 678–680.

10 Middlemast, I., Sarma, J., and Yunns, S. (1997). *Electron. Lett.* 33: 903–904.

11 Delepine, S., Gerard, F., Pinquier, A. et al. (2001). *IEEE J. Sel. Topics Quantum Electron.* 7: 111–123.

12 Noguchi, Y., Yasaka, H., Mikami, O. et al. (1990). *J. Appl. Phys.* 67: 2665–2667.

13 Chen, I.R., Zhuang, V.H., Xu, Y.I. et al. (1990). *Appl. Phys. Lett.* 56: 2502–2503.

14 Kondo, S., Yasaka, H., Noguchi, Y. et al. (1992). *Electron. Lett.* 28: 132–133.

15 Zang, Z.G., Minato, T., Navaretti, P. et al. (2010). *IEEE Photon. Technol. Lett.* 22: 721–723.

16 Tomlins, P.H. and Wang, R.K. (2005). *J. Phys. D. Appl. Phys.* 38: 2519–2535.

17 Kersey, A.D., Marrone, M.J., and Davis, M.A. (1991). *Electron. Lett.* 27: 518–520.

18 Huang, D., Swanson, E., Lin, C.P. et al. (1991). *Science* 254: 1178–1181.

19 Fujimoto, J.G. (2003). *Laser Rev.* 31: 635–642.

20 Vali, V. and Shorthill, R.W. (1976). *Appl. Opt.* 15: 1099–1100.

21 Sunada, S. and Harayama, T. (2006). *Phys. Rev. A* 74: 801–803.

22 Miki, T. and Ishio, H. (1978). *IEEE Trans. Commun.* 26: 1082–1087.

23 Zang, Z.G., Mukai, K., Navaretti, P. et al. (2011). *IEICE Trans. Electron.* E94-C: 862–864.

24 Oh, S.H., Kim, K.S., and Ju, J.J. (2009). *Opt. Express* 17: 10189–10194.

25 Tsai, C., Su, Y., Tsai, C. et al. (2004). *IEEE Photon. Technol. Lett.* 16: 2412–2417.

7
Solar Cells

7.1 Introduction

7.1.1 Overview of Chapter

Since the emergence of organisms on Earth, they have primarily existed with the heat and light provided by the Sun. Most of the energy required by humans comes directly or indirectly from the Sun. For example, humans have mostly utilized fossil fuel energy, such as coal, oil, and natural gas, which store prehistoric solar energy from the photosynthesis process. The significant increase in energy demand and consumption has led to advancements in human society and civilization, but it has also caused air pollution and the degradation of the ozone layer, seriously threatening the living environment of humanity. Therefore, changing the energy structure and developing green and renewable energy have become the shared desire of people and are highly valued by governments worldwide. Among various new clean energy sources, solar energy is becoming increasingly important in the energy structure of sustainable human development due to its independence from geographical and environmental conditions. In an era of reducing fossil fuel usage, humanity has begun to consciously develop and utilize solar energy. Solar energy is essentially a vast amount of nuclear energy released by the nuclear fusion of hydrogen atoms within the Sun. Over billions of years, solar energy has proven to be an inexhaustible source of clean energy. In particular, the share of photovoltaic (PV) power generation in total global power generation has increased annually. Currently, in the renewable clean power generation sector, it ranks second only to hydropower and has caught up with wind power generation. Some institutions predict that by around 2050, PV power generation will become the primary energy source for humanity and will have significant potential for future development [1].

7.1.2 Definition of Solar Cells

A solar cell is a kind of photoelectric device. After sunlight irradiation, the energy of light can be converted into electricity. PV power generation uses solar cells to directly convert solar energy into electricity. The basic principle of PV power generation is that semiconductor materials absorb sunlight whose photon energy is greater than the semiconductor bandgap width to produce electron–hole pairs (EHPs) and then realize photoelectric conversion through charge separation and collection. According to the different charge-separation methods, solar cells can be divided into two types of solar cells: junction-charge-separation type and interface-charge-separation type.

Junction-charge-separating solar cells are usually PN junction or *p-i-n* junction cells or thin-film cells made of inorganic semiconductor materials. The mechanism of photoelectric conversion is the PV effect, that is, sunlight with photon energy greater than the semiconductor bandgap width is absorbed by PN junction or *p-i-n* junction made of *p*-type, *n*-type, or *i*-type semiconductors to produce EHPs. The EHPs generated or diffused into the junction region separated and drifted toward the electrodes to form photogenerated current and photogenerated voltage under the action of the internal electric field.

In contrast, the interface-charge-separation solar cells are generally thin-film cells composed of organic semiconductor materials, and the mechanism of photoelectric conversion differs. After absorbing sunlight and generating EHPs, such a solar cell cannot form an effective PN junction between *p*-type and *n*-type materials through diffusion due to the issues of low carrier mobility, short diffusion length, and short exciton lifetime in organic semiconductor materials. Therefore, only the energy level difference at the interface of two organic semiconductor materials with different energy levels can separate the EHPs diffusing to the interface, and the separated electrons and holes are transmitted to the two electrodes to form photogenerated current and photogenerated voltage respectively, hereby realizing photoelectric conversion.

7.1.3 Importance of Solar Cells in Renewable Energy

Figure 7.1 shows the projected global electricity demand for various renewable energy sources through 2050 [1]. As can be seen, the use of renewable energy will experience rapid growth, among which the growth of solar power generation is more pronounced. In general, the development of alternative energy sources, such as wind power and hydropower, is affected by the geographical environment of each country and cannot be effectively applied. While biomass energy is considered the most effective and most likely alternative energy to replace oil. However, its development is greatly constrained in situations of food scarcity. Therefore, solar cells as a form of green energy are listed as one of the priorities for future research and development.

The advantages of solar cells and their modules in use are as follows:

1) **Resource sustainability:** At least for the foreseeable future, as a renewable energy, solar energy is considered inexhaustible.
2) **Economic and environmental benefits:** Solar PV systems operate without fuel consumption, rotating mechanical parts, or operational emissions, effectively eliminating operational costs while maintaining zero-noise and pollution-free power generation.

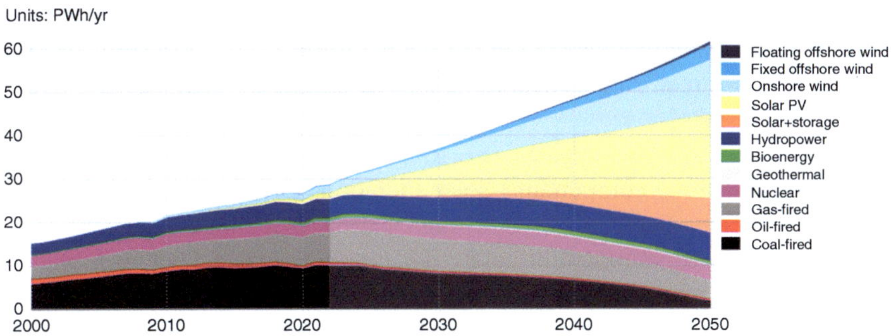

Figure 7.1 Projected global electricity demand for various renewable energy sources DNV [1].

3) **System durability:** Solar cell modules have robust modular design with minimal mechanical components, ensuring reliable performance with a service life exceeding 20 years.
4) **System flexibility:** Solar cells can convert light energy directly into direct current energy, and the scale of power generation is scalable for adaptable deployment ranging from microscale applications such as portable calculators to large power plants.
5) **Technical adaptability:** Diverse and flexible solutions of PVs can be provided through crystalline silicon, thin-film, and emerging PV technologies. Modern hybrid systems can effectively overcome the disadvantages of solar irradiance fluctuations through intelligent energy storage integration, providing reliable peak load support.
6) **Architectural integration:** Advanced thin-film solar cells can be designed to facilitate building-integrated applications through customizable light transmission and thermal management properties, enabling dual-function energy generation and architectural elements.

Globally, renewable energy legislation and financial incentives for solar cells have laid a solid foundation for the market expansion of solar cells. Currently, the primary technical problems of solar cells are further enhancing the energy conversion efficiency and reducing the production cost. Undoubtedly, the demand for solar cells will continue to grow.

7.2 Basic Principles of Solar Cells

This section provides an overview of PV principles, which are the fundamental scientific concepts governing solar cell operation.

7.2.1 Solar Spectrum and Irradiance

Solar energy serves as the fundamental energy source sustaining various activities on Earth. This energy reaches Earth through solar radiation. Solar radiation is basically electromagnetic waves, which originate from the vibrational motion of charged particles. Through radiative transfer at light speed, solar energy propagates across space without requiring a medium, enabling efficient energy delivery over vast cosmic distances.

Meteorology primarily investigates the thermal radiation characteristics on Earth consisting of three components: solar radiation, terrestrial radiation, and atmospheric radiation. Based on their distinct wavelength ranges, meteorological science classifies solar radiation as short-wave radiation (0.15–4 μm), while grouping ground and atmospheric radiation under long-wave radiation (3–100 μm). This categorization stems from their significantly different spectral properties. The visible light spectrum, constituting a critical portion of solar radiation, spans 0.4–0.7 μm. When dispersed through a prism, this range separates into a continuous spectrum of colors, commonly described as seven chromatic bands: red, orange, yellow, green, blue, indigo, and violet. It should be noted that spectral components beyond the visible range (both shorter than 0.4 μm and longer than 0.7 μm) elude direct human visual perception. However, specialized instrumentation enables their precise detection and measurement.

The Sun is a massive incandescent gaseous sphere with distinct temperature stratification. Its surface (photosphere) maintains a temperature around 5800 K, while the internal thermonuclear fusion processes at the core generate temperatures exceeding 1.5×10^7 K. Through continuous electromagnetic radiation spanning ultraviolet (UV), visible, and infrared spectra, this celestial body emits vast amounts of luminous and thermal energy into space. As the primary energy source for Earth's atmospheric dynamics, solar radiation fundamentally governs various atmospheric physical processes, including photoionization in the ionosphere, greenhouse effect in the troposphere, and photochemical reactions in the stratosphere.

Solar irradiance (denoted as I [W/m^2]) measures the electromagnetic radiation power received per unit time on a surface perpendicular to the radiation direction. As a key radiometric quantity, it specifically defines the instantaneous intensity of solar energy delivery. The solar spectrum can be roughly modeled using blackbody radiation principles.

Actual observations and measurements at the Earth's atmospheric environment, however, differ from theoretical calculations and exhibit complex spatiotemporal patterns determined by the following factors.

1) **Sun–Earth distance:** The Earth's orbit around the Sun is elliptical, so the distance between the Sun and Earth changes on an annual cycle. The intensity of solar radiation received on Earth is inversely proportional to the square of the distance between the Sun and Earth. When the Earth passes perihelion, the solar energy per unit area of the Earth's surface is 7% more than when the Earth passes aphelion.
2) **Sun altitude:** Solar radiation intensity follows the cosine law of illumination: it is maximal when the Sun is at zenith (90° elevation) and diminishes with decreasing elevation angles. The higher the height of the Sun, the greater

its radiation intensity; conversely, the lower the radiation intensity. Higher solar elevation concentrates radiant energy over smaller surface areas, while oblique angles cause energy dispersion across larger terrain surfaces, reducing per-unit-area energy reception.

3) **Insolation duration effect:** Diurnal radiation accumulation varies with seasonal and latitudinal changes: summer solstice periods exhibit extended photoperiods (14–16 hours at mid-latitudes), enabling greater daily energy accumulation. Winter conditions show reduced insolation duration (8–10 hours) coupled with lower solar elevations. Daylength variability amplifies poleward versus negligible equatorial variation.

To accommodate the solar radiation variations and establish a common method for solar cell efficiency characterization, a standard of solar radiation must be set. Several standard solar spectra exist, which suit different scenarios of solar cells. As shown in Figure 7.2, the standard solar spectra AM0 and AM1.5G represent solar irradiance under different atmospheric conditions, where air mass (AM) quantifies the optical path length through Earth's atmosphere relative to its zenith thickness. AM0 denotes extraterrestrial irradiance (1361 W/m^2) measured at the top of Earth's atmosphere—the reference spectrum for space applications. AM1.5G (1000 W/m^2) models terrestrial solar radiation when sunlight traverses 1.5 atmosphere thicknesses (48.2° zenith angle) with global tilt, accounting for both direct and diffuse components. The key difference arises from atmospheric interactions: AM1.5G shows absorption bands from atmospheric constituents (O_3, H_2O, CO_2) and Rayleigh scattering effects, particularly in UV and IR regions; AM0 differs from a theoretical 5800 K blackbody spectrum due to quantum effects in the Sun's photosphere, exhibiting characteristic Fraunhofer absorption lines from solar atmospheric elements, along with nonthermal emission components from chromospheric activity. These standardized spectra serve critical roles in PV device characterization and satellite system design.

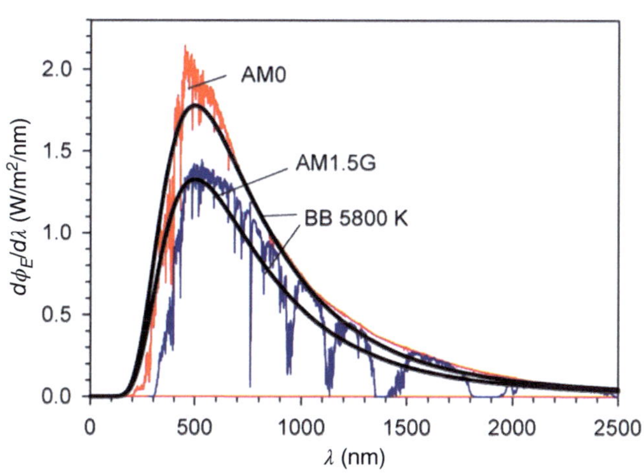

Figure 7.2 Standard solar spectra for air mass AM0, AM1.5G, and the blackbody at 5800 K. Reprinted with permission from Nayak et al. [2].

7.2.2 Photovoltaic Effect

Solar energy utilization primarily encompasses three conversion pathways: photochemical conversion, photothermal conversion, and photoelectric conversion.

- Photochemical conversion involves solar-to-chemical energy transformation through photocatalytic processes, exemplified by natural photosynthesis in plants and artificial photosynthesis systems for water splitting ($H_2O \rightarrow H_2 + O_2$).
- Photothermal conversion achieves solar-to-thermal energy transfer through radiation collection and heat absorption mechanisms, commonly realized in solar furnaces and water-heating systems.
- PV conversion represents the direct solar-to-electricity transformation through semiconductor PV effect, constituting the fundamental principle of solar cells.

This subsection will focus on the discussion of the PV effect. PV power generation leverages the semiconductor interface's PV effect to directly convert light energy into electrical energy through three essential processes: generation, separation, and collection of charge carriers. The underlying physical process will be elucidated using the PN junction solar cell as a representative system.

The PV mechanism initiates when photons impart sufficient energy to semiconductor electrons, elevating them from the valence band to the conduction band. This energy transition creates complementary charge carriers—photogenerated electrons in the conduction band and corresponding holes in the valence band, collectively known as EHPs. These carriers are generated throughout the semiconductor structure, including the *P*-type, space-charge, and *n*-type regions.

Critical charge separation occurs in the space-charge region formed at the PN junction. This area contains a built-in electric field that immediately separates the photogenerated carriers. The field exerts directional forces that drive electrons toward the *n*-type region and holes toward the *p*-type region, achieving spatial charge segregation.

In the *n*-type and *n*-type regions, carrier transport involves two mechanisms:

1) In the *n*-type region, minority-carrier holes diffuse toward the PN junction boundary, where the built-in electric field accelerates their migration into the *n*-type region, while majority electrons remain.
2) Conversely, in the *n*-type region, minority electrons diffuse to the junction boundary and are swept into the *n*-type region by the same electric field, leaving majority holes behind.

This dual transport process establishes an asymmetric charge distribution, forming a photogenerated electric field that both counteracts the intrinsic built-in field and produces an electromotive force. The resultant potential difference manifests as a measurable voltage across the semiconductor structure—the fundamental embodiment of the PV effect. Figure 7.3 schematically illustrates this PV operation.

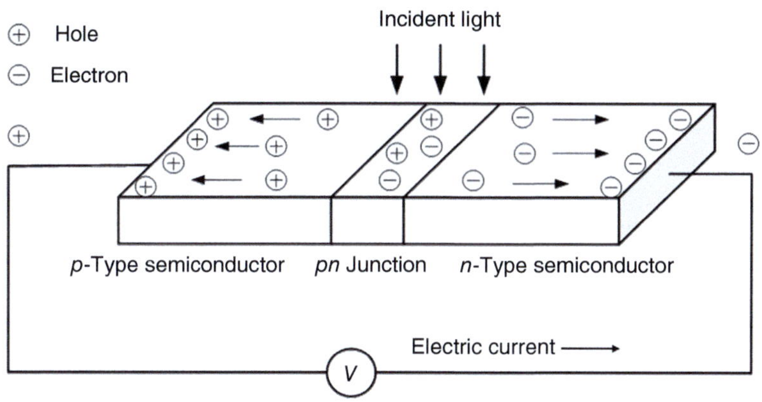

Figure 7.3 Schematic diagram of the photovoltaic effect.

7.2.3 Solar Cell Structure

At present, there are many types of solar cells. According to the material composition of solar cells, there are three primary categories: (i) silicon-based systems (monocrystalline/polycrystalline) dominating 90% of PV markets; (ii) compound semiconductors including CIGS ($CuInGaSe_2$), halogen perovskite thin-film variants with enhanced flexibility; and (iii) emerging organic/polymeric cells enabling semitransparent building-integrated PVs (BIPVs).

From technological evolution perspectives, the solar cells can be categorized into three types: (i) conventional silicon wafer devices as the first-generation solar cells; (ii) thin-film technologies emphasizing material economy as the second-generation solar cells, such as CIGS/CdTe solar cells; (iii) advanced architectures (perovskite, quantum-dot, and multi-junction architectures) as the third-generation solar cells pursuing efficiency breakthroughs beyond Shockley–Queisser limits, for example, the perovskite–silicon tandems as the third-generation solar cell can achieve high certified efficiency.

According to the substrate characteristics, solar cells can be divided into flat-panel rigid solar cells and flexible substrate solar cells. According to the classification of use, they can be divided into space solar cells, ground solar cells and PV sensors. According to the classification of working mode, they can be divided into flat-panel solar cells and concentrated solar cells.

Furthermore, based on composition and structural characteristics, solar cells are categorized into three main types: homojunction, heterojunction, and Schottky barrier solar cells. Detailed discussions of each type are provided as follows.

1) **Homojunction battery:**
   A homojunction solar cell is formed when a PN junction is created within a single semiconductor material (Figure 7.4). This structural configuration distinguishes homojunction devices, with silicon-based and gallium arsenide–based solar cells representing typical examples of this category.

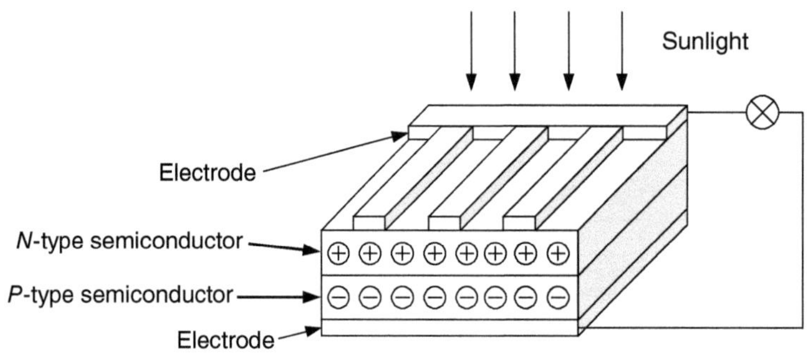

Figure 7.4 Homojunction solar cells.

2) **Heterojunction solar cells:**
   This type employs PN junctions formed between two distinct semiconductor materials with different bandgap energies. Notable examples include tin oxide/silicon, cuprous sulfide/cadmium sulfide, and cadmium sulfide/cadmium telluride configurations (Figure 7.5). When the interfacing materials exhibit compatible lattice structures, the resulting device is specifically termed a heteroface solar cell, exemplified by aluminum gallium arsenide/gallium arsenide combinations.
3) **Schottky solar cells:**
   These devices utilize the Schottky barrier phenomenon arising at metal–semiconductor interfaces (Figure 7.6). Modern variants have evolved into more sophisticated architectures such as metal–oxide–semiconductor (MOS) and metal–insulator–semiconductor (MIS) configurations through interfacial engineering.

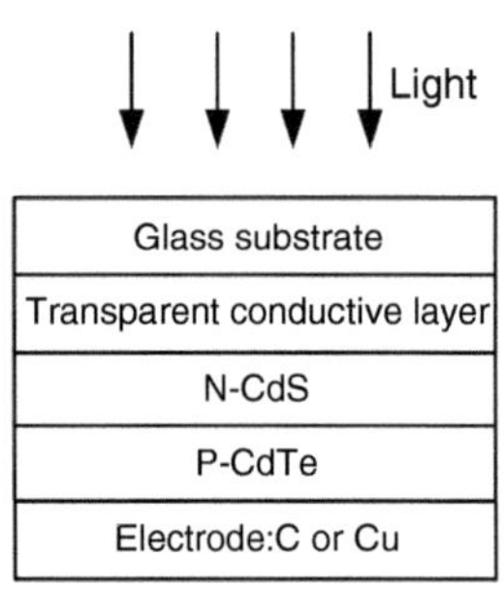

Figure 7.5 Heterojunction solar cells.

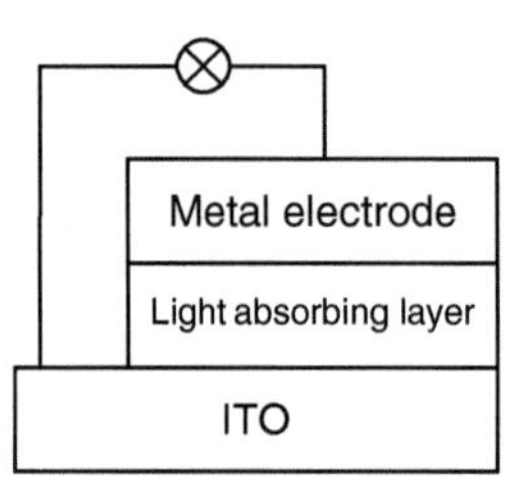

Figure 7.6 Schottky solar cells.

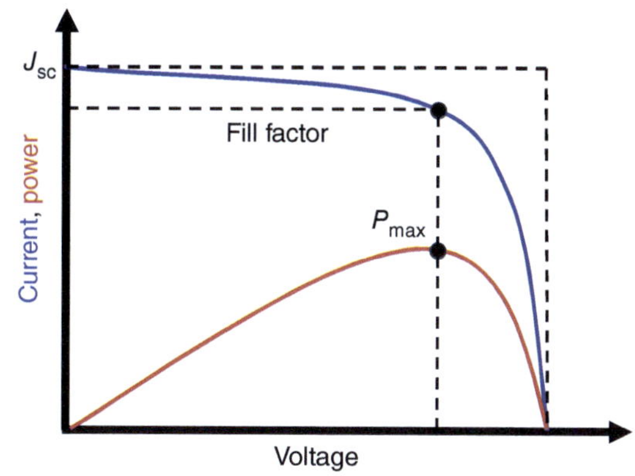

Figure 7.7 Current–voltage (I–V) cure and output power of a solar cell.

7.2.4 Parameters of Solar Cells

Under standardized testing protocols that apply AM 1.5G spectrum illumination (1000 W/m^2 irradiance) at ambient temperature (25 °C), perovskite solar cells (PSCs) employ current density–voltage (J–V) or current–voltage (I–V) characterization techniques to evaluate PV performance, as shown in Figure 7.7. Four fundamental parameters are derived from J–V curve analysis: short-circuit current density (J_{sc}), open-circuit voltage (V_{oc}), fill factor (FF), and the power conversion efficiency (PCE). The PCE is the primary quality indicator, directly determining device effectiveness and energy conversion capability. Furthermore, advanced characterization incorporates external quantum efficiency (EQE) evaluation, which measures incident-photon-to-current conversion efficiency (IPCE) across various wavelengths, providing essential spectral response data for comprehensive efficiency assessment and system optimization. The hysteresis factor is also important for solar cells to characterize the current–voltage response discrepancies under varying voltage sweep directions. These solar parameters will be explained in detail as follows.

1) **Short-circuit current density (J_{sc}):**
   Short-circuit current density (J_{sc}) represents the photocurrent density generated in PV devices under zero external bias conditions, measured in mA/cm^2. This parameter corresponds to the vertical intercept of the current–voltage (I–V) characteristic curve with the Y-axis.
   The magnitude of J_{sc} is determined by three primary factors:
   - **Material properties:** Bandgap width of the light-absorption layer and defect density in device structure
   - **Measurement conditions:** Incident light intensity and spectral distribution
   - **Device configuration:** Effective illumination area and charge separation and collection efficiency

   A narrower bandgap in the absorber material enhances light harvesting across broader spectral ranges, thereby increasing photocurrent generation. However, this improvement must be balanced with open-circuit voltage considerations for overall device performance optimization.

2) **Open-circuit voltage (V_{oc}):**
The open-circuit voltage (V_{oc}) is the voltage measured across the terminals of a device under open-circuit conditions when the current flow is zero. Its unit is volts (V), and it corresponds to the point where the current–voltage curve intercepts the horizontal (voltage) axis.
The attainment of high V_{oc} values is limited by several factors, including fundamental radiative and Auger recombination, which set the ultimate upper limit according to detailed balance theory. Additionally, non-radiative recombination pathways through trap states at defects, interfaces, or surfaces acting as recombination centers significantly reduce V_{oc} below theoretical maxima. Finally, the operating temperature of the device plays a crucial role, as an increase in temperature inherently reduces the V_{oc} due to an exponential rise in intrinsic carrier concentration and a corresponding decrease in the semiconductor bandgap energy.
3) **Fill factor (FF):**
The output power of a solar cell depends significantly on the operation point. Usually larger J_{sc} and V_{oc} can potentially create larger output power. However, as the output power is the product of current and voltage, the maximum power point tracking (MPPT) depends on the specific shape of the I–V curve, and changes in environmental factors such as sunlight intensity and temperature can cause this point to shift, requiring MPPT systems to dynamically optimize power extraction by adjusting load conditions. At the standard measurement condition (AM 1.5G illumination at 25 °C), FF characterizes the ideality of the shape of the I–V curve for maximum power output. The FF is calculated by Eq. (7.1), which is the ratio of the maximum power point to the product of the open-circuit voltage and short-circuit current.

$$\mathrm{FF} = \frac{P_{max}}{V_{oc} \times J_{sc}} \tag{7.1}$$

On an I–V characteristics (current–voltage) curve, the rectangle defined by the origin and the maximum power point represents the device's maximum achievable power output. The largest possible rectangle enclosed by the axes that also intersects this curve is formed by the outer tangent of the upper curve. Consequently, the closer the ratio between these two rectangle areas is to 1, the more ideal the device characteristics become. This indicates smaller series resistance, larger parallel resistance, more ideal Ohmic contacts, and reduced carrier recombination behavior.
4) **Power conversion efficiency (PCE):**
PCE refers to the ratio between the maximum output power (P_{max}) and incident light power (P_{in}) of PV devices, which can directly reflect the performance numerically. The calculation equation is

$$\mathrm{PCE} = \frac{P_{max}}{P_{in}} = \frac{V_{oc} \times J_{sc} \times \mathrm{FF}}{P_{in}} \tag{7.2}$$

It can be seen that the three parameters J_{sc}, V_{oc}, and FF restrict each other and jointly determine the level of efficiency. This parameter is typically expressed as a percentage and serves as a key benchmark for assessing advancements in solar energy technology, such as improvements in material design or reductions in energy losses during power generation.

5) **Hysteresis factor (HI):**
The I–V curve of solar cells can vary during working, influencing the J_{sc}, V_{oc}, FF, and PCE. For commercialization validation of solar cells, extended reliability assessments are necessary. Researchers usually evaluate the hysteresis of perovskite cells by hysteresis factors, which is another parameter to evaluate the performance of solar cells. Its calculation formula is shown as follows.

$$HI = \frac{PCE_{reverse-scan} - PCE_{forward-scan}}{PCE_{reverse-scan}} \tag{7.3}$$

where $PCE_{reverse\text{-}scan}$ is the PCE value measured under the condition of reverse scanning, and $PCE_{forward\text{-}scan}$ is the PCE value measured under the condition of forward scanning. This phenomenon arises from ion migration, capacitive effects, and charge-trapping dynamics in the photoactive layer. The study shows that improving the interface quality can reduce the hysteresis factor of perovskite cells.

7.3 Typical Solar Cells

Following our study of solar cell introductions and basic principles, this section will analyze several representative types: silicon (Si) solar cells, notable for their widespread application and high technological maturity; copper indium gallium selenide (CIGS) solar cells, distinguished by excellent absorption efficiency and potential for flexible applications; cadmium telluride (CdTe) solar cells, recognized for cost-effectiveness and simplified manufacturing processes; and perovskite solar cells, characterized by remarkably rapid efficiency improvements and ongoing research advancements. A brief introduction to each type will precede detailed analysis.

7.3.1 Si Solar Cell

Silicon solar cells are the most recognized solar cell technology. Especially the c-Si based cells are projected to dominate the market as over 90% of the global PV market. This prominence stems from c-Si's natural abundance, good stability, nontoxicity, and cost competitiveness [3].

With an indirect bandgap of 1.12 eV, c-Si is well-suited to the solar spectrum, corresponding to an absorption cutoff wavelength of approximately 1160 nm, and is close to the optimum for converting solar energy into electricity using a single absorber. However, it exhibits a relatively small absorption coefficient in the infrared region. Effective light absorption can still be achieved with thin wafers (down to 100 μm versus the typical 180 μm thickness), through optimized light-control designs such as surface texturing, antireflective coatings, and rear-surface mirrors.

Due to the indirect bandgap, radiative recombination is weak in silicon, resulting in long-lived photogenerated carriers. Consequently, the primary recombination process is Auger recombination (a nonradiative mechanism). Although this principle holds theoretically, impurities or crystallographic defects can dominate carrier recombination in practical scenarios.

Recently, the electronic quality of c-Si wafers has improved to such an extent that solar cell advancements now primarily depend on contacting systems and surface passivation. Passivation reduces carrier recombination at the silicon surface and is effectively achieved using dielectric materials like aluminum oxide, silicon nitride, silicon carbide, hydrogenated amorphous silicon, and silicon oxide [3].

Since the first Si-based solar cell with nearly 4.5% efficiency was developed at Bell Labs [4], Si-based solar cells have maintained a dominant position [5]. Key innovations emerged between the 1990s and 2000s—including passivated emitter rear-contact structures, surface texturing, firing technology, and screen printing (SP)—which significantly enhanced PCE and reduced manufacturing costs. Consequently, these advances drove the industrialization of silicon PV [6].

Figure 7.8 illustrates the schematic structure of *n*-type silicon-based solar cells [7]. Structurally, it features a PN junction with a lightly doped *p*-type base and a shallow, heavily doped *n*-type emitter. An additional layer, identified as the heavily doped *p*-type back surface field (BSF), enhances rear-side passivation. Front-side cathode contacts form finger electrodes, while the anode contact covers the entire rear surface.

Fundamentally, crystalline Si solar cells generate PV power through two physical stages: (1) photon absorption creating EHPs, followed by (2) EHP separation and transport to external electrodes. High-efficiency c-Si cell technology therefore prioritizes: unique structural design, optimized light absorption, efficient photogenerated carrier collection, minimized carrier recombination losses, and reduced electrode resistance/area. Major approaches include (1) passivated emitter rear cell (PERC),

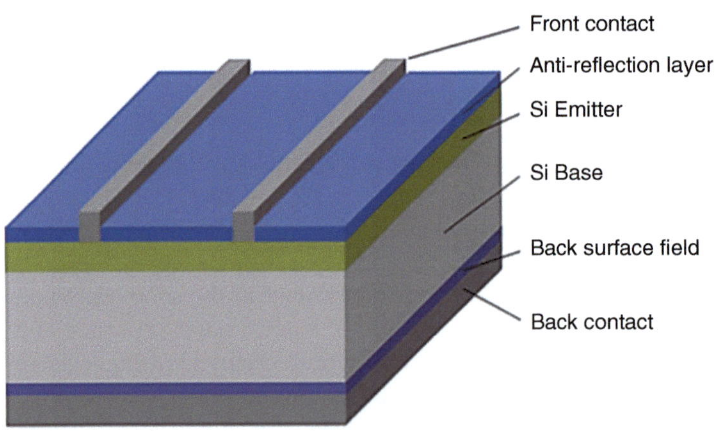

Figure 7.8 A representative structure of a Si-based solar cell showing its basic elements. Reprinted with permission from Xiao and Xu [7].

(2) interdigitated back-contact cell (IBC), (3) heterojunction with intrinsic thin-layer cell (HIT), and (4) heterojunction back-contact solar cells (HBC) [6].

Among these, PERC architecture introduces a rear passivation stack—typically combining dielectric layers with lasers creating point contacts—to significantly reduce recombination losses while still utilizing conventional front-side contacts, leading to commercially achieved efficiencies exceeding 24%. IBC cells eliminate shading losses completely by moving the grid electrodes to the rear, employing intricate interdigitated n^+ and p^+ finger diffusion regions patterned beneath the rear passivation layers. HIT cells leverage the excellent passivating properties of intrinsic amorphous silicon layers deposited on both sides of the wafer, combined with doped amorphous silicon emitters and contacts, enabling very high open-circuit voltages due to low surface recombination. HBC technology merges the back-contact advantage of IBC with the high-quality passivation of amorphous silicon heterojunctions, positioning it as one of the most potent structures currently explored. Ultimately, the pursuit of higher efficiency continues to drive innovations that synergistically integrate these principles—advancing passivation quality, optimizing light management, and minimizing electrical losses through novel contact schemes and heterojunction concepts.

7.3.2 CuInGaSe Solar Cell

Different from silicon solar cells, copper indium gallium selenide (CIGS) solar cells are thin-film PV devices. As one of three mainstream thin-film PV technologies—alongside cadmium telluride and amorphous silicon—CIGS shares key characteristics: its layers are thin enough to be deposited on flexible substrates.

This material is a I–III–VI_2 compounds consisting of copper, indium, gallium, and selenium. It forms a solid solution range between copper indium selenide (CIS, $x = 1$) and copper gallium selenide (CGS, $x = 0$), with the chemical formula $CuIn_xGa_{1-x}Se_2$. The tetrahedrally bonded chalcopyrite crystal structure exhibits a tunable bandgap from ~1.0 eV (CIS) to ~1.7 eV (CGS) with increasing gallium content.

In CIGS solar cells, a p-type CIGS layer serves as the absorber, paired with an n-type semiconductor window layer. The typical structure is illustrated in Figure 7.9. While CIGS outperforms polysilicon at the cell level, its module efficiency remains lower due to maturation challenges. CIGS thin-film technologies traditionally

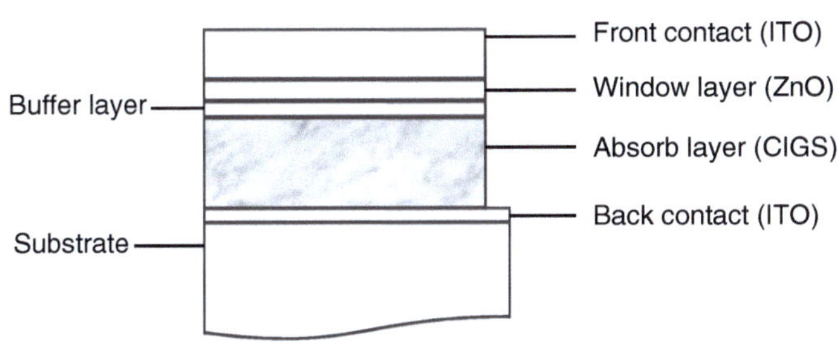

Figure 7.9 The structure of CIGS thin-film solar cell.

use high-temperature deposition and deliver optimal performance on rigid glass substrates, though advances in low-temperature CIGS deposition have significantly narrowed this performance gap.

CIGS-based solar cells can be fabricated on rigid and flexible substrates using various vacuum and non-vacuum techniques. Vacuum methods include co-evaporation [8], physical vapor deposition (PVD) [9], pulsed laser deposition (PLD) [10], chemical vapor deposition (CVD) [11], metal–organic chemical vapor deposition (MOCVD) [12], electron beam deposition (EBD) [13], molecular beam epitaxy (MBE) [14], and sputtering [15].

CIGS solar cell fabrication begins with the deposition of the electrical back contact layer on the substrate, followed by the coating of the CIGS layer, which can be applied using either vacuum or non-vacuum techniques. The process concludes with the window layer coating, primarily deposited using vacuum deposition methods.

To deposit a Cu-In(Ga) alloy film, the selenization method is commonly employed. This process begins with the preparation of a prefabricated Cu-In(Ga) alloy film, which is subsequently subjected to selenization. Common selenium sources include H_2Se gas or solid selenium, with H_2Se gas typically yielding higher-quality CIS/CIGS samples and improved energy conversion efficiency. Methods for preparing the alloy prefabricated layer, upon which this selenization relies, primarily include: the co-sputtering selenization process, the magnetron sputtering selenization process, and the electrodeposition selenization process.

Among these, the co-sputtering approach enables precise compositional control and uniformity by simultaneously depositing alloy elements, while magnetron sputtering offers scalability for industrial applications, and electrodeposition provides a cost-effective, solution-based alternative suitable for complex geometries, ensuring versatility in CIS/CIGS film fabrication for solar cell efficiencies. Regarding electrodeposition, CIS/CIGS layers can be prepared through one-step deposition [16] or multi-step deposition [17]. The one-step approach mixes all elements in a single sedimentation fluid for simultaneous deposition. In multi-step deposition, the prefabricated layer is obtained by electrodeposition, and the elemental ratio is subsequently adjusted.

7.3.3 CdTe Solar Cell

Cadmium telluride (CdTe) solar cells are Group II–VI polycrystalline thin-film solar cells, known for their low cost and suitability for large-area monolithic integration. Research on CdTe for PV applications gained significant momentum in the 1970s, building upon earlier investigations into CdTe material synthesis and photoconductivity that began around 1947.

Cadmium telluride (CdTe) is a direct bandgap semiconductor with a bandgap width of approximately 1.5 eV, well-suited for absorbing sunlight. Among common II–VI PV compounds, CdTe possesses a relatively high average atomic number, a small enthalpy of formation (indicating stability), a relatively low melting point, a large lattice constant, and significant ionic character relative to Group IV or III–V

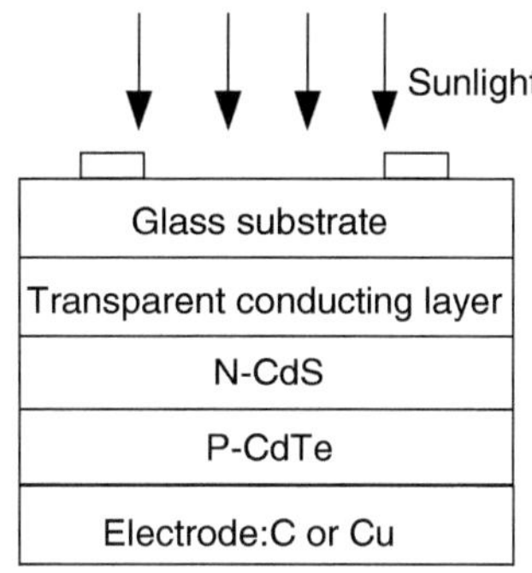

Figure 7.10 Structure of cadmium telluride battery.

semiconductors. Crucially, undoped CdTe exhibits intrinsic *p*-type conductivity due to cadmium vacancies acting as acceptors.

Figure 7.10 illustrates the structure of a typical CdTe heterojunction solar cell. Fabrication typically begins by depositing a transparent conductive oxide (TCO) film, such as SnO_2:F or ITO, onto a glass substrate. A thin *n*-type cadmium sulfide (CdS) layer is then applied, often via chemical bath deposition (CBD) or magnetron sputtering. Subsequently, the *p*-type CdTe absorber layer is deposited, usually by close-spaced sublimation (CSS) or vapor transport deposition (VTD). Finally, a back electrical contact is formed, often involving a complex stack to form an Ohmic contact with the *p*-type CdTe. In these cells, the *n*-type CdS film acts as the window layer, allowing most sunlight to pass through to the CdTe absorber layer where the majority of photons are absorbed. Due to the polycrystalline nature of the films and the processing conditions (especially the $CdCl_2$ treatment), significant interdiffusion occurs between the CdS and CdTe layers, forming a $CdTe_{1-x}S_x$ alloy region at the junction interface.

Several preparation methods exist for cadmium telluride (CdTe) polycrystalline thin films. Commonly employed techniques include VTD, CSS, and SP. Vapor-deposition techniques, such as VTD and CSS, enable the large-scale production of CdTe PV modules on flexible substrates.

Enhancing the efficiency of CdTe thin-film solar cells primarily involves increasing the open-circuit voltage (V_{OC}) and FF. One strategy is to increase the built-in potential of the PN junction by reducing the density of compensating donors (or acceptor-like defects) within the CdTe absorber layer. Additionally, reducing the density of recombination centers within the junction region improves the diode quality factor. Another approach to achieving higher efficiencies involves developing tandem cell architectures utilizing II–VI compound semiconductors. For instance, alloy systems like $CdTe_{1-x}S_x$ offer tunable bandgaps ranging from approximately 1.5 eV (CdTe) to 2.4 eV (CdS), making them suitable candidates for the wide-bandgap top cell in a tandem structure paired with CdTe.

The environmental and health implications associated with cadmium (Cd) toxicity remain a significant consideration for CdTe thin-film technology. Addressing public concerns regarding potential cadmium contamination requires robust end-of-life recycling programs and clear communication about material stability. While elemental cadmium is highly toxic, CdTe itself is a stable compound with very low solubility and bioavailability; studies indicate that its toxicity is significantly lower than that of elemental cadmium. The future success of this technology hinges on further developing low-cost, high-efficiency manufacturing processes while effectively managing environmental stewardship throughout the product lifecycle, including proven recycling strategies.

7.3.4 Organic Solar Cells

Organic solar cells (OSCs), also known as polymer or plastic solar cells, are PV devices that employ conductive organic polymers or small organic molecules, which absorb photons and transport charges to convert sunlight into electricity via photophysical processes [18]. Compared to silicon-based counterparts, OSCs—particularly advantageous for powering small autonomous sensors—offer beneficial characteristics such as mechanical flexibility, molecular tunability, potential low-cost manufacturing using solution-processed fabrication, and the possibility of semitransparent designs, facilitating integration into windows, building facades, and flexible electronics (Figure 7.11).

However, these devices also exhibit significant limitations, including susceptibility to photochemical degradation and low power conversion efficiencies compared with those achieved by inorganic solar cells. Despite challenges in efficiency and stability, OSCs remain a prominent research area due to prospects for cost reduction and performance enhancement. By 2015, efficiencies above 10% were achieved through refined device engineering, with reported efficiencies reaching 17.3% for organic PV by 2018 [19].

Current OSC efficiencies are substantially lower than those of silicon technologies. Additional hurdles include sensitivity to environmental degradation and insufficient protective encapsulation. Performance augmentation requires improved charge-carrier generation and transport. Optimizing morphology control, charge transport pathways, and interfacial engineering is crucial to mitigate recombination losses and enhance cost-effectiveness. Tandem architectures (multi-junction cells) represent a primary research avenue for efficiency improvement. Compared to single-junction devices employing narrow-bandgap materials, tandem structures reduce thermalization losses in the photon-to-electron conversion process.

Commercial deployment remains limited relative to silicon and other thin-film solar cells. Large-scale commercial manufacturing has yet to be established.

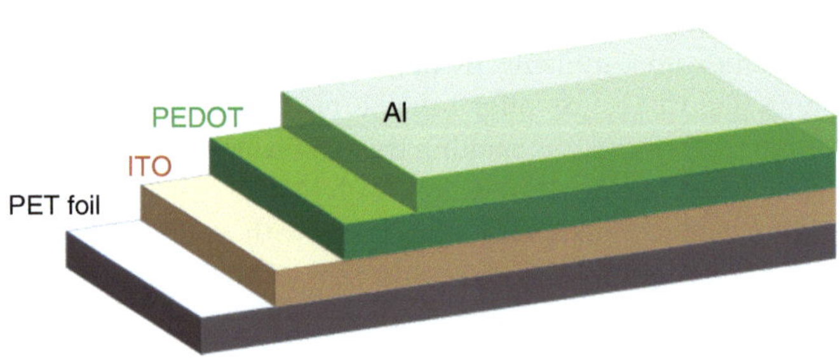

Figure 7.11 Schematic of plastic solar cells.

While Konarka Technologies pioneered roll-to-roll production utilizing polymer bulk heterojunction cells (e.g., P3HT:PCBM, Poly(3-hexylthiophene): [6,6]-Phenyl C61 butyric acid methyl ester) starting in 2008, initial module efficiencies of 3–5% were inadequate for commercial viability, ultimately contributing to the company's closure. Early encapsulation difficulties resulted in low FFs (<70%); however, developments such as inverted bulk-heterojunction structures and novel donor/acceptor combinations subsequently enabled FFs exceeding 75% by 2013 [20].

Ongoing research emphasizes scalable manufacturing platforms, particularly roll-to-roll processing of solution-based materials, to lower production costs. Stability issues, however, currently preclude applications at grid scale. Existing commercial implementations are primarily confined to niche segments like consumer electronics and low-power devices.

7.3.5 Perovskite Solar Cell

7.3.5.1 Crystal Structure of Perovskite Materials

Perovskite originally refers to the mineral $CaTiO_3$, named after the Russian mineralogist Lev Perovski. Today, the term encompasses a broad category of compounds adopting the ABX_3 crystal structure, as depicted in Figure 7.12. In conventional organic–inorganic hybrid perovskite materials, the A-site is occupied by monovalent cations (organic or inorganic) [21], such as Cs^+, methylammonium ($CH_3NH_3^+$, MA^+), or formamidinium ($CH(NH_2)_2^+$, FA^+). The B-site typically hosts divalent inorganic cations (e.g., Pb^{2+} or Sn^{2+}), while the X-site consists of monovalent halide anions (Cl^-, Br^-, or I^-). For the representative $MAPbI_3$ perovskite, the A-site corresponds to $CH_3NH_3^+$, the B-site to Pb^{2+}, and the X-site to I^-. The Pb^{2+} cations and I^- anions form $[PbI_6]^{4-}$ octahedra, with the $CH_3NH_3^+$ cations residing in the interstitial spaces between these octahedra.

Ionic radii critically govern the stability of the perovskite crystal structure. To maintain high structural stability, the ionic radii of A, B, and X must yield a tolerance factor t close to 1 [22]:

$$t = \frac{R_A + R_X}{\sqrt{2}(R_B + R_X)} \tag{7.4}$$

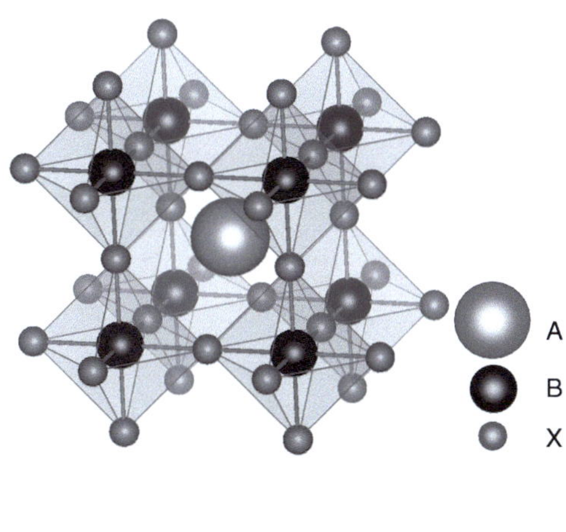

Figure 7.12 Schematic diagram of perovskite material structure.

where R_A, R_B, and R_X denote the ionic radii of the respective ions in their appropriate coordination environments. Eq. (7.4) reveals that even the largest stable monovalent alkali cation, Cs^+ (with $R_A \approx 1.67$ Å), yields $t < 1$ in lead-halide perovskites (e.g., $t \approx 0.81$ for $CsPbI_3$). Consequently, larger organic cations (methylammonium, MA^+; formamidinium, FA^+) are typically required to approach $t \approx 1$. Significant deviation from this range promotes structural distortion or symmetry reduction. Alternative strategies to achieve $t \approx 1$ include mixing cations at the A-site or partial substitution at the B-site. For example, combinations of smaller inorganic cations (e.g., Cs^+) with larger organic cations (e.g., FA^+, $R_A \approx 2.53$ Å) are employed in compositions like $Cs_{0.1}FA_{0.9}PbI_3$ or triple-cation formulations $Cs_xMA_yFA_{1-x-y}PbI_3$.

7.3.5.2 Device Structure of Perovskite Solar Cells

The structures of commonly used perovskite solar cells are depicted in Figure 7.13 [23]. Two principal device architectures for perovskite cells are prevalent: the planar structure and the mesoporous structure. The planar structure is categorized as either the regular (*n-i-p*) or inverted (*p-i-n*) configuration, based on the deposition sequence of the charge transport layers.

Mesoporous perovskite cells typically incorporate TiO_2 or Al_2O_3 nanoparticles as a mesoporous scaffold positioned between the compact layer and the light-absorbing layer to facilitate electron extraction and support perovskite film formation. However, with advancements in perovskite thin-film deposition technology, the preparation of high-quality perovskite thin films has become increasingly refined, diminishing the necessity for the mesoporous layer. Consequently, this simplifies device fabrication, and perovskite device design has progressively shifted toward planar architectures.

At present, planar perovskite cells primarily consist of *n-i-p* and *p-i-n* configurations. Devices with an *n-i-p* structure are constituted by bottom electrode/electron transport layer/perovskite layer/hole transport layer/top electrode, with common electron transport materials including SnO_2 and TiO_2, and hole transport materials including Spiro-OMeTAD (2,2′,7,7′-Tetrakis-(N,N-di-4-methoxyphenylamino)-9,9′-spirobifluorene), P3HT, and PTAA (Poly[bis(4-phenyl)(2,4,6-trimethylphenyl) amine). In contrast, *p-i-n* structured devices comprise bottom electrode/hole transport layer/perovskite layer/electron transport layer/top electrode; prevalent hole

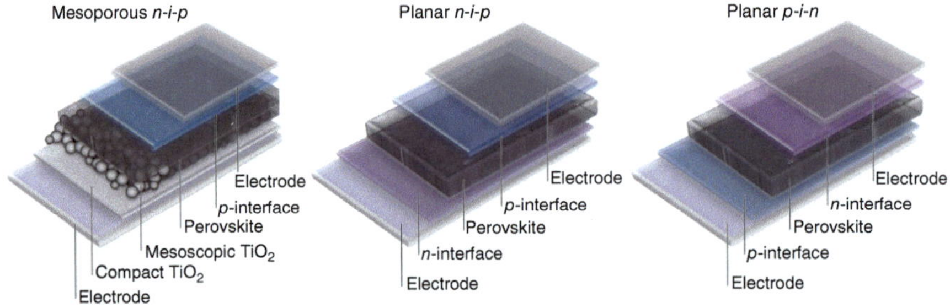

Figure 7.13 Component structure of a common perovskite battery. Reprinted with permission from Luo et al. [23].

transport materials for this configuration include PTAA and NiO_x, while common electron transport materials encompass PCBM and C_{60}. Planar *n-i-p* cells represent the most frequently adopted high-efficiency architecture due to their superior device efficiency and streamlined preparation process. Generally, the PCE of *p-i-n* devices is marginally lower than that of *n-i-p* devices, and this structure is more commonly utilized in tandem solar cells. The superior performance of *n-i-p* structures is primarily attributed to lower trap densities and more favorable band alignment at the electron transport layer/perovskite heterojunction compared to their *p-i-n* counterparts. Accordingly, research is intensifying to enhance the interfacial properties of *p-i-n* cells to narrow the performance disparity while leveraging their advantages in scalability and tandem integration for next-generation PV systems.

7.3.5.3 Carrier Transport in Perovskite Solar Cells

The working principle of perovskite solar cells is illustrated in Figure 7.14 [24]. The operational process primarily involves four sequential stages. The first stage is charge-carrier generation. During operation, incident photons absorbed by the perovskite material excite electrons from the valence band to the conduction band, creating EHPs bound by Coulomb attraction (excitons). Subsequently, in the exciton dissociation and charge-separation stage, the Coulomb-bound pairs dissociate, and the resultant free electrons and holes migrate toward the cathode and anode, respectively, driven by the internal electric field and/or energy level gradients. The third stage involves charge-carrier transfer across interfaces, where electrons are transferred across the perovskite/electron transport layer interface and holes across the perovskite/hole transport layer interface, followed by extraction and conduction through their respective transport layers. The final stage is carrier collection at the electrodes. This process establishes a steady-state circuit current.

Note that this description assumes ideal conditions; during actual device operation, phenomena such as carrier recombination, ion migration, and phase segregation may occur, significantly impacting the stability and performance of perovskite devices. These nonideal phenomena, particularly carrier recombination, represent significant efficiency bottlenecks in perovskite solar cells. Recombination encompasses multiple pathways: Shockley–Read–Hall (SRH) recombination via localized trap states within the bulk perovskite or at interfaces, bimolecular radiative recombination, and Auger recombination. Defect-induced SRH recombination,

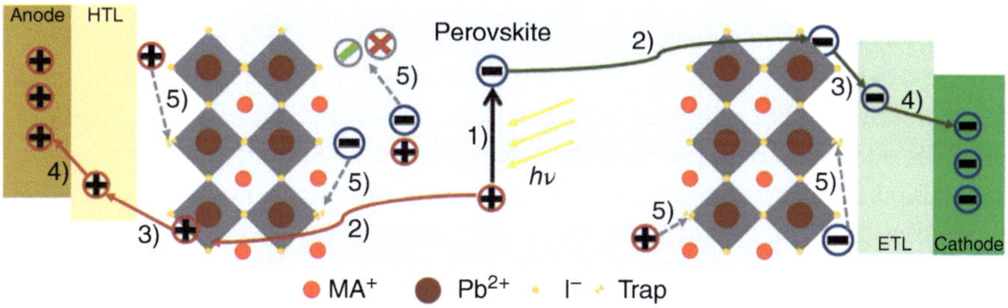

Figure 7.14 Carrier transport in perovskite solar cells. Reprinted with permission from Gao et al. [24].

arising from point defects, grain boundaries, or surface imperfections, is often the dominant non-radiative loss mechanism, causing energy dissipation as heat and reducing short-circuit current (J_{sc}), open-circuit voltage (V_{oc}), and FF. Minimizing these recombination losses through advanced material synthesis, interface engineering, and defect passivation strategies is a primary focus of ongoing research to enhance device performance and approach the theoretical efficiency limits.

7.3.5.4 Primary Characterization Instrument

The principal characterization techniques employed included scanning electron microscopy (SEM), X-ray diffraction (XRD), UV–visible spectroscopy, fluorescence spectroscopy, Fourier transform infrared spectroscopy (FTIR), Mueller matrix ellipsometry, contact angle measurement, and atomic force microscopy (AFM), among others utilized in this investigation.

Scanning electron microscopy SEM functions by directing a focused electron beam onto a sample, inducing emissions of secondary and backscattered electrons. Detection and mapping of these signals yield high-resolution images revealing sample surface topography. Such imaging enables precise characterization of microscale morphological features. Furthermore, when coupled with energy-dispersive X-ray spectroscopy (EDS), SEM enables elemental compositional analysis and spatial distribution mapping within the sample.

X-ray diffraction: XRD analysis exploits the phenomenon whereby incident X-rays undergo coherent scattering (diffraction) by crystalline lattices. Constructive interference of scattered X-rays produces a diffraction pattern diagnostic of the atomic arrangement. The fundamental relationship governing diffraction is expressed by Bragg's law:

$$2d \sin \theta = n\lambda \tag{7.5}$$

where λ is the X-ray wavelength, n is an integer representing the diffraction order, d denotes the interplanar spacing, and θ corresponds to the diffraction angle. Primary applications of XRD include determination of interplanar spacings (d-values), unit cell parameters, crystal system classification, quantitative phase analysis through relative peak intensities, average crystallite size via Scherrer analysis of peak broadening, and microstrain assessment from peak profile analysis. For instance, Feng et al. observed that $MAPbI_3$–dimethyl sulfide (DS) films exhibited significantly enhanced diffraction peak intensities compared with standard $MAPbI_3$ films in XRD analysis, indicating improved crystallinity upon DS addition to the perovskite precursor [25]. Similarly, XRD analysis has enabled investigations of lattice parameter evolution in PuGa alloys under self-irradiation [26], quantified perovskite nanoparticle size variations with temperature [27], and characterized strain-induced non-radiative losses in halide perovskite films [28].

Ultraviolet–visible spectroscopy: UV–visible spectroscopy operates on the principle that molecules exhibit characteristic absorption spectra within the UV and visible wavelength regions. This facilitates the identification of material composition and quantification of concentrations. Thus, this technique characterizes materials through their absorption profiles. In solar cell applications, both the absorption intensity and spectral range of the light-absorbing layer critically influence the short-circuit current density, a key parameter for evaluating perovskite solar cell performance.

Fluorescence spectroscopy: Fluorescence spectroscopy is based on the photoluminescence (PL) phenomenon. Upon absorption of incident radiation, molecules or atoms undergo electronic excitation from the ground state (valence band) to an excited state (conduction band). Subsequent relaxation to the ground state results in the emission of radiation with energy characteristic of the electronic transition. As different substances exhibit distinct energy level distributions, their fluorescence excitation and emission spectra provide qualitative identification. Experimentally, initial excitation spectrum testing identifies the optimal excitation wavelength—determined by the wavelength yielding maximum fluorescence intensity for the sample. Subsequently, fluorescence emission spectra are acquired under this optimal excitation wavelength. Within the perovskite solar cell domain, PL analysis qualitatively characterizes film defects and assesses carrier transport/extraction dynamics at interfaces between the perovskite layer and charge transport layers. In addition to steady-state PL measurements, fluorescence spectrophotometry enables transient fluorescence spectroscopy (TRPL, time-resolved photoluminescence), frequently implemented via the time-correlated single-photon counting (TCSPC) method. Analysis of the fluorescence decay curve yields the excited-state lifetime, commonly employed to evaluate carrier lifetimes in perovskite films.

Fourier transform infrared spectroscopy: FTIR spectroscopy functions on the principle that molecular vibrations absorb infrared radiation at frequencies characteristic of specific chemical bonds and functional groups. This dependence on molecular structure makes FTIR invaluable for elucidating molecular structure and chemical composition. Analysis of infrared absorption spectra facilitates molecular spatial configuration determination and calculation of key chemical parameters [29, 30].

Mueller matrix ellipsometry: Mueller matrix ellipsometry operates by analyzing the change in the polarization state—specifically the amplitude ratio (Ψ) and phase difference (Δ)—between incident and reflected light upon interaction with a sample. This technique is primarily employed to measure thickness and optical constants of single-layer or multilayer thin films and can also characterize the optical properties of bulk materials. In perovskite solar cell research, Mueller matrix ellipsometry is extensively utilized for layer thickness determination and measurement of thin-film optical constants. Characterization involves establishing an optical model based on the transfer matrix method, followed by fitting to experimental data [31].

Contact angle measurement: Contact angle measurement quantifies the angle formed at the interface between a liquid droplet and a solid surface, governed by the interplay of the liquid's cohesive forces and adhesive forces between the liquid and the solid. Within perovskite solar cell research, contact angle measurements evaluate film surface hydrophobicity. Enhanced hydrophobicity correlates with superior air stability, while hydrophilicity is often associated with improved film-forming properties [32].

Atomic force microscopy: AFM operates based on the measurement of forces—primarily van der Waals interactions—between a sharp tip and the sample surface. The tip, mounted on a sensitive cantilever, is raster-scanned across the sample surface. Deflection of the cantilever due to tip–sample interactions is monitored to generate a topographical map representing surface features [33].

7.4 Manufacturing of Solar Cells

7.4.1 Si Solar Cell

7.4.1.1 Atomic and Band Structures of Crystalline Silicon

Silicon is a Group IV element in the periodic table with atomic number 14. The ground-state electron configuration of an isolated silicon atom is $1s^2\ 2s^2\ 2p^6\ 3s^2\ 3p^2$. However, when silicon atoms approach one another to form a crystal lattice, orbital hybridization occurs. According to hybridization theory, the 3s and 3p orbitals hybridize to form degenerate sp^3 hybrid orbitals, resulting in new energy bands. Figure 7.15 illustrates that as the interatomic distance decreases, the initially discrete 3s and 3p energy levels broaden into nascent energy bands.

Consider a crystalline silicon structure comprising N silicon atoms. The degeneracies of the atomic 3s and 3p energy levels are 1 and 3, respectively. Consequently, the nascent 3s and 3p bands contain 2N and 6N states, respectively, where electrons can occupy specific discrete states (though treated as quasi-continuous due to the large N). As the interatomic spacing continues to diminish further, s-p orbital hybridization commences, merging the initial s and p bands into a single hybrid band encompassing 8N states capable of accommodating 8N electrons. When the atomic distance approaches the equilibrium lattice constant (d_0), this merged band

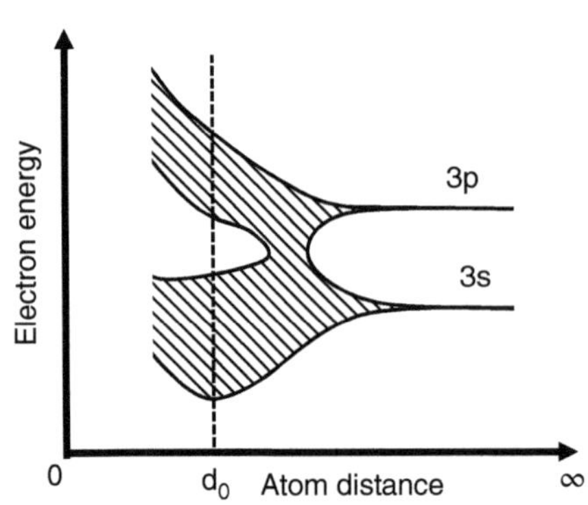

Figure 7.15 Schematic representation of band formation when silicon atoms form a crystal.

undergoes splitting into two distinct bands: the valence band and the conduction band. Each of these bands contains 4N states capable of holding 4N electrons. At low temperatures, all 4N valence electrons occupy the lower-energy valence band, leaving the conduction band unoccupied.

The band formation process confirms that upon crystal formation via orbital hybridization, silicon atoms adopt an effective electron configuration with four equivalent sp^3 hybrid orbitals in the outermost shell, each possessing one unpaired electron. Within the actual crystal structure, these sp^3 hybrid orbitals are oriented at an angle of $109°28'$ relative to each other, with their symmetry axes directed toward the vertices of a tetrahedron. Each of the four unpaired electrons forms a covalent bond with an electron from an adjacent silicon atom, also directed along the tetrahedral bonds, establishing the stable diamond cubic crystal structure. This diamond structure consists of two interpenetrating face-centered cubic (FCC) sublattices displaced relative to each other by one-quarter of the body diagonal length. Silicon atoms reside on both FCC sublattices.

Silicon crystallizes in the diamond cubic structure, which belongs to the cubic crystal system. The unit cell contains 8 atoms, with a lattice constant a = 5.4395 Å. The nearest-neighbor atomic distance is 2.35 Å, and the atomic density is $5\times10^{22}/\text{cm}^3$. The most technologically significant crystallographic planes in silicon are the low-index facets: {111}, {110}, and {100}, with corresponding zone axes <111>, <110>, and <100>.

The {111} planes are bilayers (diatomic planes). The largest interplanar spacing within the {111} family is between adjacent bilayers (d_{111} = 3.135 Å). The smaller spacing exists between the two atom layers within each bilayer (0.78 Å). The areal density refers to the number of atoms per unit area within a given {111} bilayer plane. The bond density represents the density of covalent bonds crossing between adjacent {111} bilayers. Among the primary planes: The {111} planes are close-packed, exhibiting the largest interplanar spacing and the lowest bond density. The {110} planes possess intermediate areal density, interplanar spacing, and bond density. The {100} planes are the least densely packed, possessing the smallest interplanar spacing and highest bond density.

Due to the large {111} interplanar spacing and low bond density, silicon exhibits a strong propensity to cleave along {111} planes, which are therefore the primary cleavage planes. Cleavage also occurs, though less readily, along {110} planes, which are utilized in certain device fabrication processes. The <110> direction represents the close-packed atomic row direction. The <110> direction within the {111} plane corresponds to the primary slip system in silicon.

Silicon wafers oriented along the <110> and <100> directions exhibit a high tendency for fracture (warpage or breakage) because the angle between the wafer surface and the prevalent {111} cleavage planes is significant. Furthermore, the etching behavior differs markedly: the densely packed {111} planes have a high surface atom density, leading to a low chemical etching rate. Conversely, the less densely packed {110} and {100} planes possess lower surface atom densities and consequently exhibit higher etching rates.

7.4.1.2 Properties of Crystalline Silicon

The relative atomic mass of silicon is 28.085, and it exhibits a diamond cubic structure under ambient pressure. At room temperature, silicon exists as a solid with a gray metallic luster, high hardness, and brittle fracture behavior, characterized by tensile strength significantly exceeding shear strength and negligible ductility. However, when the heat-treatment temperature exceeds 750 °C, silicon undergoes a brittle-to-ductile transition, enabling slip dislocation and plastic deformation under applied stress.

Silicon possesses high chemical stability and is usually insoluble in single acids but can be solubilized by hydrofluoric acid. It is prone to oxidation; consequently, silicon in nature predominantly exists as oxides. In its crystalline form, silicon spontaneously develops a native oxide layer ranging from several to several tens of nanometers in thickness. Thermal oxidation produces a dense insulating silicon dioxide (SiO_2) layer on the surface of crystalline silicon, which serves as a protective barrier, diffusion source, or insulating layer in silicon-based devices.

Crystalline silicon is an indirect bandgap semiconductor material with superior semiconductor properties. Its electrical and optical characteristics can be summarized as follows:

1) **Electrical conduction characteristics:** The resistivity of silicon can be modulated over a broad range (approximately 10^{-5}–10 Ω cm) through doping with small concentrations of *n*-type (Group V elements) or *p*-type (Group III elements) impurities. At room temperature, intrinsic silicon exhibits a carrier concentration of approximately $1.5 \times 10^{10}/\text{cm}^3$ and a corresponding resistivity of about 2.3×10^3 Ω cm. Silicon demonstrates high electron mobility (≈1350 $\text{cm}^2/(\text{V s})$) and relatively lower hole mobility (≈480 $\text{cm}^2/(\text{V s})$).
2) **PN junction characteristics:** PN junctions fabricated in crystalline silicon exhibit rectifying (unidirectional) electrical characteristics. Semiconductor devices and solar cells based on crystalline silicon PN junctions typically display low leakage currents and operational tolerance up to 150 °C. However, silicon's electron mobility is lower than that of materials like germanium (Ge) or gallium arsenide (GaAs). Consequently, silicon-based devices exhibit less effective performance compared to GaAs devices under high-temperature, high-frequency, or intense radiation conditions.
3) **Optoelectronic characteristics:** Silicon possesses a bandgap of 1.12 eV. As an indirect bandgap semiconductor, its optical absorption primarily occurs within the infrared spectrum, with an absorption coefficient typically 10^2–10^3 times lower than direct bandgap compound semiconductors. This transmissivity permits infrared light with photon energies below 1.12 eV to pass through the material. When subjected to illumination, crystalline silicon PN junctions generate photocurrent—a principle fundamental to silicon solar cell technology. Conversely, while silicon PN junctions can also produce electroluminescence, luminous efficiency is extremely limited ($/ < 0.01\%$) due to the requirement for phonon mediation in exciton recombination inherent to its indirect bandgap structure. Strategies to enhance the luminous efficiency of silicon remain an area of active research.

7.4.1.3 Synthesis Method of Crystalline Silicon

Crystalline silicon (c-Si) is categorized into two principal crystal types: monocrystalline silicon (mono-Si) and polycrystalline silicon (poly-Si). Monocrystalline silicon is further distinguished based on the fabrication process into floating-zone grown (FZ) and Czochralski grown (CZ). Polycrystalline silicon is predominantly produced as cast polycrystalline silicon (cast mc-Si) via the directional solidification casting method.

FZ monocrystalline silicon exhibits superior purity, electrical characteristics, and PCE compared to CZ monocrystalline silicon. However, CZ silicon holds advantages in terms of achievable wafer diameter, mechanical processing attributes, and manufacturing cost. Consequently, the vast majority of commercially produced monocrystalline silicon solar cells utilize CZ material. FZ silicon is employed only in specialized applications demanding exceptionally high cell efficiencies and where cost considerations are secondary.

Historically, cast mc-Si surpassed mono-Si in market share for large-scale commercial solar cell deployment due to its significantly lower preparation cost and long-standing established manufacturing. However, recent advances in cell production technology have substantially improved the tolerance of monocrystalline silicon solar cells to material impurities and defects while simultaneously reducing manufacturing costs. These improvements, coupled with the inherent higher conversion efficiency of monocrystalline silicon, have revitalized its market position. Consequently, numerous new domestic and international solar cell manufacturing facilities have adopted monocrystalline silicon as the substrate material of choice when implementing advanced cell architectures.

7.4.2 Perovskite Solar Cells

The primary preparation methods for perovskite polycrystalline films include the one-step method [34], two-step method [35], and various other techniques [36–38]. Nucleation behaviors and film quality vary with different preparation processes. Consequently, precise control of film growth based on process characteristics is essential for optimizing device performance.

7.4.2.1 One-Step Method

The one-step method involves dissolving the requisite perovskite precursor materials concurrently, followed by fabrication of the perovskite film through a single spin-coating process. Initially, this approach employed *N,N*-dimethylformamide (DMF) and dimethyl sulfoxide (DMSO) to dissolve lead(II) iodide (PbI_2) and methylammonium iodide (MAI), yielding a homogeneous perovskite precursor solution. However, owing to the disparate solubilities of PbI_2 and MAI within the solvent system, their precipitation kinetics diverged significantly. During DMF solvent evaporation, dendritic perovskite crystallites formed readily, generating numerous voids within the film. To address this challenge, Seok et al. introduced a solution-based synthesis method for organolead trihalide perovskite solar cells ($CH_3NH_3Pb(I_{1-x}Br_x)_3$, $x = 0.1$–0.15), achieving a breakthrough efficiency of

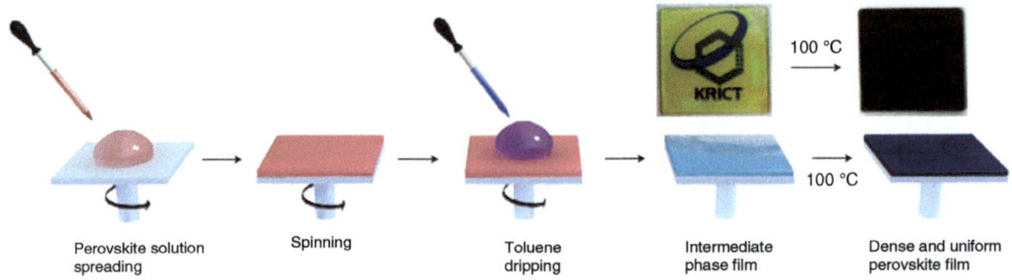

Figure 7.16 Flow chart of one-step solution method to prepare perovskite polycrystalline thin film. Reprinted with permission from Jeon et al. [39].

16.2% [39]. As shown in Figure 7.16, this was realized by dissolving precursors in a mixed γ-butyrolactone (GBL)/DMSO solvent, spin-coating the solution onto a TiO_2 substrate, and inducing rapid crystallization via toluene drop-casting. This process formed a uniform, dense perovskite layer through a key CH_3NH_3I–PbI_2–DMSO intermediate phase. The resulting solar cell architecture combined advantages of mesoscopic and planar structures, delivering a certified PCE of 16.2% with no hysteresis and demonstrating the critical role of solvent engineering in fabricating efficient, low-cost solution-processed PVs.

7.4.2.2 Two-Step Method

As illustrated in Figure 7.17, the two-step method entails fabricating high-quality perovskite films through sequential deposition: initially forming a PbI_2 layer followed by reaction with a cationic precursor solution [40]. Pioneered by Michael Grätzel and colleagues, this approach first deposits PbI_2 onto mesoporous TiO_2 substrates. Subsequent immersion in methylammonium iodide (MAI) solution generates $MAPbI_3$ perovskite films, achieving power conversion efficiencies of 15% [35]. As perovskite device efficiencies advance, the bandgap characteristics and film quality have become increasingly critical. Research advancements have refined this two-step process through the incorporation of additives such as formamidinium iodide (FAI), methylammonium bromide (MABr), and methylammonium chloride (MACl) into the cationic solution, enabling fabrication of high-quality mixed-composition perovskite films. Building upon studies of PbI_2 passivation in two-step processed films, You et al. demonstrated planar perovskite

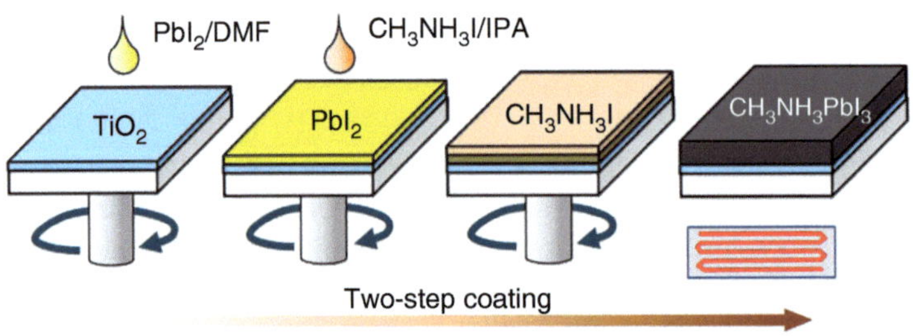

Figure 7.17 Flowchart of perovskite polycrystalline films prepared by two-step continuous deposition. Reprinted with permission from Im et al. [40].

solar cells exceeding 21% efficiency, establishing a foundational methodology for high-performance perovskite device fabrication [41].

7.4.2.3 Vapor-Deposition Method

Vapor deposition enables precise, solvent-free fabrication of high-purity perovskite films with controlled morphology and composition. In dual-source deposition (Figure 7.18), separate organic (CH_3NH_3I) and inorganic ($PbCl_2$) precursors are co-evaporated in vacuum, allowing stoichiometric tuning of mixed-halide perovskites like $CH_3NH_3PbI_{3-x}Cl_x$ [42]. XRD confirms such films adopt an orthorhombic crystal structure identical to solution-processed counterparts, with characteristic peaks at 14.12° (110), 28.44° (220), and 43.23° (330), while minimal PbI_2 (12.65°) or $CH_3NH_3PbCl_3$ (15.68°) signals indicate exceptional phase purity. These uniform films can be integrated into planar heterojunction solar cells, achieving solar power conversion efficiencies exceeding 15%, with simplified device architectures.

7.4.2.4 Vapor-Assisted Solution Method

The vapor-assisted solution method was developed to overcome the limitations inherent in both conventional solution-based processes and vapor-deposition techniques [42]. This approach represents a modification of the two-step deposition method, wherein MAI/FAI deposition is achieved via vapor deposition instead of a solution process. As illustrated in Figure 7.19, this gas–solid phase crystallization technique enhances control over film topography and grain size, while effectively mitigating delamination issues commonly associated with liquid–solid interactions. Perovskite films fabricated using the vapor-assisted solution method exhibit

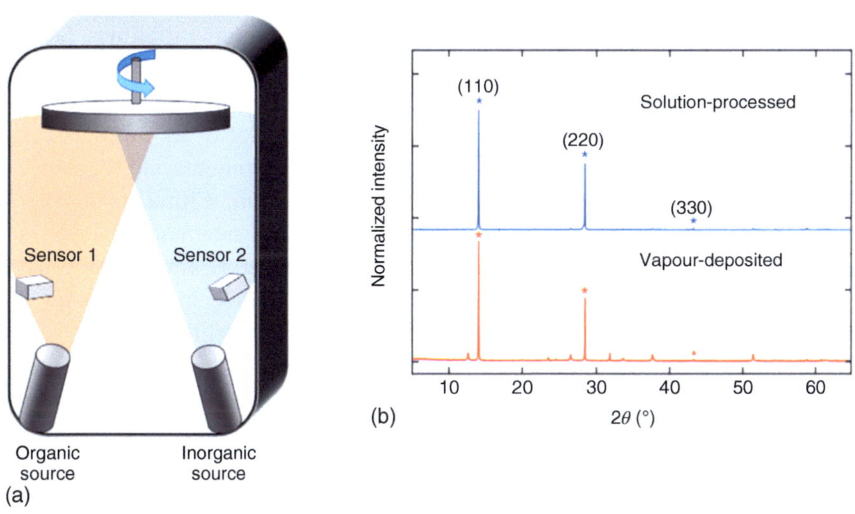

Figure 7.18 (a) Schematic draw of dual-source thermal evaporation. The organic source was methylammonium iodide (MAI) and $PbCl_2$. (b) XRD spectra of a solution-processed perovskite film (blue) and vapor-deposited perovskite film (red). Reprinted with permission from Liu et al. [42].

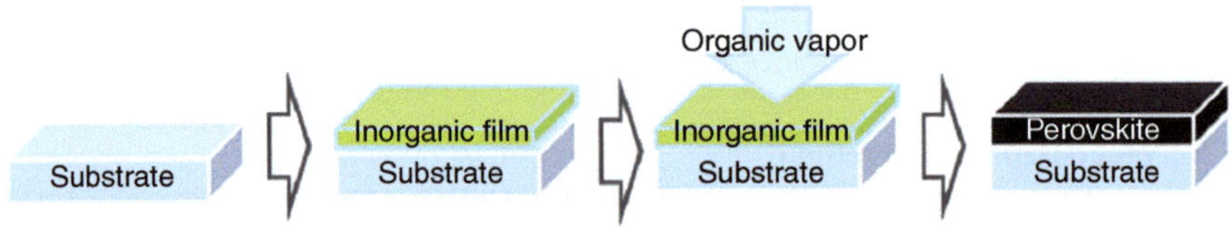

Figure 7.19 Schematic illustration of perovskite film formation through vapor-assisted solution process. Reprinted with permission from. Chen et al. [43].

uniform coverage, large grain size, and complete conversion to the perovskite phase [43]. However, the practical application of this strategy remains limited. Gas–solid reactions typically require extended durations to reach full conversion, and solar cells produced with this technology have demonstrated efficiencies primarily in the 10–12% range. Furthermore, the dependence on vacuum processes results in significant energy consumption, hindering its scalability for large-scale manufacturing.

7.5 Applications of Solar Cells

Aerospace: The earliest application of solar cells for power generation occurred in the aerospace sector. Solar cells were first employed in space aboard the Pioneer 1 satellite in 1958. Capable of directly converting solar energy into electricity, solar PVs are now extensively utilized to power space stations and satellites. Equipped with batteries and energy storage systems, solar arrays ensure spacecraft functionality during orbital periods without direct solar irradiance. Furthermore, deploying solar cells on spacecraft significantly reduces mass, lowers power system demands, and offers extended operational lifetimes. Solar cells will inevitably be the primary power source for missions to Mars and future planned human outposts on the Moon and Mars.

Solar-powered vehicles: Research into applying solar cells to automobiles commenced around the 1980s, with significant developments subsequently undertaken in the United States, the Netherlands, Japan, China, and other nations. Solar vehicles represent a prominent category of new energy vehicles attracting considerable attention and are widely regarded as among the most environmentally sustainable options within the automotive industry. Recent years have witnessed substantial progress in related solar vehicle technologies due to rapid global economic and technological advancement. A primary limitation for electric vehicles (EVs) remains driving range, restricting long-distance travel. Should solar cell technology overcome the specific challenge of generating sufficient power for extended EV range, vehicle endurance limitations could be significantly mitigated. A solar vehicle, designed primarily for terrestrial transport, relies on solar power as its main energy source. The first recorded solar-powered car model, developed by General Motors engineer William G. Cobb in 1955, was a 15 inch miniature vehicle powered by 12 selenium PV cells driving a small electric motor. While solar radiation is the primary energy source, some vehicle designs integrate batteries to supplement power or utilize solar panels to recharge batteries and power auxiliary systems.

Communications: In the communications sector, solar cells find application in solar-powered signal towers and wireless emergency telephones, among other uses. China's extensive deployment of 5G infrastructure presents a significant energy demand, as a single 5G base station consumes approximately 18 times more power than its 4G counterpart. Furthermore, the reduced coverage radius of a 5G base station necessitates a much denser network than 4G. Deploying conventional power lines to remote or mountainous locations is often prohibitively expensive. Solar cell power systems offer a viable alternative due to their operational reliability, minimal susceptibility to topographical constraints, and absence of grid distance limitations.

Transportation infrastructure: Transportation infrastructure applications predominantly include solar-powered street lights and traffic signals, increasingly adopted in urban areas and scenic locations. A standard solar PV street light equipped with a 100 W PV module generates approximately 100 kWh annually. Data from the Energy Research Institute of the National Development and Reform Commission indicate that China's installed capacity of approximately 25 million solar street lights annually conserves 2.5 billion kWh of electricity, equivalent to saving 875,000 tons of standard coal. This translates to significant reductions in pollutant emissions, including 2.18 million tons of CO_2, 66,000 tons of SO_2, and 595,000 tons of carbon particulates annually. Additionally, PV modules can be integrated into highway noise barriers. Transforming conventional noise barriers into PV noise barriers delivers dual benefits: effective sound attenuation alongside substantial electricity generation.

Building-integrated photovoltaics (BIPV): BIPV utilizes PV materials to replace conventional building elements within the outer envelope, such as roof tiles, skylights, or facade cladding. While existing structures can be retrofitted, BIPV is increasingly incorporated as a primary or supplementary power source in new construction. The primary advantage of integrated PV lies in offsetting initial costs through savings on the building materials and labor required for the conventional components the BIPV modules supplant. These benefits position BIPV as one of the fastest-growing segments within the PV industry. BIPV applications began in the 1970s with aluminum-framed PV modules installed on remote buildings without grid access. The 1980s saw rooftop PV demonstrations on grid-connected structures near centralized power stations. By the 1990s, dedicated BIPV products designed for seamless integration into building envelopes emerged commercially. A 2011 economic assessment by the National Renewable Energy Laboratory, alongside a historical overview, indicated that significant technical hurdles needed to be addressed before BIPV panel installation costs achieved competitiveness with conventional systems. However, consensus is growing that widespread commercialization of BIPV is essential for achieving the European target of net-zero energy buildings by 2020.

Photovoltaic power station: Large-scale integration of solar PV technology enables the direct conversion of sunlight into electrical energy, facilitating deployments ranging from utility-scale solar farms to distributed rooftop systems interconnected with the electricity grid. This technology confers substantial advantages, including significant reductions in greenhouse gas emissions by displacing fossil fuel–based generation, enhanced energy security through decentralized power production, and long-term cost reductions driven by declining PV module prices and minimal operational expenditures post-installation. However, several challenges persist. The inherent intermittency of solar irradiation, stemming from meteorological variability and diurnal solar cycles, leads to variable power output that can induce grid instability. Grid integration complexities arise from temporal disparities between peak solar generation periods and electricity demand profiles, necessitating considerable infrastructure investments. Further limitations encompass the high initial capital investment required for installation and associated energy storage systems, land-use conflicts arising from utility-scale installations, and the constrained conversion efficiencies characteristic of commercial PV modules. Remediation strategies are advancing rapidly. For instance, lithium-ion battery storage systems capture surplus daytime energy for release during nocturnal or overcast conditions, thereby flattening the supply profile. Smart grid technologies facilitate real-time monitoring, demand–response management, and bidirectional energy transfer, optimizing distribution efficiency. Advanced grid-forming inverters bolster grid stability by actively regulating voltage and frequency during solar intermittencies. AI-driven forecasting algorithms will enhance grid management by predicting generation and demand patterns, enabling proactive balancing. Future integration of hybrid systems combining solar PV with complementary renewable sources (e.g., wind, hydro) and next-generation storage technologies such as flow batteries promises further improvement in system reliability.

7.6 Conclusion

Solar cells have evolved from laboratory innovations to become a cornerstone of the global renewable energy infrastructure. The current technological landscape is characterized by the continued dominance of silicon-based cells, commanding over 90% market share. Concurrently, thin-film technologies such as CIGS and perovskite exhibit disruptive potential: perovskite cells achieve theoretical efficiencies exceeding 30% with significantly lower costs, while flexible CIGS modules facilitate the advancement of BIPVs. Table 7.1 compares key PV cell types, analyzing their conversion efficiency, advantages, disadvantages, costs, and primary application scenarios. The comparison reveals that each PV cell technology occupies distinct application niches, with no single solution simultaneously achieving high efficiency and low cost.

In the realm of PV cell material advancements, silicon-based technologies remain predominant. Monocrystalline silicon cells exhibit superior conversion efficiency, and recent technological progress has reduced their manufacturing costs below those of polycrystalline silicon cells. Consequently, monocrystalline silicon is

Table 7.1 Comparison of various photovoltaic cells [44–47].

Cell types	Conversion efficiency (%)	Advantages	Disadvantages	Applicable scenario	Cost (cells)
Monocrystalline silicon (c-Si)	24.2	Established, high-efficiency technology	Relatively high manufacturing expense	Large-scale, field-based photovoltaic power plants	1.04–1.12 CNY/W
Polycrystalline silicon (multi-Si)	22.8	Mature and widely adopted manufacturing process	Lower conversion efficiency compared to monocrystalline silicon	Utility-scale photovoltaic farm implementation	0.73–0.83 CNY/W
Silicon-based thin film (a-Si/μc-Si)	11.9	Semitransparent; compatible with flexible substrates	Significantly lower efficiency than crystalline silicon counterparts	Building-integrated photovoltaics (BIPV)	4.54 CNY/W (e.g., Mitsubishi Heavy Industries)
Copper indium gallium selenide (CIGS)	23.4	Low weight; superior low-light response under diffuse conditions	Relies on indium, a relatively scarce element	Potential in BIPV applications	~5.00 CNY/W
Cadmium telluride (CdTe)	22.1	Demonstrated operational longevity	Utilizes tellurium and cadmium, elements facing supply constraints	Development in BIPV configurations	~4.60 CNY/W (e.g., First Solar)
Gallium arsenide (III–V: GaAs)	35.5	Exceptionally high efficiency; tolerance for elevated temperatures and radiation environments	Prohibitively expensive for terrestrial use	Aerospace systems requiring maximum power-to-weight ratios (e.g., satellites)	~38.00 CNY/W
Dye-sensitized solar cell (DSSC)	11.7	Extended operational lifetime potential; relatively uncomplicated architecture	Volatilization risk associated with liquid electrolytes; limited long-term stability under operating conditions	Primarily under laboratory investigation	N/A
Perovskite solar cell (PSC)	26.7	Low predicted manufacturing cost; utilization of abundant precursor materials	Current instability and insufficient operational lifespan	Experimental stage for BIPV integration	~1.00 CNY/W (e.g., Golden Concord Nano Pilot Estimates)

progressively replacing polycrystalline silicon within the PV market. Thin-film and emerging PV cell materials provide benefits including thinness, flexibility, and translucency, demonstrating significant potential for PV power generation and BIPV applications.

Concerning PV power generation applications, this technology presents promising opportunities not only for centralized power supply but also for distributed systems integrated within sectors such as construction, transportation, and agriculture. High-power monocrystalline modules (>600 W) have reduced utility-scale levelized costs of electricity, while thin-film technologies enable novel market applications: CIGS leverages low-light performance for curved architectural surfaces, and lightweight perovskite or GaAs cells energize satellites, drones, and vehicle-integrated systems. Distributed PV power stations can achieve synergistic integration with buildings, vehicles, and land resources. Such integration fosters complementary advantages that concurrently reduce carbon emissions, thereby contributing to climate change mitigation, and stimulate industrial innovation in these sectors.

However, persistent challenges impede sustainable growth. Intrinsic efficiency limitations confront mainstream technologies, with silicon approaching its theoretical maximum of 29.4%, necessitating complex tandem architectures such as perovskite/silicon assemblies aimed at efficiencies surpassing 30%.

The path forward depends on balancing the efficiency–cost–sustainability triad. Next-generation development concentrates on multi-junction cells, particularly perovskite/silicon tandems approaching commercial viability, alongside material innovations designed to reduce reliance on precious metals. Integration is equally critical: AI-driven energy management integrates solar generation with smart grids and storage infrastructure. Crucially, enhancement of lifecycle sustainability must prioritize further reduction in manufacturing carbon footprints.

References

1 DNV (2024). Energy Transition Outlook 2024. https://www.dnv.com/energy-transition-outlook/ (accessed 1 June 2025).
2 Nayak, P.K., Garcia-Belmonte, G., Kahn, A. et al. (2012). *Energy Environ. Sci.* 5: 6022–6039.
3 Schmidt, J., Peibst, R., and Brendel, R. (2018). *Sol. Energy Mater. Sol. Cells* 187: 39–54.
4 Chapin, D.M., Fuller, C.S., and Pearson, G.L. (1954). *J. Appl. Phys.* 25: 676–677.
5 Battaglia, C., Cuevas, A., and De Wolf, S. (2016). *Energy Environ. Sci.* 9: 1552–1576.
6 Liu, J., Yao, Y., Xiao, S. et al. (2018). *J. Phys. D. Appl. Phys.* 51: 123001.
7 Xiao, S. and Xu, S. (2014). *Crit. Rev. Solid State Mater. Sci.* 39: 277–317.
8 Repins, I., Contreras, M.A., Egaas, B. et al. (2008). *Prog. Photovolt. Res. Appl.* 16: 235–239.
9 He, X., Ercius, P., Varley, J. et al. (2019). *Prog. Photovolt. Res. Appl.* 27: 255–263.

10 Tsai, M.G., Tung, H.T., Chen, I.G. et al. (2013). *J. Am. Ceram. Soc.* 96: 2419–2423.

11 Park, J.H., Afzaal, M., Kemmler, M. et al. (2003). *J. Mater. Chem.* 13: 1942.

12 Choi, I.H. and Lee, D.H. (2007). *Thin Solid Films* 515: 4778–4782.

13 Venkatachalam, M., Kannan, M.D., Jayakumar, S. et al. (2008). *Thin Solid Films* 516: 6848–6852.

14 Nakada, T., Furumi, K., and Kunioka, A. (1999). *IEEE Trans. Electron Devices* 46: 2093–2097.

15 Delahoy, A.E., Chen, L., Akhtar, M. et al. (2004). *Sol. Energy* 77: 785–793.

16 Chraibi, F., Fahoume, M., Ennaoui, A. et al. (2004). *Moroccan J. Condens. Matter* 5: 88–96.

17 Bhattacharya, R.N., Hiltner, J.F., Batchelor, W. et al. (2000). *Thin Solid Films* 361–362: 396–399.

18 Jo, J., Kim, S.S., Na, S.I. et al. (2009). *Adv. Funct. Mater.* 19: 866–874.

19 Meng, L., Zhang, Y., Wan, X. et al. (2018). *Science* 361: 1094–1098.

20 Guo, X., Zhou, N., Lou, S.J. et al. (2013). *Nat. Photonics* 7: 825–833.

21 Chen, Q., De Marco, N., Yang, Y.M. et al. (2015). *Nano Today* 10: 355–396.

22 Goldschmidt, V.M. (1926). *Naturwissenschaften* 14: 477–485.

23 Luo, D., Su, R., Zhang, W. et al. (2020). *Nat. Rev. Mater.* 5: 44–60.

24 Gao, F., Zhao, Y., Zhang, X. et al. (2020). *Adv. Energy Mater.* 10: 1902650.

25 Feng, J., Zhu, X., Yang, Z. et al. (2018). *Adv. Mater.* 30: 1801418.

26 Ravat, B., Oudot, B., and Baclet, N. (2007). *J. Nucl. Mater.* 366: 288–296.

27 Tsai, H., Nie, W., Blancon, J.C. et al. (2016). *Nature* 536: 312–316.

28 Jones, T.W., Osherov, A., Alsari, M. et al. (2019). *Energy Environ. Sci.* 12: 596–606.

29 Gan, Z., Yu, Z., Meng, M. et al. (2019). *APL Mater.* 7: 031107.

30 Shao, J., Lu, W., Yue, F. et al. (2007). *Rev. Sci. Instrum.* 78: 013111.

31 Jiang, Y., Soufiani, A.M., Gentle, A. et al. (2016). *Appl. Phys. Lett.* 108: 061905.

32 Yazdi, M.R.A., Rahimzadeh, A., Chougi, Z. et al. (2018). *AIP Adv.* 8: 025109.

33 Chen, B.X., Rao, H.S., Li, W.G. et al. (2016). *J. Mater. Chem. A* 4: 5647–5653.

34 Xiao, M., Huang, F., Huang, W. et al. (2014). *Angew. Chem. Int. Ed.* 53: 9898–9903.

35 Burschka, J., Pellet, N., Moon, S. et al. (2013). *Nature* 499: 316–319.

36 Choi, W., Na, S., Park, C. et al. (2019). *Sol. Energy* 178: 56–60.

37 Li, J., Munir, R., Fan, Y. et al. (2018). *Joule* 2: 1313–1330.

38 Park, M., Cho, W., Lee, G. et al. (2019). *Small* 15: 1804005.

39 Jeon, N.J., Noh, J.H., Kim, Y.C. et al. (2014). *Nat. Mater.* 13: 897–903.

40 Im, J.H., Kim, H.S., and Park, N.G. (2014). *APL Mater.* 2: 081510.

41 Jiang, Q., Chu, Z., Wang, P. et al. (2017). *Adv. Mater.* 29: 1703852.

42 Liu, M., Johnston, M.B., and Snaith, H.J. (2013). *Nature* 501: 395–398.

43 Chen, Q., Zhou, H., Hong, Z. et al. (2014). *J. Am. Chem. Soc.* 136: 622–625.

44 China Photovoltaic Society (2020). Report on 2019 China PV technology development—research progress of new solar cells (1). *Sol. Energy* 5: 5–15.

45 China Photovoltaic Society (2020). Report on 2019 China PV technology development—research progress of new solar cells (2). *Sol. Energy* 6: 5–16.

46 China Photovoltaic Society (2020). Report on 2019 China PV technology development—research progress of new solar cells (3). *Sol. Energy* 7: 5–9.

47 China Photovoltaic Society (2020). Report on 2019 China PV technology development—research progress of new solar cells (4). *Sol. Energy* 8: 5–9.

8

Photodetectors

8.1 Introduction

Photodetectors are devices that convert photons into electrical signals for the detection of optical radiation; they constitute fundamental components within optoelectronics and serve as critical elements in information science and technology. The performance of photodetectors critically influences advancements in diverse fields, including optical communication, optical storage, optical computing, optical imaging, optical sensing, and spectral analysis.

Photodetectors can be classified according to several criteria. Based on detection mechanism, they are categorized into two primary types: photonic detectors and thermal detectors, as illustrated in Figure 8.1. Photonic detectors are further subdivided based on the fundamental photoelectric effect involved. Specifically, detectors relying on the internal photoelectric effect include photoconductive detectors, photovoltaic detectors (photodiodes), and phototransistors, among others. Detectors based on the external photoelectric effect comprise phototubes, photomultiplier tubes (PMTs), and camera tubes. Meanwhile, thermal detectors, operating via thermoelectric phenomena, include thermocouples, thermopiles (thermal radiometers), and pyroelectric detectors.

Classification according to wavelength sensitivity, depicted in Figure 8.2, encompasses detectors operating in the infrared (IR), visible (vis), ultraviolet (UV), X-ray, and gamma-ray regions. Detected photon energy possesses units of electron volts (eV), covering a range spanning seven orders of magnitude. Alternative classification schemes exist. For instance, device architecture distinguishes between vertical and lateral photodetector structures. Classification by material composition yields categories such as inorganic semiconductor photodetectors, organic semiconductor photodetectors, metal halide perovskite photodetectors, low-dimensional nanomaterial photodetectors, and composite photodetectors. Based on spectral response range, photodetectors are categorized as broadband or narrowband detectors.

Photodetectors based on the principle of the photoelectric effect originated with Alexandre-Edmond Becquerel's initial observation of this phenomenon in 1839. Substantial development and application, however, commenced during the

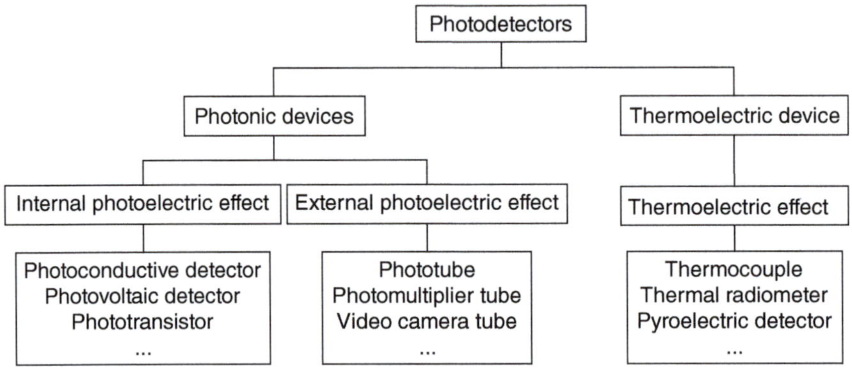

Figure 8.1 Classification of detectors by detection mechanism.

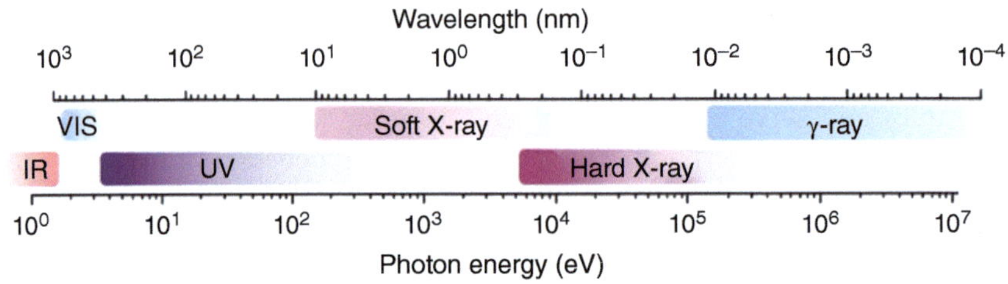

Figure 8.2 Classification of detectors by wavelength sensitivity.

twentieth century alongside advancements in quantum physics. Einstein's formulation of the photon hypothesis provided the critical theoretical framework for explaining the photoelectric effect. Since the early twentieth century, photodetectors have been employed in industrial and scientific laboratories for quantitative light intensity measurement. Early devices utilized selenium (Se) as the photosensitive material for photometric purposes. Modern photodetectors predominantly feature silicon-based PN junction architectures, a fundamental structure discovered in 1940 by Russel Ohl at Bell Laboratories. Subsequent progress in optoelectronics has substantially advanced photodetector technology.

Accelerated by technological innovation, optoelectronic materials have undergone rapid development, fostering the creation of increasingly efficient photodetectors and photoelectric conversion devices. Systems incorporating emerging materials such as organic semiconductors, perovskites, and topological insulators exhibit advantages including cost-effectiveness, flexibility, and tunable optoelectronic characteristics. Progress in novel materials and micro-nanotechnology has diversified photodetector research, enhancing capabilities across UV, vis, IR, and terahertz spectral regimes [1]. Devices based on low-dimensional nanomaterials (e.g., carbon nanotubes, graphene, transition metal dichalcogenides) demonstrate superior sensitivity, broadband spectral response, and rapid temporal response. Additionally, photodetectors leveraging artificial microstructures, band engineering, and surface plasmon resonance exhibit enhanced spectral selectivity, high-gain characteristics, accelerated response dynamics, and improved integration capacity.

Functioning as transducers converting optical signals to electrical outputs, photodetectors constitute fundamental components of intelligent optoelectronic systems. Applications span diverse military and civilian domains, including satellite reconnaissance, missile guidance, meteorological surveillance, and automated monitoring systems. While broadband detectors exhibit sensitivity across extended spectral ranges, encompassing X-ray, UV, vis, and IR detection, narrowband detectors target specific wavelengths, facilitating biological sensing, secure optical communications, and other applications requiring spectral discrimination. Currently, photodetectors find extensive implementation in noncontact detection fields including optical communications, military technologies, industrial automation, environmental analysis, security systems, and wearable electronics.

8.2 Principle of Photodetectors

8.2.1 Photoconductive Type

Photoconductors (or photoresistors) represent the simplest photodetector configuration structurally, serving as fundamental components in various technological and practical applications. A photoconductive detector is a semiconductor-based photodetector that operates on the photoconductive effect. This device modulates its electrical conductivity in response to absorbed photon energy. This section introduces the operational principle, structure, performance parameters, and several representative types of photoconductive detectors.

8.2.1.1 Photoconduction Effect

A photoconductive detector (photoresistor) utilizes the photoconductive effect in semiconductors, defined as the phenomenon where the electrical conductivity of a semiconductor material changes due to variations in incident light intensity. Upon illumination, semiconductor materials absorb photon energy. This absorption elevates the concentration of charge carriers, thereby enhancing the material's conductivity [2].

Photoconductors are classified based on their operating principles and material properties:

Intrinsic Photoconduction The intrinsic photoconductive effect occurs when the incident photon energy ($h\nu$) exceeds the material's bandgap energy. This promotes electrons from the valence band to the conduction band, generating photogenerated free electrons in the conduction band and photogenerated free holes in the valence band. The resultant increase in free charge-carrier density elevates the material's electrical conductivity. This process corresponds to intrinsic absorption, as depicted in Figure 8.3.

Extrinsic Photoconduction The extrinsic photoconductive effect arises in doped semiconductors, where photons excite bound carriers associated with impurity energy levels. This comprises two mechanisms (Figure 8.4): donor-level excitation, where photons promote electrons from neutral donor impurity levels into the

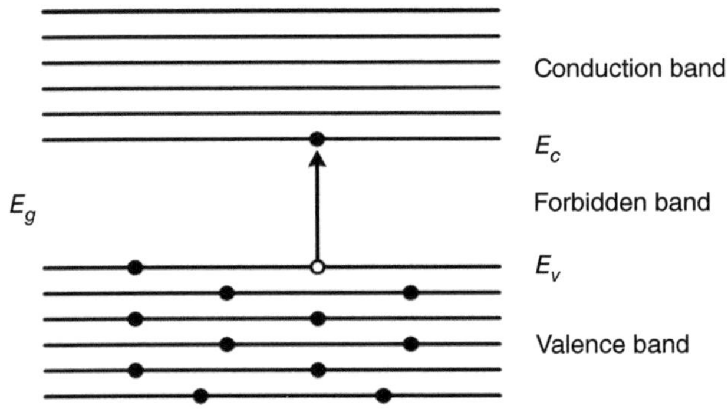

Figure 8.3 Intrinsic absorption of semiconductors.

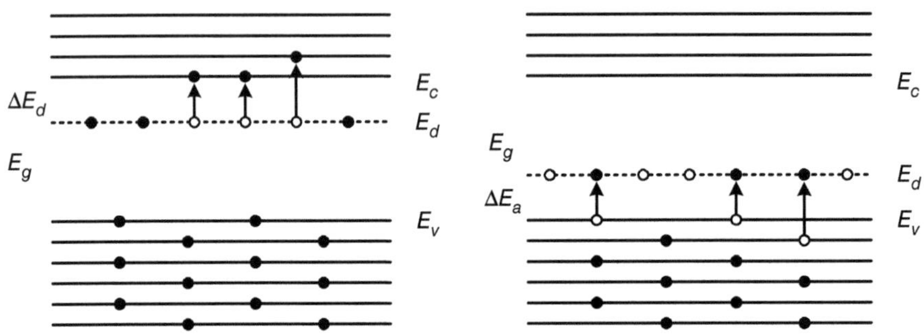

Figure 8.4 Impurity absorption of (a) *n*-type donor and (b) *p*-type acceptor.

conduction band; acceptor-level excitation, where photons excite electrons from the valence band to neutral acceptor impurity levels, effectively generating free holes in the valence band. Both excitation processes generate photogenerated free carriers (electrons or holes), leading to increased electrical conductivity. This phenomenon is termed impurity absorption.

The concentration of impurity atoms within a semiconductor crystal is typically several orders of magnitude lower than the intrinsic atom density. Consequently, the extrinsic photoconductive effect is significantly weaker than its intrinsic counterpart. However, due to the lower-ionization energies of impurities (relative to the bandgap), extrinsic photoconduction facilitates the development of IR detectors. Both photoconductive effects share core operational mechanics. Given the significantly weaker response of the extrinsic effect and the predominance of intrinsic effects in most materials, subsequent sections will collectively refer to the photoconductive effect without explicit distinction unless otherwise specified.

8.2.1.2 Theory of Photoresistor

Materials exhibiting the photoconductive effect, including intrinsic and extrinsic semiconductors (such as silicon and germanium), as well as cadmium sulfide, cadmium selenide, and lead sulfide, can be utilized as photosensors whose conductivity varies with incident light intensity. These devices are termed photoconductive detectors. Photoresistors, the most common photoconductive devices, exhibit a rapid decrease in electrical resistance upon illumination at wavelengths corresponding to their bandgap energies.

Under dark conditions, photoresistors maintain a high resistance. When light with photon energy exceeding the material's bandgap irradiates the device, valence band electrons absorb photons and transition to the conduction band, generating electron–hole pairs (EHPs). This photogeneration process increases the population of free charge carriers, thereby reducing resistivity and consequently lowering resistance. Enhanced light intensity generates more carriers, resulting in proportionally lower resistance values. After illumination ceases, carrier recombination occurs over time, gradually restoring the resistance to its initial state.

A photoresistor structure typically consists of photoconductive material with electrodes deposited at both ends; the working principle and schematic symbol are depicted in Figure 8.5. Upon application of a bias voltage U_{bb}, photogenerated carriers undergo directional drift under the imposed electric field, creating a measurable photocurrent I_p in the external circuit.

A photoresistor is fabricated by mounting a photoconductive material onto an insulating substrate—such as hard glass, mica, or high-frequency ceramic—followed by electrode deposition at opposite termini of the photoconductor. Lead wires are subsequently connected to the electrodes, and the assembly is finally encapsulated within a metal or plastic housing incorporating an optical window. The typical structural configuration is illustrated in Figure 8.6.

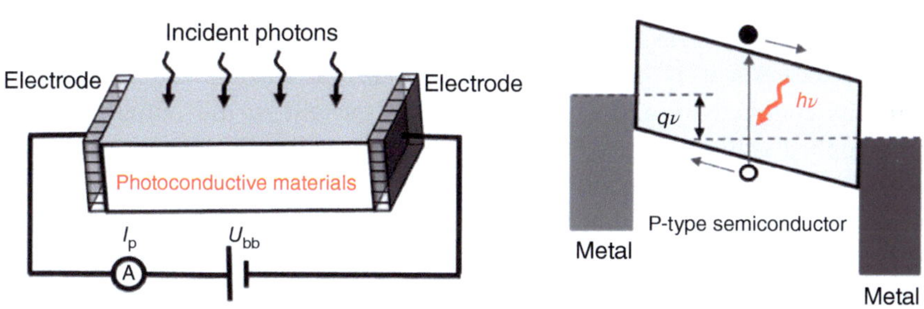

Figure 8.5 Schematic of photoresistor and the band structure.

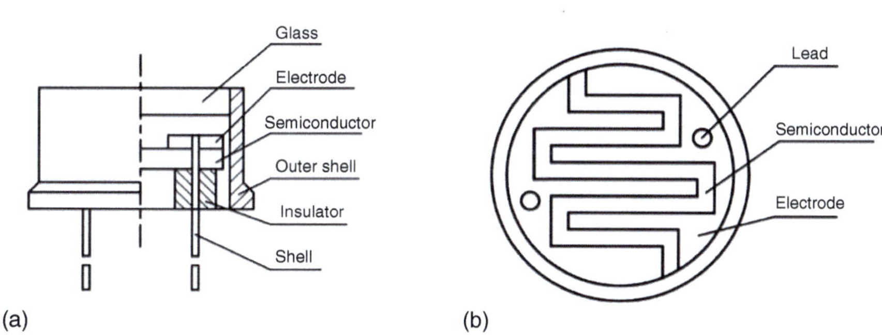

Figure 8.6 Photoresistor structure: (a) side view and (b) top view.

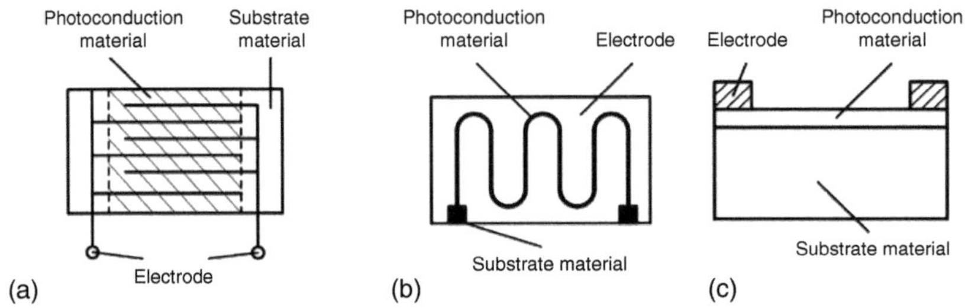

Figure 8.7 Three typical structures of photoresistors: (a) comblike structure, (b) serpentine structure, and (c) engraved-line structure.

Under light illumination, the photocurrent of a photoresistor is inversely proportional to the interelectrode length. To enhance sensitivity, the distance between the electrodes is minimized during device design. Consequently, common photoresistor configurations include comblike, serpentine, and engraved-line structures, as illustrated in Figure 8.7. The comblike structure entails etching interdigitated comb-shaped grooves into the substrate material (typically glass), filling these grooves with an electrode material such as gold or graphite, and coating the surface with a photoconductive layer. In the serpentine structure, the photoconductive material is patterned into a meandering strip, flanked by metal conductive layers functioning as electrodes on both sides and encapsulated within an outer insulating substrate. The engraved-line structure is fabricated by etching narrow strips of photoconductive material onto a prepared substrate, followed by evaporative deposition of electrode material onto these strips.

When a photoresistor is illuminated, nonequilibrium carriers are generated, increasing its conductivity and producing a photocurrent in the output circuit under an applied voltage. Taking the photoresistor structure in Figure 8.8 as an example, assume its photoconductive material is an n-type semiconductor with length L, width w, height d, and light irradiation with power P, which is uniformly incident along the x-direction.

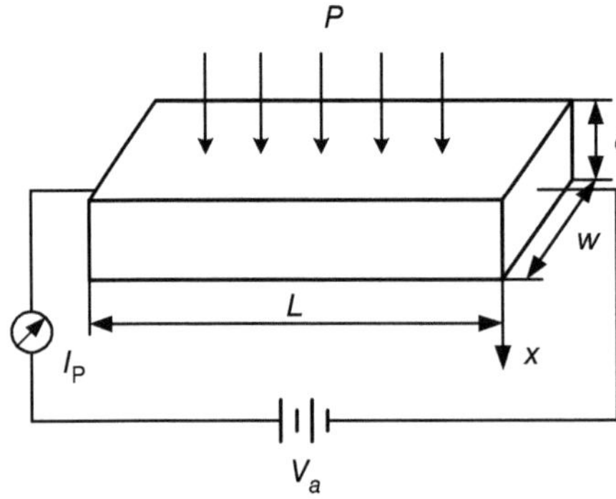

Figure 8.8 Geometric parameters of photoresistor.

If the absorption coefficient of the photoconductive material is α and its surface reflectivity is R, the incident light power $P(x)$ varies along the vertical direction within the material by

$$P(x) = P(1 - R)e^{-\alpha x} \tag{8.1}$$

The statistical average of photogenerated carriers should also be a function of the incident depth x. Let $n(x)$ denote the photogenerated carrier concentration at position x. Then, the drift current density $J(x)$ at that position under the applied electric field is

$$J(x) = en(x)v \tag{8.2}$$

where v represents the drift velocity of photogenerated carriers under the applied electric field E. The drift velocity can be further expressed as

$$v = \mu_n E = \frac{\mu_n V_a}{L} \tag{8.3}$$

where μ_n is the electron mobility, L is the distance between the electrodes, and V_a is the applied voltage. Then the average output photocurrent is

$$I_P = \int_0^{\mathrm{d}} J(x)w \cdot \mathrm{d}x \tag{8.4}$$

In the equation, $A = w \cdot \mathrm{d}$ is the detector electrode area and $\mathrm{d}A = w \cdot \mathrm{d}x$ is the differential area element. Under uniform illumination of the photosensitive area $(w \cdot L)$, we can obtain

$$I_P = w \cdot e \cdot v \int_0^{\mathrm{d}} n(x)\mathrm{d}x \tag{8.5}$$

where $n(x)$ depends on the generation and recombination rates of photogenerated carriers in the semiconductor material under the external electric field.

Using the steady-state condition where the electron generation rate equals the recombination rate, $n(x)$ can be determined. If the optical power absorbed per unit length in the x-direction by the photoconductive material is $\alpha P(x)$, the optical power absorbed per unit volume becomes $\alpha P(x)/wL$. Thus, the number of photogenerated carriers produced per unit time and unit volume (i.e., the generation rate) is

$$G_0 = \frac{\alpha P(x)}{wLhv} \tag{8.6}$$

If the average carrier lifetime is τ_0, the number of recombined carriers (i.e., the recombination rate) can be expressed as

$$R_0 = \frac{n(x)}{\tau_0} \tag{8.7}$$

Under steady-state conditions, $G_0 = R_0$:

$$\frac{\alpha P(x)}{wLhv} = \frac{n(x)}{\tau_0} \tag{8.8}$$

Using Eq. (8.8), the photogenerated carrier concentration $n(x)$ is

$$n(x) = \frac{\alpha P(x)\tau_0}{wLh\nu} \tag{8.9}$$

Substituting Eq. (8.1) into Eq. (8.9), it yields

$$n(x) = \frac{\alpha P(1-R)e^{-\alpha x}\tau_0}{wLh\nu} \tag{8.10}$$

By substituting Eq. (8.10) into Eq. (8.5), the average output photocurrent of the photoconductive detector is obtained as

$$I_P = \frac{\alpha P(1-R)\tau_0 e\mu_n V_a}{h\nu L^2}\int_0^{\mathrm{d}} e^{-\alpha x}\mathrm{d}x \tag{8.11}$$

At the extreme cases where the light has no reflection ($R = 0$) and is completely absorbed by the photoconductive materials ($\alpha d \gg 1$), the photocurrent I_{P0} is

$$I_{\mathrm{P0}} = \frac{e\mathrm{P}\tau_0\mu_n V_a}{h\nu L^2} \tag{8.12}$$

The quantum efficiency (QE) can be obtained using Eqs. (8.11) and (8.12):

$$\eta = \frac{I_P}{I_{P0}} = \alpha(1-R)\int_0^{\mathrm{d}} e^{-\alpha x}\mathrm{d}x \tag{8.13}$$

which represents the efficiency of generating EHPs in the photoresistor, calculated by the intensity ratio between the absorbed light and the incident light.

Substituting Eq. (8.13) into Eq. (8.11) simplifies it to

$$I_P = \frac{\eta eP}{h\nu}\left(\frac{\mu_n V_a \tau_0}{L^2}\right) = \frac{\eta eP}{h\nu}\left(\frac{v\tau_0}{L}\right) \tag{8.14}$$

Since the transit time of carriers between the electrodes under the external electric field is

$$t_d = \frac{L}{v} \tag{8.15}$$

Substituting the aforementioned expression into Eq. (8.14) yields

$$I_P = \frac{\eta eP}{h\nu}\left(\frac{\tau_0}{t_d}\right) \tag{8.16}$$

Letting $G = \tau_0/t_d$, the equation can be rewritten as

$$I_P = G\frac{\eta e}{h\nu}P \tag{8.17}$$

where $G = \tau_0/t_d$ is the internal gain, representing the effective contribution of photogenerated carriers to the external circuit current of the detector.

When $G = 1$, it indicates that the average carrier lifetime t_0 exactly equals the transit time t_d between the electrodes, resulting in each photogenerated EHP contributing precisely one electron charge to the external circuit current. When $G < 1$, t_0 is shorter than t_d, indicating that the average contribution per photogenerated EHP is less than one electron charge. Conversely, when $G > 1$, t_0 exceeds t_d, implying that a single photon excitation drives multiple electrons through the external circuit, thereby amplifying the photocurrent contribution beyond

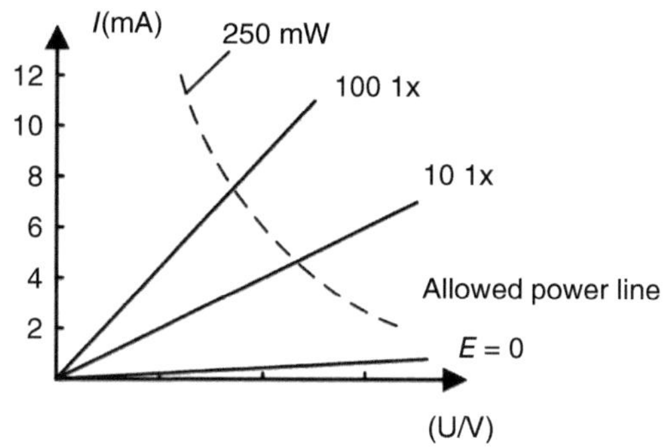

Figure 8.9 Current–voltage characteristics of photoresistor.

one electron charge per photon. To achieve higher internal gain, semiconductor materials with long carrier lifetimes and high mobility are typically selected, while simultaneously reducing the interelectrode spacing and increasing the bias voltage. Under optimized conditions, the internal gain G can attain values on the order of $\sim 10^2$. Photoconductive detectors inherently possess internal gain, represented by G, which depends on material properties, structural dimensions, and applied bias voltage.

8.2.1.3 Current–Voltage Characteristics

The current–voltage characteristics of a photoconductive detector describe the relationship curve between the voltage applied across the photoresistor and the resulting photocurrent under constant illumination. As the photoresistor behaves as an Ohmic device, it adheres to Ohm's law. Consequently, its volt-ampere characteristic is depicted as a straight line passing through the origin, illustrated in Figure 8.9 and supported by Eq. (8.14).

When utilizing a photoconductive detector, the actual power dissipated must remain below the maximum permissible value indicated by the rated power curve. To meet this requirement, the load resistance must be designed such that the load line does not intersect with the rated power curve.

8.2.1.4 Photoelectric Characteristics

At room temperature, a photoconductive detector in darkness (i.e., no illumination) exhibits a measurable conductance due to thermally excited nonequilibrium carriers. This conductance is termed dark conductance, and its reciprocal represents the dark resistance. Typically, the dark conductance is very low—meaning dark resistance is extremely high—with practical devices often achieving dark resistances exceeding 1 MΩ, sometimes reaching 100 MΩ. When a bias voltage is applied, the resulting current under these conditions is defined as the dark current (I_{dark}).

Under illumination, the resistance of the photodetector at a specific light level is called the light resistance (R_{light}), and the corresponding current is the light current (I_{light}). The photocurrent (I_{ph}), which drives detection sensitivity, is calculated from their difference ($I_P = I_{light} - I_{dark}$). For optimal performance, a photoconductive detector should exhibit high dark resistance (to minimize I_{dark}) and low light resistance (to maximize I_P). Correspondingly, reduced dark current and increased photocurrent directly enhance the detector's sensitivity.

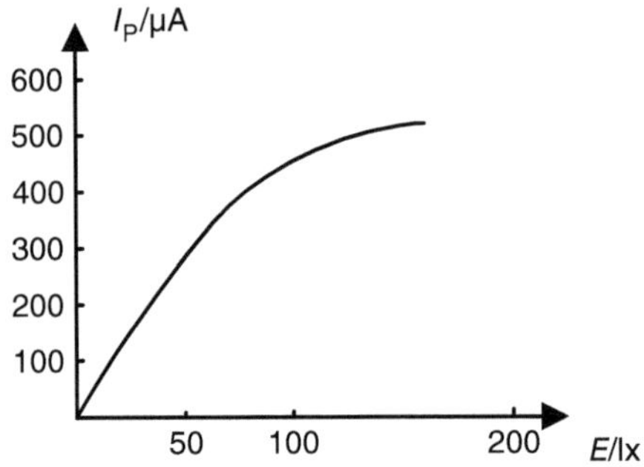

Figure 8.10 Optoelectronic characteristics of cadmium sulfide (CdS) photoresistor.

The photoelectric characteristics describe how the current through the device—under fixed voltage—relates to incident luminous flux (illuminance). Importantly, under weak illumination, τ_0 and τ_d remain constant, resulting in direct proportionality: photocurrent (I_P) $\propto$ luminous flux (P), i.e., $I_P \propto P$. In contrast, under strong illumination, τ_0 becomes dependent on photogenerated carrier concentration, and τ_d varies due to higher carrier densities or temperature rises; thus, I_P exhibits nonlinear behavior with P.

The general equation describing the photoelectric characteristics of a photoresistor is

$$I_P(\lambda) = S_g U^a E^\gamma \tag{8.18}$$

where S_g is the photoconductive sensitivity, dependent on the detector material, U is the applied voltage, E denotes illuminance, I_P is the photocurrent (defined as $I_{light}-I_{dark}$), a is the voltage exponent ($a \approx 1$ for Ohmic contact, $a \approx 1.1$–1.2 for non-Ohmic contact); and γ is the illumination exponent ($\gamma = 1$ for linear photoconductivity under weak light, $\gamma = 0.5$ for nonlinear behavior under strong light). As illustrated in Figure 8.10, the photoelectric characteristic curve shows I_P versus illuminance E. This curve transitions from linear to nonlinear response as illuminance increases.

8.2.1.5 Sensitivity and Gain

Generally, photoconductive detectors operate under weak illumination conditions, where the voltage exponent $a = 1$ and the illumination exponent $\gamma = 1$. Thus, Eq. (8.18) can be rewritten as

$$I_P = S_g UE = g_P U \tag{8.19}$$

where $g_P = S_g E$ is termed the photoconductance of the photoconductive detector, with units of siemens (S).

By the definition of sensitivity, the photoconductive sensitivity S_g is the ratio of the response quantity to the input quantity:

$$S_g = \frac{g_P}{E} \tag{8.20}$$

When considering the current generated by dark conductivity, the total current flow through the photoconductive detector is

$$I_{light} = I_P + I_D = g_P U + g_D U = g_B U \tag{8.21}$$

where I_P is the photocurrent; I_{light} is the bright current, I_D is the dark current, g_D represents dark conductivity, g_B is the bright conductivity. Therefore, the photoconductivity of the photoconductive detector is

$$g_P = g_B - g_D \tag{8.22}$$

A key characteristic parameter is the photoconductive gain G, defined as the ratio of the external photocurrent (formed by photogenerated carriers under an electric field in a photoconductor of length L) to the internal photocurrent generated by photons:

$$G = \frac{I_P}{qN} \tag{8.23}$$

Under illumination, assuming N EHPs are generated per unit time with lifetimes τ_n and τ_p, the photogenerated free electron and hole concentrations are

$$\Delta n = \frac{N\tau_n}{AL}; \Delta p = \frac{N\tau_p}{AL} \tag{8.24}$$

Consequently, the material's conductivity increases by $\Delta\sigma$, termed the photoconductivity, given by

$$\Delta\sigma = q(\Delta n\mu_n + \Delta p\mu_p) \tag{8.25}$$

The photocurrent induced by the photoconductivity $\Delta\sigma$

$$I_P = \frac{U\Delta\sigma A}{L} = \frac{qAU(\Delta n\mu_n + \Delta p\mu_p)}{L} = \frac{qNU}{L^2}(\tau_n\mu_n + \tau_p\mu_p) \tag{8.26}$$

Substituting Eq. (8.26) into Eq. (8.23) yields

$$G = \frac{U}{L^2}(\tau_n\mu_n + \tau_p\mu_p) = \frac{U}{L^2}\tau_n\mu_n + \frac{U}{L^2}\tau_p\mu_p = G_n + G_p \tag{8.27}$$

where G_n is termed the electron gain coefficient and G_p is the hole gain coefficient of the photoconductive detector.

Furthermore, G_n and G_p can be simplified as

$$G_n = \frac{U}{L^2}\tau_n\mu_n = \frac{\tau_n}{t_n}; \quad G_p = \frac{U}{L^2}\tau_p\mu_p = \frac{\tau_p}{t_p} \tag{8.28}$$

where t_n represents the electron drift time, and t_p the hole drift time. For semiconductors where electron and hole lifetimes are equal ($t_n = t_p = \tau_0$), the gain of the intrinsic photoresistor is

$$G = G_n + G_p = \frac{\tau_n}{t_n} + \frac{\tau_p}{t_p} = \tau_0\left(\frac{1}{t_n} + \frac{1}{t_p}\right) \tag{8.29}$$

Defining $1/t_n + 1/t_p = 1/t_d$ gives

$$G = \frac{\tau_0}{t_d} \tag{8.30}$$

where t_d is the transit time required for charge carriers to traverse the electrode spacing L.

8.2.1.6 Prehistory Effect

The prehistory effect refers to a phenomenon where the response characteristics of a photoconductive detector depend on its prior operational state (e.g., illumination or darkness) before use. This phenomenon is categorized into the dark-state prehistory effect and the bright-state prehistory effect.

Dark-state prehistory effect: This occurs when the detector is kept in darkness prior to testing or operation. Upon sudden illumination, a longer duration in the dark state results in a slower rise time of the photocurrent. Generally, both lower operating voltages and lower illuminance conditions enhance this effect. The dark-state-prehistory-effect curve for a CdS photoresistor is shown in Figure 8.11. Under dark conditions, charge carriers within the material (e.g., EHPs) are gradually captured by defect states or deep-level traps, forming a "dark state." Upon illumination, these trapped carriers must first be released before they can contribute to conduction, leading to a delay in current response.

Bright-state prehistory effect: This arises when the detector is pre-illuminated before testing or operation. If the pre-illumination intensity differs from the target illuminance during operation, a hysteresis phenomenon is observed. The bright-state prehistory effect curve for a CdS photoresistor is illustrated in Figure 8.12. These phenomena are caused by the characteristics of semiconductor materials like cadmium sulfide (CdS): pre-illumination alters the carrier concentration and trap-state filling within the material, resulting in a required relaxation time for conductivity changes. When switching from strong to weak light, excess carriers require time to recombine; when switching from weak to strong light, carrier generation also requires time to accumulate.

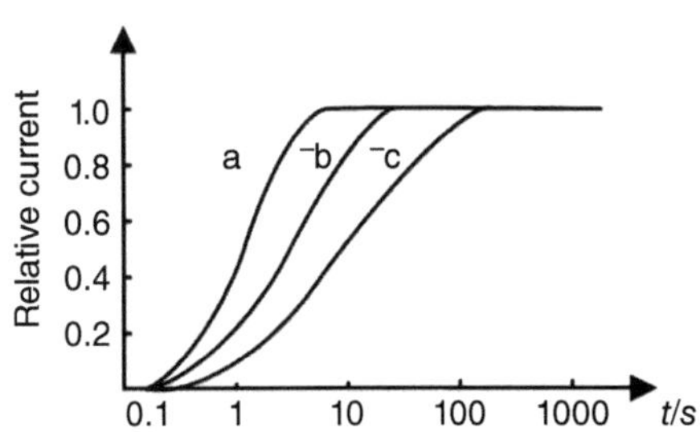

Figure 8.11 Dark-state-prehistory-effect curves of cadmium sulfide photoresistor (a–c correspond to device stored in dark from short to long time).

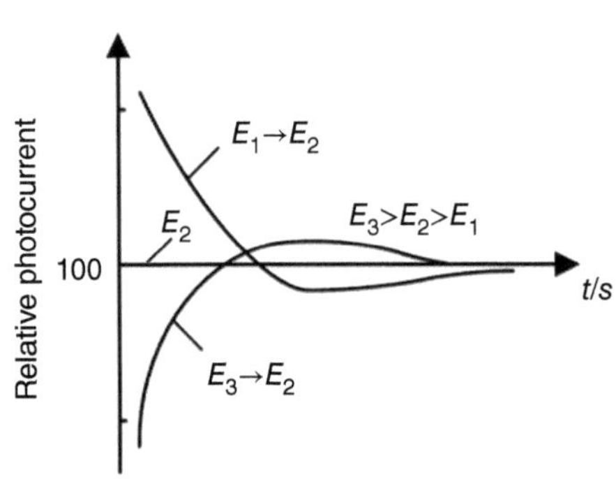

Figure 8.12 Bright-state-prehistory-effect curve of cadmium sulfide photoresistor.

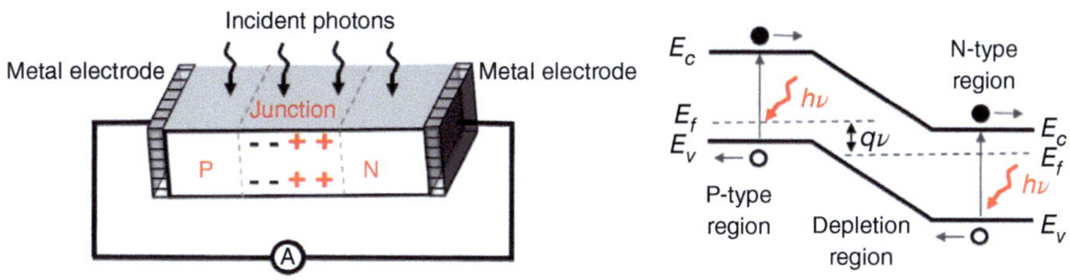

Figure 8.13 Schematics of a PN junction photodiode detector and its energy band diagram.

8.2.2 Photodiode Type

Compared to photoconductors, photodiodes exhibit self-powering characteristics, requiring no external bias voltage. Based on structural and operational principle differences, photodiodes can be classified into several distinct types, each possessing unique technical attributes and application suitability.

8.2.2.1 PN Junction Photodiode

The PN junction photodiode is a fundamental photodetector type. As illustrated in Figure 8.13, it comprises *p*-type and *n*-type semiconductor regions. Photons with sufficient energy incident upon the intrinsic region of the device generate EHPs. These pairs are generated near, or diffuse toward, the PN junction. Under the influence of the junction's strong built-in electric field, electrons are swept toward the *n*-type region and holes toward the *p*-type region. This charge separation results in a measurable photocurrent when the device is integrated into a circuit or an open-circuit photovoltage. Thus, optical signals are converted into electrical signals through this mechanism.

8.2.2.2 Schottky Barrier Photodiode

A Schottky barrier photodiode is formed at the interface between a metal and a semiconductor. As depicted in Figure 8.14, the junction between the metal and

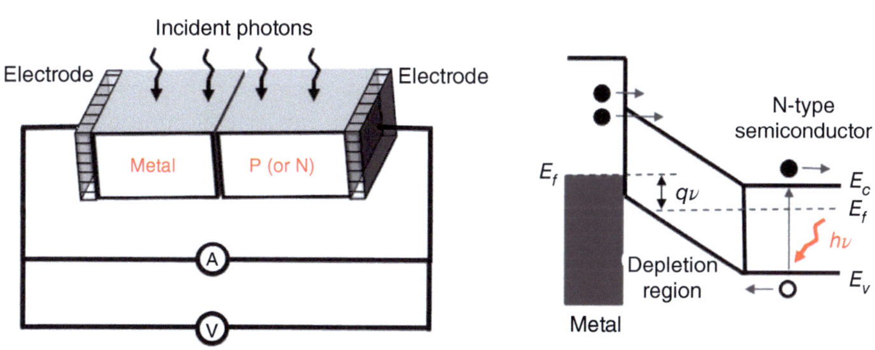

Figure 8.14 Schematics of a Schottky barrier photodiode detector and its energy band diagram.

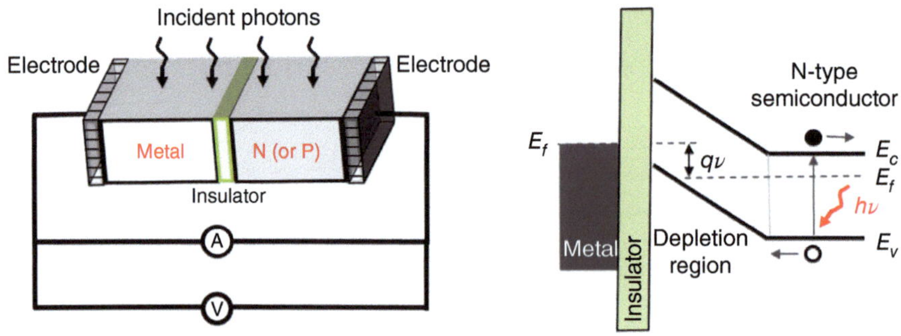

Figure 8.15 Schematics of a metal–insulator–semiconductor (MIS) detector and its energy band diagram.

the semiconductor establishes a distinctive barrier potential. Analogous to a PN junction, this Schottky barrier facilitates the separation of photogenerated EHPs. Photo-carrier generation occurs both within the semiconductor bulk and the metal–semiconductor interface region. Compared to PN junction photodiodes, Schottky barrier photodiodes offer significant fabrication advantages: a simple metal deposition process onto the semiconductor suffices, eliminating the need for high-temperature diffusion or epitaxy. Furthermore, these devices exhibit exceptionally fast response speeds. A key drawback, however, is a relatively higher dark current compared to PN junction diodes.

8.2.2.3 Metal–Insulator–Semiconductor Photodiode

A metal–insulator–semiconductor (MIS) photodiode comprises a metal electrode, a thin insulating layer, and a semiconductor substrate (Figure 8.15). The insulator physically isolates the metal gate from the semiconductor. This insulator thickness is a critical parameter, typically maintained below 10 nm to facilitate quantum mechanical tunneling. In an *n*-type semiconductor substrate, the application of a negative bias voltage to the metal gate repels majority carriers (electrons) from the semiconductor–insulator interface, inducing a depletion region. The operational principle relies on photon absorption within the semiconductor to generate EHPs. Photogenerated carriers tunnel through the insulator and are subsequently swept across the junction by the electric field, thereby generating a measurable photocurrent. This configuration enables internal signal amplification through the insulator's capacitive coupling, which enhances carrier collection efficiency without the need for external gain circuitry. Furthermore, passivation provided by the insulator suppresses dark current, thereby enhancing signal-to-noise (SNRs) ratios in low-light applications.

8.2.2.4 Avalanche Photodiode

As shown in Figure 8.16, the avalanche photodiode (APD) exploits carrier avalanche multiplication to detect weak optical signals. Under a sufficiently high reverse bias, the depletion region extends through the intrinsic layer. Photon absorption occurs predominantly within the *p*-type region. Photogenerated minority carriers

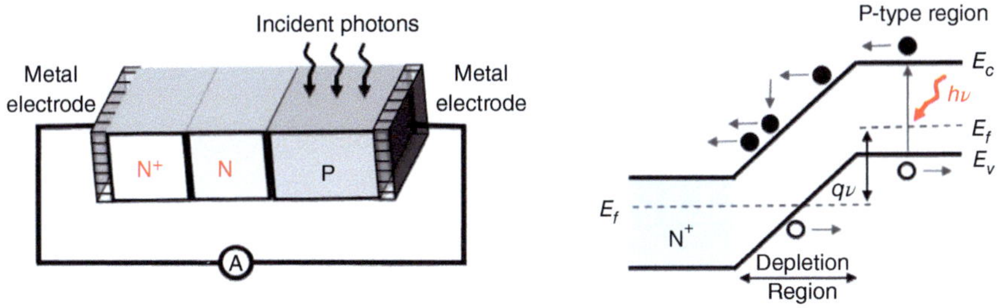

Figure 8.16 Schematics of an avalanche photodiode detector and its energy band diagram.

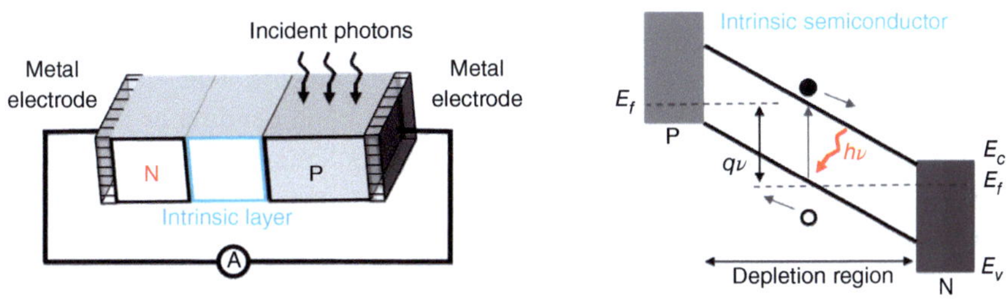

Figure 8.17 Schematics of a PIN photodiode detector and its energy band diagram.

(e.g., electrons generated in the *p*-region) diffusing to the depletion edge undergo rapid acceleration by the high electric field. These carriers gain sufficient kinetic energy to initiate impact ionization upon collision with lattice atoms, creating secondary EHPs—hence the term "avalanche." This internal gain mechanism endows APDs with high speed, enhanced sensitivity, and high QE.

8.2.2.5 PIN Photodiode

As illustrated in Figure 8.17, a PIN photodiode features an undoped, intrinsic semiconductor layer sandwiched between *p*-type– and *n*-type–doped regions (not "n-type and n-type"). When reverse biased, the intrinsic (I) layer becomes fully depleted of free charge carriers, forming a wide depletion region. If the p^+ and n^+ contact layers utilize wide-bandgap semiconductors, incident photons are primarily absorbed within the wide intrinsic depletion region, generating EHPs directly within this active zone. These pairs are promptly separated by the applied electric field. The absence of free carriers in the intrinsic region minimizes diffusion-based current components. Consequently, the dominant dark current component arises from generation–recombination (G–R) centers within the depletion region. If the contact layers possess sufficient bandgap width to inhibit minority carrier injection and thermal generation at typical operating temperatures, their contribution to dark current becomes negligible.

8.2.3 Photovoltaic Type

A photovoltaic photodetector is a photoconversion device that utilizes the photovoltaic effect. This effect occurs when photons incident on a semiconductor material

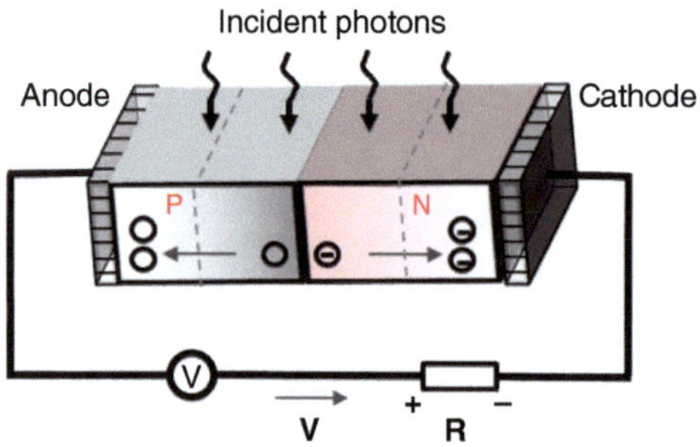

Figure 8.18 Photovoltaic photodetector.

generate EHPs, which are then separated by a built-in electric field, creating an electrical signal. The internal electric field arises from band bending at junctions between dissimilar materials, such as PN junctions or Schottky junctions. Devices exploiting the photovoltaic effect are termed photovoltaic devices, with an operating principle analogous to that of solar cells (Figure 8.18).

The fundamental photovoltaic photodetector structure consists of a single PN junction. More complex device architectures include the introduction of an additional intrinsic (undoped) semiconductor layer between the p-type and n-type regions, forming a P–I–N photodetector. This design enhances device sensitivity. These structures facilitate the separation of photogenerated charge carriers within or near their depletion region; this spatial separation is essential for generating the electrical signal. The width of the charged depletion region depends on the doping concentration within the semiconductor. Applying a reverse bias voltage widens this region, thereby increasing the efficiency of photogenerated carrier separation.

Enhancing the detector's temporal response (speed) can be achieved by increasing the material's doping density to reduce its series resistance. Alternatively, incorporating an intrinsic layer between the p and n regions largely determines the resulting width of the space-charge region (SCR). The resulting P–I–N structure exhibits lower resistivity and a wider SCR, resulting in lower capacitance. These characteristics render the P–I–N configuration particularly suitable for high-speed integrated circuit applications.

8.2.4 Photomultiplier Tube Type

The PMT is a photodetector capable of detecting single photons, operating on the principle of electron multiplication via secondary emission. Figure 8.19 depicts the structure and working principle of a conventional PMT. It comprises a vacuum glass envelope containing a photocathode, an anode, and a series of intermediate dynodes. Incident light striking the photocathode induces the emission of photoelectrons. These primary electrons are accelerated toward the first dynode by an applied electric field. Upon impact, each incident electron stimulates the dynode to emit multiple secondary electrons through secondary electron emission. The secondary electrons are sequentially accelerated toward subsequent dynodes by the

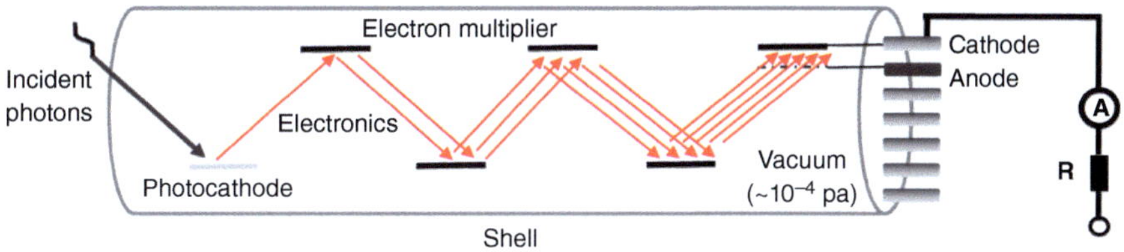

Figure 8.19 Schematic diagram of photomultiplier tube.

electric field. This process repeats at each dynode stage, resulting in a cascade multiplication of electrons. The progressively amplified electron current is ultimately collected at the anode, achieving typical gain factors ranging from 10^4 to 10^8.

This substantial gain confers upon PMTs significantly higher sensitivity than other photodetectors, enabling the detection of extremely weak optical signals [3]. Owing to their high gain, rapid temporal response, and output current proportional to the incident photon flux, PMTs are extensively utilized in astrophotometry and precision scientific instrumentation.

8.3 General Metrics of Photodetectors

Photodetectors, serving as core components in optical communications, imaging systems, and quantum sensing, critically determine the precision of signal conversion and system sensitivity. Understanding the fundamental parameters of detectors and their correlation mechanisms constitutes the foundation for optimizing device design and enhancing performance. Later, we systematically analyze several key performance parameters of photodetectors.

8.3.1 Dark Current

Dark current refers to the electric current generated within a photodetector under an applied bias voltage in the absence of optical illumination. This current constitutes a primary noise source, and its magnitude correlates with the device's effective active area and operating temperature. Lower dark current generally indicates superior device performance.

Common photodetectors, including photodiodes, phototransistors, and charge-coupled device (CCD) sensors, exhibit a finite dark current even under zero illumination. The magnitude of this current varies significantly depending on the detector's operating principle and conditions, ranging from sub-nanoampere levels to values orders of magnitude higher. While negligible in many applications, dark current critically impacts scenarios involving weak-light detection. Furthermore, dark current directly contributes to shot noise, which is proportional to its square root. The ratio of the photocurrent (I_p) to the dark current (I_d), known as the on–off ratio, quantifies the detector's responsivity to incident light. Both dark current and photocurrent are fundamental determinants of SNR. Consequently, excessive dark current degrades the SNR by increasing overall device noise.

8.3.2 Responsivity

The responsivity (R) characterizes a photodetector's capacity to convert incident light into an electrical signal. It is defined as the detector's output signal per unit incident optical power. Responsivity generally includes current responsivity (R_I, in A/W) and voltage responsivity (R_V, in V/W), quantifying the photocurrent and photovoltage generated per unit optical power, respectively. Higher responsivity indicates a stronger efficiency of photon-to-electron conversion within the device. The mathematical expression for current responsivity is

$$R = \frac{I_{\text{light}} - I_{\text{dark}}}{P_{\text{incident}}} \tag{8.31}$$

where I_{light} is the output photocurrent, I_{dark} is the output dark current, and P_{incident} is the incident optical power. Higher responsivity signifies superior device performance. An alternative expression of responsivity incorporating quantum efficiency is

$$R = \frac{\eta q}{h\nu} \tag{8.32}$$

where η is the quantum efficiency, q is the elementary charge (1.602×10^{-19} C), $h\nu$ is the energy of the incident photon. This equation directly relates responsivity to the fundamental photoconversion process.

The responsivity value depends significantly on the detector's material properties, structural design, and operating conditions. Enhancing the photocurrent generated can improve responsivity. Furthermore, the responsivity has spectral dependence at a specific photon energy, reflecting its photoelectric conversion capability at that wavelength. Crucially, the responsivity of a device varies considerably with the wavelength of incident light; therefore, its spectral response curve is typically the most informative characterization.

8.3.3 Detectivity

Detectivity (D), is defined as the reciprocal of the noise-equivalent power (NEP). It corresponds to the reciprocal of the optical power required to produce a signal-to-noise ratio of unity. The general expression for detectivity is given by

$$D = \frac{R}{I_{\text{noise}}} = \frac{1}{\text{NEP}} \tag{8.33}$$

where R is the responsivity, I_{noise} is the noise current. Detectivity quantifies sensitivity to weak optical signals. However, since the detectivity depends significantly on the bandwidth and the detector area, specific detectivity, or D^*, is more often used as a figure of merit for a photodetector to characterize performance, which is equal to the reciprocal of noise-equivalent power (NEP), normalized per square root of the sensor's area and frequency bandwidth.

$$D^* = \frac{\sqrt{A \cdot \Delta f}}{\text{NEP}} \tag{8.34}$$

where Δf is the bandwidth, and A is the detector area. The unit of specific detectivity is Jones (cm $Hz^{1}/^{2}$/W). State-of-the-art devices exhibit D values exceeding 10^{12} Jones. For the case the noise is dominated by the short noise of dark current, the specific detectivity is calculated by

$$D^* = \frac{R\sqrt{A}}{\sqrt{2qI_{\text{dark}}}} \tag{8.35}$$

8.3.4 Quantum Efficiency

Quantum efficiency (QE) is categorized into internal quantum efficiency (IQE) and external quantum efficiency (EQE). IQE is defined as the ratio of the number of collected charge carriers to the number of photons absorbed by the material. EQE is defined as the ratio of the number of collected charge carriers to the number of incident photons. Henceforth in this text, quantum efficiency refers specifically to EQE, denoted by the symbol η. Based on its definition, η is expressed by the relationship:

$$\eta = \frac{Rhc}{q\lambda} \times 100\% \tag{8.36}$$

where h is the Planck constant, c is the speed of light in vacuum, R is the detector's responsivity, q is the elementary charge (1.602×10^{-19} C), and λ is the wavelength of the incident light.

As a dimensionless quantity typically expressed as a percentage, EQE quantifies the photon-to-charge-carrier conversion efficiency at a specific wavelength. Equation (8.36) demonstrates that EQE is directly proportional to responsivity R. Consequently, detectors with high EQE exhibit enhanced photoresponse sensitivity, establishing this parameter as a critical performance metric for optoelectronic devices.

8.3.5 Spectral Range

The spectral response of a photodetector characterizes the dependency of its relative responsivity on the wavelength of incident optical radiation. This relationship is typically depicted by a graphical curve illustrating the photocurrent generated under optical excitation across a wavelength range, thereby establishing the operational spectral window. Conventionally, the x-axis denotes wavelength, whereas the y-axis corresponds to quantum efficiency, current responsivity, or normalized arbitrary units.

8.3.6 Response Time

Response time characterizes the rapidity of a device's reaction to incident optical signals. Upon illumination, a finite interval is required for the photodetector's output signal to reach a quasi-stable state; conversely, when illumination ceases, a comparable duration is necessary for response recovery to the initial state. For PN junction and Schottky junction photodetectors, response time is primarily governed by three

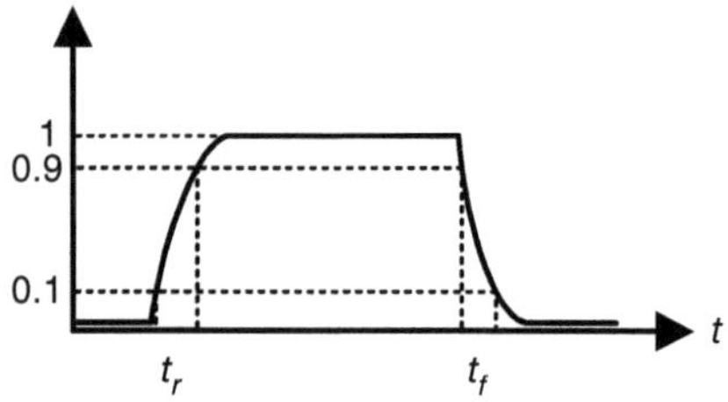

Figure 8.20 The rise (t_{rise}) and fall times (t_{fall}) of photodetector.

components: the carrier transit time (τ_{drift}) in the depletion region, the diffusion time (τ_{diff}) in the quasi-neutral region, and the *RC* time constant (τ_{RC}). These time constants span nanoseconds to milliseconds, modulated by charge transport dynamics and defect densities. Specifically

$$\tau_{drift} = \frac{w}{v_d} \tag{8.37}$$

$$\tau_{diff} = \frac{d^2}{D} \tag{8.38}$$

$$\tau_{RC} = 2.2RC \tag{8.39}$$

where w denotes depletion region width, v_d represents the saturation drift velocity, d reflects the carrier diffusion length in the quasi-neutral region, D is the minority carrier diffusion coefficient, R comprises the sum of device series resistance and load resistance, and C signifies the total device capacitance.

As carrier diffusion velocities are typically lower than drift velocities, extended diffusion times limit overall response speed. Consequently, maximizing depletion layer width is advantageous. Standard characterization defines rise time (t_{rise}) as the interval for signal ascent from 10% to 90% of peak amplitude, while fall time (t_{fall}) represents descent from 90% to 10%, as illustrated in Figure 8.20. Short response times (typically $t_{rise} < 1$ μs and $t_{fall} < 10$ μs) are prerequisite for high-frequency applications including optical communications and high-speed imaging.

Under pulsed optical excitation, photoconductive detectors exhibit distinct illumination-dependent temporal responses. Under weak illumination, the rise time (t_{rise}) and fall time (t_{fall}) are symmetric and coincide with the carrier lifetime τ_0 ($t_{rise} = t_{fall} = \tau_0$). However, intense illumination induces a condition where t_{fall} significantly exceeds t_{rise} (Figure 8.21). Upon termination of excitation, the restoration of pre-irradiation equilibrium requires recombination of photogenerated carriers. As the carrier density decreases during recombination, the reduced recombination probability prolongs the recovery duration. The elevated free carrier concentration generated under strong irradiation extends the time required for the resistance to recover to its dark-state value, leading to illumination-dependent temporal evolution of the dark resistance recovery.

8.3.7 Noise

Noise in photodetectors consists primarily of external noise and internal noise. External noise originates from fluctuations in the output current or voltage induced

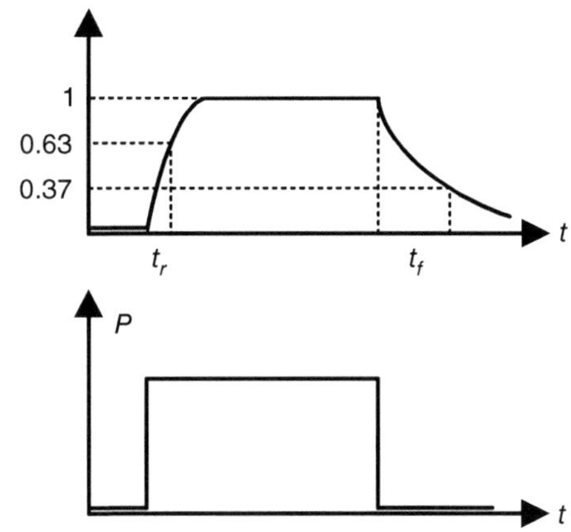

Figure 8.21 Photodetector responses upon strong light excitation.

by extraneous light outside the signal spectrum. Prevalent sources encompass industrial interference, radio-frequency disturbances, and cosmic/atmospheric phenomena. Internal noise comprises four principal types: $1/f$ noise (flicker noise), thermal noise, G–R noise, and shot noise. 1/f noise, dominant at low operational frequencies, arises from surface defects or material imperfections. Thermal noise stems from the random thermal motion of charge carriers and exhibits a dependence on temperature. G–R noise results from statistical variations in the generation and recombination lifetimes of photogenerated carriers. Shot noise, attributed to the inherent statistical fluctuation of discrete charge carriers traversing a junction, is frequency-independent within the low-to-mid frequency range but becomes frequency-dependent at higher frequencies.

Additional critical performance metrics for photodetectors include linearity, operational temperature range, and SNR.

8.4 Perovskite Photodetectors

A variety of materials are employed in photodetectors, including silicon, indium gallium arsenide (InGaAs), mercury cadmium telluride (HgCdTe), perovskite, and other semiconductor materials. These materials are integral to the photoelectric conversion process owing to their unique physical and chemical properties. Among them, perovskite semiconductor materials represent prominent candidates for radiation detection due to their high light absorption capacity, elevated carrier mobility, extended carrier lifetime, and straightforward fabrication process. The original perovskite was the compound $CaTiO_3$, composed of calcium, titanium, and oxygen, discovered by Gustav Rose in Russia in 1839 [4]. The crystal structure comprises octahedra formed by titanium and oxygen atoms, separated by calcium atoms. Subsequent investigations established that this structure remains stable upon substitution of the constituent elements; for instance, in cesium lead bromide, cesium replaces calcium, lead replaces titanium, and bromine replaces oxygen. Consequently, compounds exhibiting this crystal structure are termed "perovskites," with the structural formula denoted as ABX_3. Currently, perovskite is no longer exclusively defined as the mineral $CaTiO_3$ but encompasses a broad class of compounds. When the X site is occupied by halogen elements, the structure is categorized as a halide perovskite. If metal elements occupy the A or B sites, it is specifically termed a metal halide perovskite.

A perovskite photodetector is defined as a photodetector in which perovskite serves as the photoresponsive material. These devices constitute a novel category of detectors characterized by high efficiency, rapid response speed, superior sensitivity, low noise, low cost, and exceptional photoelectric conversion performance. Compared to traditional silicon-based counterparts, perovskite photodetectors exhibit broader application prospects, particularly in domains such as vis-light detection, UV-light detection, X-ray detection, gamma-ray detection, neutron detection, and related fields. Specifically, perovskite materials display low carrier recombination probabilities, high carrier mobility, extended carrier diffusion distances, and prolonged lifetimes, rendering them promising candidates for X-ray detection materials.

For perovskite detectors, three common device architectures are photoconductors, photodiodes, and phototransistors (Figure 8.22) [5]. Based on the direction of current flow, the photoelectric conversion component can be classified into vertical and lateral structures. Vertical configurations typically include vertical photoconductors and photodiodes, whereas lateral photoconductors and phototransistors belong to lateral structures. Vertical geometries are widely adopted due to their compatibility with pixelated matrices. However, for lateral structures, the electric field strength depends solely on electrode spacing and is independent of film thickness, conferring lateral structures an advantage of low driving voltage requirements.

Furthermore, phototransistors feature a three-terminal configuration comprising a gate, source, and drain. In this arrangement, the channel current between the source and drain can be modulated not only by gate polarization but also through charge generation upon interaction with radiation. Moreover, when the gate voltage is biased below the threshold, the charge-carrier density in the channel is reduced, and the resultant photocurrent originates from the separation and collection of charges ionized by high-energy radiation photons. This structure benefits from gate-mediated signal amplification, thereby enhancing the detected photocurrent. Nevertheless, research on perovskite phototransistors remains relatively scarce, potentially attributable to performance degradation caused by ion migration.

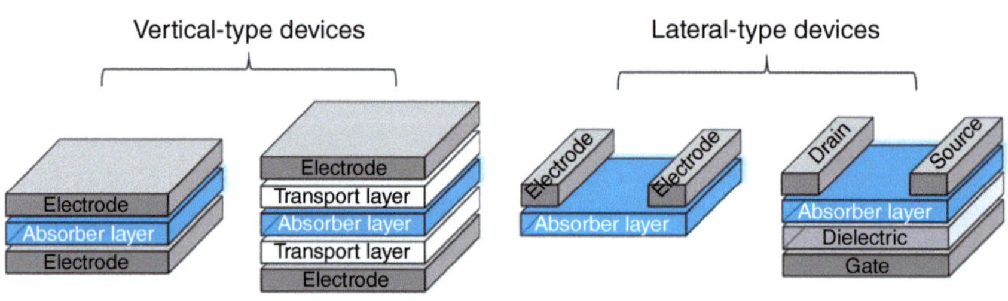

Figure 8.22 Device structure of photoconductors, photodiodes, and phototransistors based on perovskites.

Current research in perovskite photodetectors prioritizes three primary domains: (i) device architecture: structural design constitutes a critical research focus, such as investigations concentrating on bipolar junctions and PN junctions. Among these configurations, bipolar junctions demonstrate superior response speed and sensitivity. (ii) Material engineering: optimizing perovskite materials represents a decisive factor governing optoelectronic performance. Contemporary research emphasizes chemical composition refinement, surface modification, and epitaxial growth techniques to enhance material stability and photoelectric characteristics. (iii) Charge transport and doping: carrier dynamics and dopant integration profoundly influence device performance. Current investigations center on dopant selection, concentration optimization, and fundamental charge transport mechanisms.

Perovskite detectors can be categorized based on their detection wavelength, including IR, vis-light, UV, X-ray, and gamma-ray detection systems. The spectral response range of perovskite photodetectors is tunable through varying the halide composition (e.g., the Br/I ratio) or integrating quantum dots. Photodetectors based on perovskite material for detecting different wavelengths will be discussed in detail in the following subsections.

8.4.1 Infrared Detectors

Perovskite IR photodetectors utilize the photoconductive or photovoltaic effect to convert incident IR photons into electrical signals. Absorbed photons generate EHPs, which subsequently separate and are collected at electrodes, yielding photocurrent. Composed of materials such as hybrid organic–inorganic lead halides or formamidinium-based compositions, perovskites exhibit tunable bandgaps (0.5–1.7 eV) via compositional engineering. This enables efficient photon absorption spanning the near-IR (NIR, 700–1500 nm) to short-wave infrared (SWIR, 1.5–3 μm) spectral ranges. State-of-the-art devices demonstrate responsivities exceeding 10 A/W within the NIR regime. Hybrid systems incorporating graphene or colloidal quantum dots offer an extended spectral response into the mid-wave infrared (MWIR, 3–8 μm). Self-powered detectors represent a novel architecture exploiting the bulk photovoltaic effect to obviate external bias requirements.

Material engineering strategies, including the development of 2D/3D heterostructures (e.g., $(PEA)_2PbI_4$/3D perovskite), have been investigated to enhance stability and suppress ion migration [6]. Encapsulation techniques (e.g., atomic layer deposition [ALD]) and additive engineering (e.g., zwitterionic molecules) are employed to improve stability and prolong operational lifetimes under humidity/heat stress [7]. Despite persistent challenges in long-term operational stability and scalable fabrication, perovskite IR photodetectors demonstrate significant potential to supplant conventional technologies (e.g., InGaAs, HgCdTe) owing to their tunable optoelectronic properties, solution processability, and cost efficiency. Ongoing research focuses on extending spectral coverage into the long-wave infrared (LWIR, 8–15 μm) and integrating AI-driven readout circuits to enable advanced smart sensing systems [8].

Perovskite IR detectors can enable diverse novel applications. Examples include low-cost thermal imaging for surveillance, autonomous vehicles, and nondestructive testing; portable NIR analyzers for agricultural crop monitoring and pharmaceutical chemical identification; SWIR detectors relevant for dark current (I_{dark}) characterization and optical fiber communication; and noninvasive medical diagnostics such as glucose monitoring and blood oxygenation sensing. Furthermore, perovskite–quantum dot hybrid detectors facilitate environmental sensing, notably methane/CO_2 detection within the MWIR band.

8.4.2 Visible-Light Detectors

Visible-light photodetectors based on perovskites operate through the photoelectric effect to convert vis light (380–780 nm) into electrical signals. The active layer, typically composed of organic–inorganic hybrid perovskites such as $MAPbI_3$ and $CsPbBr_3$, absorbs photons to produce EHPs. Owing to the high carrier mobility (>10 $cm^2/V/s$) and extended diffusion lengths (1–10 μm) of these materials, these charge carriers undergo rapid separation, enabling efficient collection at the electrodes. Under the influence of a built-in or applied electric field, photogenerated electrons and holes are transported via drift or diffusion, resulting in a measurable photocurrent. This photoelectric conversion process transforms optical signals into electrical signals, thereby forming the fundamental operational principle of photodetectors.

Perovskite materials exclusively absorb photons possessing energies greater than or equal to the material's bandgap energy. Consequently, the spectral sensitivity limit (long-wavelength cutoff) of a photodetector is determined by the bandgap relationship.

$$\lambda_{max} = \frac{1240}{E_g} \tag{8.40}$$

where λ_{max} denotes the cutoff wavelength (in nanometers, nm), beyond which photon detection ceases, and E_g represents the optical bandgap (in electron volts, eV) of the semiconductor material. Incident photons with wavelengths $\lambda \leq \lambda_{max}$ provide sufficient energy to excite electrons across the bandgap from the valence band to the conduction band, generating charge carriers. The tunable bandgap (1.5–2.3 eV) of perovskites permits precise spectral selectivity across the vis spectrum.

Research advances in vis-light photodetectors based on perovskites have recently been achieved. Significant progress has been realized in hybrid photodetectors based on graphene/$CsPbBr_{3-x}I_x$ perovskite NCs (nanocrystals), which exhibit a high photoresponsivity of $\sim 10^8$ A/W and a detectivity of $\sim 10^{16}$ Jones at an irradiance of 0.07 $\mu W/cm^2$ under 405 nm illumination [9].

8.4.3 Ultraviolet-Light Detectors

UV photodetectors constitute critical components across diverse applications spanning environmental monitoring, flame detection, biomedical imaging, and space exploration. Traditional UV photodetectors, particularly those fabricated from silicon (Si) or gallium nitride (GaN), are constrained by limitations pertaining to cost, mechanical flexibility, and spectral selectivity. In recent years, perovskite materials have emerged as a revolutionary alternative, attributable to their exceptional optoelectronic properties, including tunable bandgaps, high absorption coefficients, and extended carrier diffusion lengths [10]. This section elucidates the fundamental principles governing perovskite-based UV photodetectors, with particular emphasis on their operational mechanisms, inherent material advantages, and prevalent device architectures.

Perovskite UV photodetectors primarily function via two distinct mechanisms: the photoconductive and photovoltaic effects. Within the photoconductive mode, often utilized in lateral device configurations, incident UV photons generate EHPs within the perovskite active layer. An applied external bias—or the material's inherent internal electric field—facilitates the separation and directional transport of these charge carriers. The extended carrier lifetimes characteristic of perovskites permit carriers to traverse the material multiple times before recombination, thereby inducing photoconductive gain. This photocurrent gain exhibits direct proportionality to both illumination intensity and the product of carrier lifetime (τ) and mobility (μ). Conversely, the photovoltaic mode governs vertically structured devices, exemplified by photodiodes. In this configuration, the built-in electric field generated at the interface between the perovskite and charge transport layers (e.g., perovskite/electron transport layer or perovskite/hole transport layer) spontaneously separates photogenerated carriers without requiring an external bias. The resultant separated carriers are subsequently collected at the respective electrodes, generating a detectable photovoltage or photocurrent. While this mode confers distinct advantages, notably low dark current noise and superior temporal response characteristics, its effectiveness is critically contingent upon precise energetic band alignment at the perovskite heterojunction interfaces.

Perovskite UV photodetectors are implemented in several distinct architectures tailored to specific performance goals: photoconductor-type devices utilize a simple lateral electrode–perovskite–electrode structure, exhibiting high responsivity (>10^3 A/W) due to incident power-proportional conductivity changes, but they suffer from slow response times caused by persistent photoconductivity. Photodiode-type configurations employ vertical structures incorporating electron and hole transport layers (e.g., TiO_2, Spiro-OMeTAD), which create an internal electric field for rapid carrier extraction, achieving fast response times (<1 ms) and high detectivity (>10^{12} Jones), although interfacial defects necessitate passivation to manage dark current. Finally, self-powered photodetectors leverage ferroelectric polarization or asymmetric contacts to operate without an external power source, enabling portable applications; this capability is exemplified by devices like those using $CH_3NH_3PbCl_3$, which generate an open-circuit voltage under UV illumination through spontaneous carrier migration [1].

8.4.4 X-Ray Detectors

X-rays are electromagnetic radiation characterized by extremely short wavelengths, high energy, and strong penetrating power. These properties facilitate

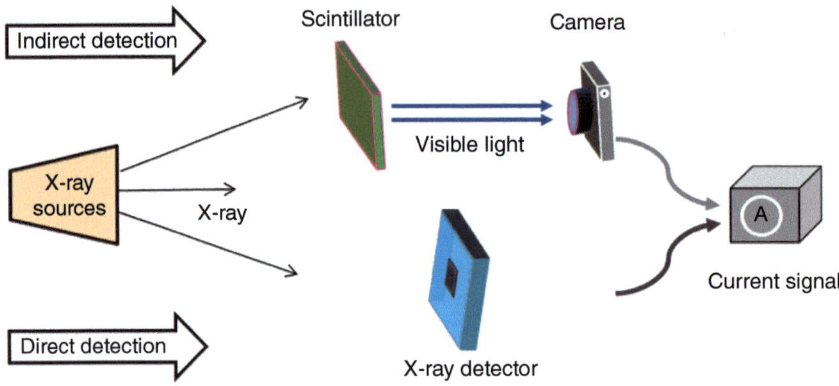

Figure 8.23 Schematic diagram of indirect and direct X-ray detection.

their widespread application in diverse fields, including security screening, medical diagnostics, crystallography, and industrial nondestructive inspection. Consequently, X-ray detection represents a critical technological approach within disciplines such as space science and nuclear science.

Detection of X-rays, particularly due to their inherent penetrability, presents greater challenges than their generation. Traditional detector materials, such as amorphous selenium (α-Se) and silicon (Si), are constrained by limitations in sensitivity, radiation tolerance, and cost. Perovskites, composed of high-atomic-number elements, exhibit a fundamental advantage in attenuating high-energy radiation. Furthermore, perovskite materials offer tunable electronic properties, high luminescence efficiency, and large carrier mobility–lifetime ($\mu\tau$) products. Crucially, bandgap and electronic dimensionality (3D/2D/1D/0D) can be engineered through compositional modulation [11].

Perovskite X-ray detectors are categorized based on their operating principle: direct detection and indirect detection (Figure 8.23). The classification is further refined by material chemistry into lead-based halides, bismuth-based halides, and copper-based halides. Key material examples, underlying physical mechanisms (e.g., photoconductive gain, scintillation), and representative applications are systematically reviewed.

8.4.4.1 Direct-Type X-Ray Detectors

Direct detection relies on perovskites absorbing X-ray photons and generating electrical signals directly, without intermediate energy conversion. The primary conversion mechanisms are the photoelectric effect and Compton scattering. The photoelectric effect dominates at lower X-ray energies (<100 keV), particularly in perovskites containing high-atomic-number (Z) elements (e.g., Pb, Bi, I), where enhanced photon absorption occurs through electron ejection. Compton scattering becomes significant at higher energies (>1 MeV), producing secondary electrons.

The carrier generation and collection processes involve the following steps: (i) incident X-ray photons ionize perovskite atoms, creating high-energy secondary electrons; (ii) these secondary electrons generate electron–hole (e–h) pairs via impact ionization, typically yielding approximately 1 e–h pair per 5 eV of deposited energy (the theoretical limit); and (iii) an external electric field (1–100 V/μm) separates the e–h pairs, generating a measurable photocurrent. This direct charge collection process efficiently converts incident radiation into an electrical signal.

The operational mode depends on the energy deposition dynamics. There are two primary operational modes: current mode and voltage mode. In current mode, the device measures the continuous photocurrent under an applied bias, making it suitable for high-intensity radiation detection. In voltage mode, the device measures the open-circuit voltage induced by carrier accumulation, which is ideal for low-dose detection.

The magnitude of the maximum photocurrent, I_{CC}, accounting for carrier generation and collection, is described by

$$I_{CC} = \Phi n q \tag{8.41}$$

where Φ is the photon absorption rate, n is the number of the generated EHPs per absorbed photon, and q is the elementary charge. The influencing factors are charge collection efficiency dependent on trap density (ideal: $<10^{10}/\text{cm}^3$) and the mobility-lifetime product (ideal: $\mu\tau > 10^{-4}\ \text{cm}^2/\text{V}$).

The key parameters used to characterize detector performance are as follows:

(a) **Absorption coefficient:** This parameter quantifies a material's ability to attenuate X-rays. The absorption capability depends on the effective atomic number (Z_{eff}) and density (ρ) of the material, where higher atomic numbers result in greater X-ray attenuation. The absorption coefficient (α) is calculated by

$$\alpha = \frac{\rho Z_{\text{eff}}^4}{AE^3} \tag{8.42}$$

where A is the atomic mass and E is the X-ray photon energy.

(b) **Spatial resolution:** This refers to the ability to distinguish adjacent details in an object. In digital imaging systems, resolution is primarily determined by pixel size—higher resolution yields better image quality. Standard measurement methods include slanted-edge, slit, and line-pair test patterns. The spatial resolution limit for visual discernment is characterized by the spatial frequency at which the modulation transfer function (MTF) equals 0.2.

(c) **Detection limit:** Defined by IUPAC as the dose rate producing an SNR of 3:1. Lower detection limits are essential for medical radiation imaging. This parameter depends on signal sensitivity and system noise; thus, precise measurement of small currents improves detector performance.

(d) **Dark current:** Critical for direct X-ray detection, this leakage current affects noise levels. Lower dark currents reduce electronic noise and improve SNR ratios. Detectors typically require dark current densities below $10^{-10}\ \text{A/cm}^2$.

(e) **Sensitivity:** Expressed in unit of μC·Gy^{-1}/cm^2, measuring charge collected per unit X-ray dose. It reflects a detector's efficiency in converting photons to collectable charge. High sensitivity enables better image contrast at lower doses, crucial for medical applications. Sensitivity (S) is calculated by

$$S = \frac{Q}{AD} \tag{8.43}$$

where Q is collected charge, A is effective detector area, and D is radiation exposure.

(f) **Response time:** Ranging from milliseconds (polycrystalline films) to nanoseconds (single crystals), this parameter is vital for dynamic imaging.

(g) **$\mu\tau$ product:** This determines charge-carrier collection efficiency in direct detectors. Higher $\mu\tau$ values correspond to greater collection efficiency.

(h) **Stability:** X-ray photodetectors require both radiation hardness and environmental stability for long-term, reliable operation. For instance, $CsPbBr_3$ demonstrates high radiation hardness, retaining 90% of its initial sensitivity after exposure to 10^6 Gy total dose, a significant improvement compared to α-Se detectors, which degrade significantly at 10^3 Gy [12]. Regarding environmental stability, encapsulated $CsPbBr_3$ devices maintain their performance for over 1000 hours under accelerated aging conditions of 85 °C and 85% relative humidity [13].

Some recent research advances of X-ray detectors based on perovskite materials are discussed as follows:

Lead-based Halide Perovskite X-ray Detectors Lead-based perovskites are front-runners in direct X-ray detection due to lead's high atomic number (Pb, $Z = 82$), which significantly enhances X-ray absorption through the strong photoelectric effect. Their direct bandgap values (1.5–2.3 eV) facilitate efficient charge-carrier generation. When high-energy X-rays (keV–MeV) interact with these materials, they ionize lead atoms, generating secondary electrons. These secondary electrons then collide with the crystal lattice, causing impact ionization, which produces e–h pairs (approximately 1 pair formed per 5 eV of deposited energy). An externally applied electric field (1–50 V/μm) drives these e–h pairs toward the electrodes, enabling charge collection and generating the measurable photocurrent.

Typical examples of these materials include $CsPbBr_3$, $MAPbI_3$ ($CH_3NH_3PbI_3$), and $FAPbBr_3$ ($HC(NH_2)_2PbBr_3$). $CsPbBr_3$ is a cubic-phase perovskite with a bandgap of approximately 2.3 eV. Its rigid inorganic framework contributes to a sensitivity exceeding 30,000 μC/Gy/cm^2, a detection limit below 1 nGy/s, and excellent radiation resistance (tolerating doses greater than 10^6 Gy). $MAPbI_3$ ($CH_3NH_3PbI_3$) is a hybrid perovskite with a bandgap of approximately 1.6 eV. It exhibits promising properties for low-dose dental imaging applications, including a high μτ product (~2×10^{-3} cm^2/V) and strong X-ray stopping power (attenuation coefficient ~10^4/cm at 50 keV). The highest-performing $MAPbI_3$ X-ray detectors demonstrate sensitivities reaching up to 5,200,000 μC/Gy/cm^2, detection limits as low as 0.1 nGy/s, and radiation stability up to 11 Gy. $FAPbBr_3$ ($HC(NH_2)_2PbBr_3$) is a formamidinium-based perovskite offering improved thermal stability and sensitivity around 25,000 μC/Gy/cm^2. Key advantages of these lead-based perovskite detectors are their ultrahigh sensitivity (10^4–10^5 μC/Gy/cm^2), arising from strong X-ray absorption, and fast response times (nanoseconds to microseconds), crucial for real-time imaging applications.

Bismuth-based Halide Perovskite X-ray Detectors Although the performance of three-dimensional lead-based halide X-ray detectors is exceptional, their poor stability caused by ion migration hinders further development. Studies indicate that reducing the crystal dimensions of metal halide perovskites to two-dimensional, one-dimensional, or zero-dimensional structures can significantly suppress ion migration and enhance stability. Consequently, research has focused on replacing lead with metals such as bismuth, silver, and copper through compositional and dimensional engineering.

Bismuth-based halides ($ABiX_n$, where A = Rb, Cs, MA, BA; X = Cl, Br, I) are promising low-toxicity alternatives to lead-based perovskites, leveraging Bi^{3+}'s $6s^2$ lone–pair electrons. This enables tunable bandgaps (1.8–3.0 eV) achieved through Bi^{3+} 6s-6p orbital hybridization and halide p-orbital contributions. They exhibit high stability, with strong Bi–X covalent bonds resisting moisture/oxygen degradation—$Cs_3Bi_2I_9$, for example, retains >90% efficiency after 1000 hours of aging. Additionally, they display defect tolerance, maintaining low trap density ($<10^{15}/cm^3$) due to Bi^{3+}'s lower electronegativity compared to Pb^{2+}, which reduces non-radiative recombination.

Recent material advances include 3D $Cs_3Bi_2I_9$, synthesized as solution-processed single crystals with <2% strain; it achieves a high X-ray sensitivity of 2.1×10^4 μC/Gy/cm^2, surpassing α-Se (0.3×10^4), though its wide bandgap (2.2 eV) limits vis-light harvesting. Another advance is 2D $(BA)_2BiI_7$, noted for breakthrough air stability (>6 months) and bandgap reduction to 1.9 eV via quantum confinement, enabling flexible photodetectors with 10^4 on/off ratios. The double perovskite $Cs_2AgBiBr_6$ is also notable, demonstrating defect suppression through Bi^{3+}/Ag^+ cation ordering that lowers Urbach energy to 15 meV and achieves 11.3% PCE in solar cells.

These Bi-based materials provide RoHS-compliant, eco-friendly alternatives to Pb-based counterparts but exhibit limitations including lower stopping power and carrier mobility. Key challenges for these materials comprise a low carrier mobility ($\sim 10^{-4}$ cm^2/V·s), a relatively wide bandgap (2.0–3.0 eV) that reduces X-ray absorption efficiency, and high defect density ($>10^{17}/cm^3$); these challenges necessitate passivation strategies such as phenylethylammonium iodide (PEAI). Critical optimization approaches encompass bandgap engineering (e.g., narrowing bandgaps to approximately 1.4 eV by alloying with Sb^{3+} in $Cs_3Bi_{1-X}Sb_XI_9$), dimensionality control (using 2D/3D heterostructures like $Cs_3Bi_2I_9/(PEA)_2BiBr_4$ to improve charge extraction), and defect passivation (employing thiocyanate /SCN^-/ ligands to reduce

Bi vacancies by up to 80% or employing ionic liquids like phenylethylammonium bromide /PEABr/ to suppress halide ion migration).

Copper-based Halide Perovskite X-ray Detectors Copper-based perovskite radiation detectors operate primarily through two mechanisms: defect-assisted absorption and photon recycling. During defect-assisted absorption, incident X-rays excite deep-level defects such as copper vacancies, generating localized e–h pairs. Simultaneously, photon recycling occurs as low-energy photons emitted from radiative recombination are reabsorbed by the material, enhancing the overall QE.

Key challenges associated with these materials include their relatively low stopping power for ionizing radiation, which necessitates the use of thick absorber films to ensure efficient charge collection; additionally, high intrinsic defect densities exceeding 10^{17} cm^{-3} demand effective interface passivation strategies, for instance, using PEAI.

Applications of copper-based halide perovskite X-ray detectors are leveraged by specific material advances. For example, $CsCu_2I_3$, a one-dimensional perovskite possessing a bandgap of approximately 3.2 eV, is used for UV–X-ray dual-band detection [14, 15].

8.4.4.2 Indirect X-Ray Detection

While direct detection operates by X-rays being absorbed directly within the perovskite material, converting photons into charge carriers (e.g., $CsPbBr_3$ detectors), indirect detection functions differently. In this approach, the perovskite materials function as scintillator materials; after being excited by X-rays, perovskite materials emit vis light, which is detected by cameras (Figure 8.23) [16].

Other than parameters similar to the direct X-ray detection, some more key parameters characterizing indirect detection performance are as follows:

(a) **Light yield:** Quantifying the efficiency with which scintillators convert absorbed high-energy radiation into low-energy photons. This parameter can be measured using an integrating sphere. The light yield can also be estimated according to the following equation:

$$\text{Light yield} = 1{,}000{,}000\frac{SQ}{\beta E_g} \tag{8.44}$$

where S is the transport efficiency of EHPs to emissive centers, Q is the luminescence QE, β is a dimensionless constant (typically ~2.5), and E_g is the energy bandgap of the scintillator material.

(b) **Radioluminescence wavelength:** Radioluminescence (RL) refers to the emission of vis or near-vis light from scintillators upon excitation by ionizing radiation. Alignment between the central wavelengths of RL and photoluminescence (PL) indicates emission originating from identical energy levels. RL intensity typically demonstrates a linear increase with rising dose rate. Crucially, the RL emission spectrum must correspond effectively with the absorption spectra of the photodetectors employed, such as CCDs, CMOS sensors, PMTs, or TFT arrays.

(c) **Afterglow:** The persistent RL intensity measured at a specific time following cessation of X-ray irradiation, primarily attributable to material defects (e.g., point defects, dislocations). Scintillators for high-speed imaging generally require the RL intensity at a defined interval post-excitation (e.g., 3 ms) to decay to less than 0.1% of its initial value. Protracted afterglow induces imaging artifacts due to signal superposition from preceding exposures.

Examples of indirect X-ray detection utilizing perovskite materials are selected and discussed as follows:

(a) **Lead-Based Halide Perovskites**
Lead-based halide perovskites, such as $CsPbBr_3$, are highly promising as X-ray scintillators owing to their superior radiation absorption and efficient light emission. The high atomic number of lead ($Z = 82$) ensures strong X-ray attenuation, while the tunable bandgap (~2.3 eV) allows for optimal vis-light emission that matches photodetector sensitivity. These materials exhibit fast decay times (<20 ns) and high light yields (>50,000 photons/MeV) [17], and they outperform conventional scintillators like LYSO (lutetium–yttrium oxyorthosilicate) in temporal resolution [18]. Furthermore, solution-processable synthesis enables cost-effective fabrication of flexible or pixelated scintillator screens. Recent advances in defect passivation further enhance radiation hardness, making these materials ideal for applications such as medical imaging and high-flux X-ray detection.

(b) **Copper-Based Halide Perovskites**
Copper-based halide perovskites, including $Cs_3Cu_2I_5$, are emerging as high-performance X-ray scintillators due to their unique advantages. With heavy copper ($Z = 29$) and iodine ($Z = 53$) atoms, they exhibit strong X-ray stopping power, comparable to lead-based materials. Their self-trapped exciton emission mechanism enables large Stokes shifts (~200 nm) and high light yields (>40,000 photons/MeV), effectively minimizing self-absorption losses [19]. These materials demonstrate decay times (<100 μs)—critical for high-frame-rate X-ray imaging—while maintaining excellent radiation stability under prolonged exposure [20]. Solution-processable synthesis allows low-cost fabrication of flexible scintillator screens, suitable for flexible detector configurations.

8.4.5 Gamma-Ray Detectors

Perovskite gamma-ray detectors are emerging as promising alternatives to conventional semiconductor-based detectors (e.g., CdTe, CZT) owing to their constituent high-atomic-number elements (e.g., Pb, I), tunable bandgap, and cost-effective solution processability. These detectors utilize the strong gamma-ray absorption properties of perovskites to convert high-energy photons into measurable electrical signals, enabling applications in medical imaging, nuclear safety monitoring, and astrophysics.

The working mechanism of metal halide perovskite detectors for gamma rays comprises two primary stages. First, gamma-ray photons interact with perovskite crystals predominantly through the photoelectric effect and Compton scattering.

The high-density lead (Pb) in perovskites enhances photoelectric absorption efficiency, generating EHPs proportional to the incident photon energy. Second, under an applied bias voltage, photogenerated carriers are separated and collected, generating a measurable current or voltage pulse.

Key performance parameters include response time and energy resolution (ER). Response time denotes the time interval between gamma-ray photon absorption and the establishment of a stable electrical output signal. The experimental approach to obtain response includes using pulsed radiation sources (e.g., δ-pulse) to excite the detector and analyzing the output waveform to extract temporal characteristics. Numerically, Monte Carlo (M.C.) modeling can be employed to simulate the detector's sensitive volume under pulsed radiation, coupled with equivalent circuit models (e.g., RLC circuits, i.e. resistor-inductor-capacitor circuits) to compute time-dependent current waveforms and response characteristics.

ER, a critical parameter quantifying a detector's ability to distinguish photons of similar energies, profoundly impacts applications such as nuclear spectroscopy, medical imaging, and astrophysics. It is defined as the full width at half maximum (FWHM) of a gamma-ray photopeak divided by the peak energy, conventionally expressed as a percentage:

$$\mathrm{ER} = \left(\frac{\mathrm{FWHM}}{E_\gamma}\right) \times 100\% \tag{8.45}$$

where E_γ represents the energy corresponding to the maximum of the photopeak channel in the energy spectrum.

Perovskite materials demonstrate exceptional promise for gamma-ray detection, particularly lead-based halides (e.g., $CsPbBr_3$) [21] and copper-based halides (e.g., $Cs_3Cu_2I_5$) [22]. Their composition of heavy elements (Pb, $Z = 82$; I, $Z = 53$) ensures strong gamma-ray attenuation, while their tunable bandgaps enable efficient radiation-to-charge conversion. Solution processability facilitates the cost-effective fabrication of large-area detectors, and their high defect tolerance contributes to radiation hardness. Recent advances in single-crystal growth have produced mobility-lifetime products exceeding 10^{-3} cm^2/V, comparable to traditional semiconductors such as cadmium zinc telluride (CZT) [23]. Therefore, these materials are highly suitable for nuclear medicine and homeland security applications, where high sensitivity and portability are critical requirements [24].

8.5 Metal–Semiconductor–Metal Photodetectors

8.5.1 Introduction

Metal–semiconductor–metal photodetectors (MSM-PDs), characterized by their simple structure, low fabrication complexity, and superior high-frequency response, represent a fundamental component of modern photonics, constituting a class of optoelectronic devices extensively employed for high-speed light detection in applications such as optical communications, imaging, and sensing. In current research and applications, MSM-PDs are widely utilized and investigated owing to

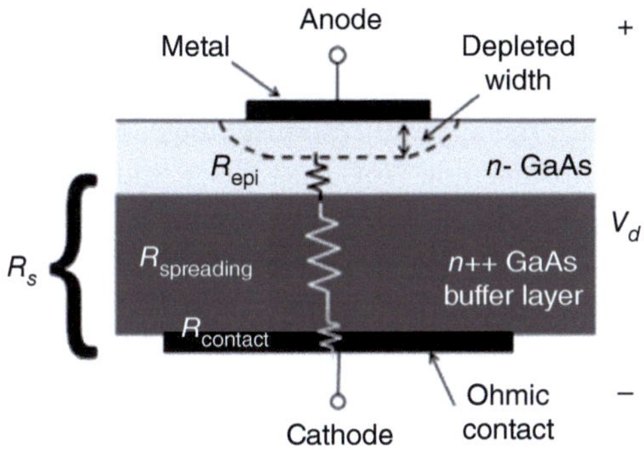

Figure 8.24 Schematic of a typical Schottky barrier diode.

their straightforward architecture, stable operation, and exemplary performance. These photodetectors operate primarily by forming junctions between metals and a doped semiconductor, with conduction mechanisms analogous to those in Schottky barrier photodetectors. Distinct from Schottky barrier photodiodes, MSM photodetectors feature two metal electrodes separated by a semiconductor material, which may be configured as one-dimensional, two-dimensional, three-dimensional structures, or a thin film. Among these contacts, both can be Schottky contact or one can be an Ohmic contact that permits the transfer of excess electrons from the metal to the semiconductor under specific conditions, resulting in modulated electrical and optical properties [25]. The schematic structure of a standard MSM configuration functioning as a Schottky barrier diode is depicted later (Figure 8.24). This section explores the operating principles, design methodologies, and technological advancements in MSM-PDs.

A common structure for an MSM-PD can be fabricated by depositing two interdigitated metallic electrodes (e.g., gold, aluminum, or titanium) on a semiconductor substrate (e.g., silicon or InGaAs). These electrodes form Schottky contacts with the semiconductor, establishing an array of alternating Schottky barriers [26]. The gaps between the electrodes define the active region where optical absorption occurs.

The photodetection mechanism can be conceptualized in several sequential steps. First, during photon absorption, incident photons with energy $h\nu > E_g$ (the semiconductor bandgap) generate EHPs within the substrate. Subsequently, under an electric field produced by an external bias voltage applied across the electrodes, carrier separation occurs, driving electrons and holes toward the anode and cathode, respectively. Finally, the resulting current generated by the collected carriers manifests as a photocurrent proportional to the incident optical power.

Key advantages of MSM photodetectors include high operational speed, inherently low dark current, and fabrication simplicity. Specifically, the interdigitated electrode geometry minimizes device capacitance, enabling bandwidths exceeding 100 GHz. Furthermore, Schottky barriers intrinsically suppress dark currents more effectively than PN junction photodetectors. Fabrication simplicity arises because doping or the growth of complex epitaxial layers is not required.

8.5.2 Design Considerations

The selection of semiconductor materials for MSM-PDs involves critical considerations. The bandgap (E_g) determines the spectral response range, such as silicon (Si) for vis light and indium gallium arsenide (InGaAs) for IR radiation. High carrier mobility (μ), exemplified by gallium arsenide (GaAs), reduces carrier transit time and enhances operational speed. A higher absorption coefficient (α) permits thinner active layers, thereby lowering capacitance. Electrode metal selection necessitates low-resistance materials like gold (Au) or titanium/gold (Ti/Au) to minimize series resistance; critically, the electrode metal influences the Schottky barrier height, which must align with the semiconductor to optimize the responsivity–dark current trade-off.

Electrode geometry optimization requires balancing finger width (W) and spacing (S). Narrower spacing increases electric field strength ($E = V/S$), enhancing carrier drift velocity, though excessively small spacing compromises light absorption. Wider finger width reduces electrode resistance but obstructs incident photons. Larger detector areas improve responsivity but elevate capacitance, ultimately limiting bandwidth.

Bias voltage significantly influences device performance. In photovoltaic mode, low-voltage bias minimizes dark current but constrains speed due to weak electric fields. Conversely, photoconductive mode employs high bias to intensify electric fields, reducing carrier transit time and maximizing bandwidth.

8.5.3 Advancements and Applications

The selection key performance metrics of MSM photodetectors include responsivity (R), bandwidth (f_{3dB}), and dark current (I_{dark}). Bandwidth is constrained by the RC time constant and carrier transit time. Dark current (I_{dark}), governed by thermionic emission over Schottky barriers, degrades the SNR when elevated; lower I_{dark} values are thus advantageous.

Technological advancements in MSM-PDs have progressed through distinct phases. Early developments (1980s–1990s) established MSM-PDs as alternatives to PIN photodiodes, utilizing GaAs and InP substrates for fiber-optic communications. During the 2000s, integration with complementary metal–oxide–semiconductor (CMOS) electronics enabled cost-effective mass production for consumer devices. Contemporary innovations leverage wide-bandgap semiconductors like gallium nitride (GaN) and silicon carbide (SiC) to extend UV detection for flame sensing and environmental monitoring. Nanostructured electrodes—including plasmonic architectures and graphene—enhance light absorption while mitigating shadowing effects. Flexible MSM-PDs incorporating organic semiconductors (e.g., poly(3-hexylthiophene):[6,6]-phenyl-C61-butyric acid methyl ester, P3HT:PCBM) facilitate bendable detectors for wearable electronics.

MSM-PDs have transitioned from specialized laboratory devices to ubiquitous elements in modern optoelectronics. Their design flexibility, compatibility with semiconductor processing, and scalability ensure enduring relevance in emerging technologies such as 6G networks, quantum computing, and the Internet of Things. Future developments will prioritize overcoming material constraints, improving QE, and enabling multifunctional operation.

8.6 Organic Photomultiplication Detectors

8.6.1 Introduction

PMTs are photodetectors characterized by exceptionally high sensitivity and ultrafast temporal response, rendering them indispensable in low-light-level detection applications. Capable of detecting extremely weak radiation, including single photons, across UV, vis, and NIR spectral regimes, these devices exhibit remarkably low noise, large photosensitive areas, and relatively low costs. Such attributes facilitate their widespread deployment in diverse fields, including nondestructive testing, quantum imaging, biophotonics, medical imaging (encompassing positron emission tomography and fluorescence lifetime imaging), radiation detection, high-energy physics, nuclear medicine, DNA sequencing, dark current characterization, and laser-induced fluorescence spectroscopy.

Traditional inorganic PMTs operate through photoelectron emission. Upon light irradiation, the photocathode emits photoelectrons. These electrons are accelerated toward the first dynode under an applied electric field, inducing secondary electron emission. The resulting increased electron flux undergoes successive acceleration toward subsequent dynodes. Via this cascade multiplication process across multiple dynodes, the initial electron population is amplified by factors of 10^4–10^8. Consequently, electrons accumulated at the anode generate a substantially amplified signal. This mechanism endows PMTs with significantly higher sensitivity than conventional photodetectors, enabling the detection of faint optical signals.

In contrast to PMTs, organic photomultiplication detectors (OPMDs) represent a significant advancement in optoelectronics. Unlike traditional inorganic detectors, OPMDs leverage the unique properties of organic semiconductors, including tunable bandgaps, solution processability, and intrinsic flexibility. Crucially, OPMDs achieve high internal gain through trap-assisted tunneling injection, wherein photogenerated carriers are captured by deliberately engineered trap states. This process induces the quantum tunneling injection of opposite carriers from the electrodes. By utilizing these organic semiconductors, OPMDs achieve high internal gain while offering advantages such as mechanical flexibility and cost-effective fabrication. Consequently, these devices have progressed from laboratory prototypes to commercially viable high-sensitivity, low-cost alternatives to conventional photodetectors. Their distinctive combination of flexibility, spectral tunability, and manufacturability via solution processing positions them at the forefront of next-generation optoelectronic technologies.

8.6.2 Design Considerations

OPMDs employ organic semiconductors, predominantly π-conjugated polymers or small molecules, to absorb incident photons and generate excitons, which are defined as bound EHPs. A defining characteristic of OPMDs is their remarkably high internal gain, typically exceeding 1000 ($G > 1000$). The operational mechanism of these devices encompasses several critical stages. Initially, photon absorption occurs when incident light of energy $h\nu$ greater than the semiconductor bandgap (E_g) leads to exciton generation. Subsequently, exciton migration transports these entities toward interfaces between donors and acceptors, such as those present in bulk heterojunction (BHJ) structures. At these interfaces, exciton dissociation takes place, resulting in the liberation of free electrons and holes. Additionally, trap-assisted multiplication contributes significantly to photocurrent amplification; defects or traps within organic layers facilitate iterative carrier injection and multiplication. Furthermore, bias-induced electrochemical doping substantially enhances electrical conductivity and sustains the gain under illuminated conditions.

OPMDs exhibit key advantages over inorganic PMTs, including mechanical flexibility through compatibility with bendable polymeric substrates such as polyethylene terephthalate (PET) or polydimethylsiloxane (PDMS). A significant operational benefit is their low-voltage requirement, functioning effectively below 10 V, in stark contrast to the kV-level biases necessary for PMTs. Additionally, they offer superior spectral tunability, as their optical bandgaps can be readily adjusted via molecular engineering, enabling coverage across the vis to NIR regions.

Material selection for OPMDs encompasses deliberate designs for the active layer and charge transport layers. For the active layer, donor–acceptor blend systems are essential; examples include P3HT:PCBM for vis-light detection and diketopyrrolopyrrole (DPP)-based polymers for NIR absorption. It is crucial that this layer possesses a low trap density to minimize charge-carrier recombination and associated noise. Regarding charge transport layers, common electron transport layers (ETLs) include materials like zinc oxide (ZnO) or titanium dioxide (TiO_2), while typical hole transport layers (HTLs) consist of poly(3,4-ethylenedioxythiophene):polystyrene sulfonate (PEDOT:PSS) or molybdenum trioxide (MoO_3).

Device architecture design significantly impacts performance. The BHJ configuration utilizes intimately blended donor and acceptor phases to maximize the interfacial area for efficient exciton dissociation. Alternatively, the planar heterojunction architecture employs sequentially deposited layers to achieve more controlled charge transport pathways.

Electrode design considerations involve selecting transparent conductors such as indium tin oxide (ITO) or silver nanowires to facilitate efficient top or bottom illumination. Implementing interdigitated electrode structures can enhance carrier collection efficiency by minimizing the distance carriers must travel before being collected.

Optimization of the applied bias is critical for balancing gain and device stability. Operating at a relatively low bias places the device in the photoconductive mode, providing moderate gain. Conversely, a high bias regime induces significant electrochemical doping within the organic layers. While this enables the high gain associated with the dominant trap-assisted multiplication mechanism, it simultaneously increases the risk of device breakdown due to excessive current flow.

8.6.3 Advancements and Applications

During the early developments phase (2000s–2010s), pioneering research reported that initial OPMDs demonstrated modest performance gains ($G \sim 100$). However, these early devices suffered from instability and high dark current. Modern innovations (2020s) focus on high-gain architectures. For example, tandem structures employ vertically stacked active layers to achieve multiplicative gain, while nanostructured electrodes utilize plasmonic nanoparticles to enhance light trapping.

Numerous challenges still persist, leading to dedicated solutions. Dark current suppression utilizes wide-bandgap charge-blocking layers, such as C60 derivatives. To increase response speed, strategies focus on optimizing carrier mobility; this can be achieved through aligned polymer chains or specific additive engineering. Stability enhancements have been achieved through barrier layers, such as those grown by ALD, including Al_2O_3 encapsulation, which mitigates degradation caused by oxygen and water ingress. Furthermore, cross-linked polymers improve thermal and mechanical resilience. Hybrid systems leverage organic–inorganic perovskites, combining the high-gain characteristic of organic materials with the superior carrier mobility of perovskites.

OPMDs find applications in diverse fields, notably biomedical imaging. Examples include their use in fluorescence-guided surgery for real-time detection of tumor-targeted fluorophores. Flexible OPMDs facilitate wearable health monitors for applications such as pulse oximetry. In low-light photography, they are poised to replace silicon photodiodes in smartphone cameras, enhancing night-mode capabilities. Furthermore, ultralow-noise OPMD architectures enable single-photon detection for quantum sensing applications. Future research directions entail on-chip integration, which may merge OPMDs with organic LEDs (OLEDs) to create compact imaging systems. Neuromorphic photonics aims to mimic synaptic plasticity for bio-inspired optical computing. Developing biodegradable eco-friendly semiconductor materials is crucial for sustainable electronics. Overall, future development aims to address performance disparities relative to inorganic counterparts, enhance operational stability, and unlock novel applications across sectors including healthcare, robotics, and environmental monitoring.

References

1 Cai, W., Li, H., Li, M. et al. (2021). *J. Phys. D. Appl. Phys.* 54: 293002.
2 Qiu, Q. and Huang, Z. (2021). *Adv. Mater.* 33: 2008126.
3 Guo, L., Chen, P., Li, L. et al. (2020). *Vacuum Electronics.* 4: 1–13.
4 Quan, L., Arquer, F.P.G., Sabatini, R.P. et al. (2018). *Adv. Mater.* 30: 1801996.
5 Wang, B., Yang, X., Chen, S. et al. (2022). *iScience* 25: 105593.
6 He, Y., Pan, W., Guo, C. et al. (2021). *Adv. Funct. Mater.* 31: 2104880.
7 Li, C., Zhou, S., Nie, J. et al. (2021). *Nano Lett.* 21: 10279–10283.
8 Lu, F.Y., Liang, Y.Y., Wang, N.N. et al. (2024). *Adv. Photonics* 6: 054001.
9 Kwak, D.-H., Lim, D.-H., Ra, H.-S. et al. (2016). *RSC Adv.* 6: 65252–65256.
10 Wang, B., Liang, D., Qaid, S.M.H. et al. (2024). *Appl. Phys. Lett.* 124: 243302.
11 Wang, B., Yang, X., Li, R. et al. (2023). *ACS Energy Lett.* 8: 4406–4413.
12 Gao, L., Sun, J.-L., Li, Q. et al. (2022). *ACS Appl. Mater. Interfaces* 14: 37904–37915.

13 Zhang, Y., Jiang, Z., Li, J. et al. (2024). *Coatings* 14(7): 918. https://doi.org/10.3390/coatings14070918.

14 Zhang, Z.-H., Yan, S.-S., Chen, Y.-L. et al. (2024). *ACS Appl. Mater. Interfaces* 16: 10398–10406.

15 Zhang, M.Y., Zhu, J.S., Yang, B. et al. (2021). *Nano Lett.* 21: 1392–1399.

16 Wang, B., Peng, J., Yang, X. et al. (2022). *Laser Photonics Rev.* 16: 2100736.

17 Wu, H.D., Ge, Y.S., Niu, G.D. et al. (2021). *Matter* 4: 144–163.

18 Nikl, M. and Yoshikawa, A. (2015). *Adv. Opt. Mater.* 3: 463–481.

19 Lian, L.Y., Zheng, M.Y., Zhang, W.Z. et al. (2020). *Adv. Sci.* 7: 2000195.

20 Ma, Z.Z., Ji, X.Z., Lin, S.L. et al. (2023). *Adv. Mater.* 35: 2300731.

21 He, Y.H., Matei, L., Jung, H.J. et al. (2018). *Nat. Commun.* 9: 1609.

22 Cheng, S., Beitlerova, A., Kucerkova, R. et al. (2020). *Phys. Status Solidi (RRL)* 14: 2000374.

23 Liu, F.Z., Wu, R., Wei, J. et al. (2022). *ACS Energy Lett.* 7: 1066–1085.

24 Maddalena, F., Tjahjana, L., Xie, A. et al. (2019). *Crystals* 9: 88.

25 Bardeen, J. (1947). *Phys. Rev.* 71: 717–727.

26 Newman, T., Bishop, W.L., Ng, K.T. et al. (1991). *IEEE Trans. Microw. Theory Tech.* 39: 1964–1971.

9

Graphene and Graphene-Based Devices

9.1 Introduction

The 2010 Nobel Prize in Physics was awarded to Dr. Andre Geim and Konstantin Novoselov for their "groundbreaking experiments concerning the two-dimensional material graphene." While graphite has been utilized by humankind for millennia, its fundamental building block, the single-layer carbon atom crystal later designated as graphene, was experimentally isolated relatively recently. Subsequently, it was discovered that isolating high-quality graphene is remarkably straightforward. Consequently, research and applications related to graphene expanded dramatically following the Nobel Prize recognition.

As an allotrope of carbon distinct from graphite, graphene exhibits outstanding electrical, optical, mechanical, thermal, magnetic, chemical, and biological properties. The relationship between its structure and these properties will be detailed in Section 9.2. Section 9.3 compares methodologies for fabricating high-quality single-layer graphene, including mechanical exfoliation, epitaxial growth, chemical vapor deposition (CVD), the unzipping of single-walled carbon nanotubes, and carbon dioxide reduction, among others. Benefitting from graphene's extraordinary properties, it has been integrated into diverse electrical and optical devices. These include high-mobility field-effect transistors (FETs), high-speed radio-frequency circuits, highly sensitive gas sensors, photodetectors, nonvolatile memories, solar cells, light-emitting diodes (LEDs), broadband modulators, optical antennas, and plasmonic components. The working mechanisms, device architectures, characterization techniques, and advantages of these graphene-based devices are explored in Section 9.4. Finally, Section 9.5 envisions the future potential of graphene-based devices and outlines current research challenges associated with their industrial implementation.

9.2 Structure and Properties of Graphene

9.2.1 Crystal Structure of Graphene

A ball and stick model of the single-layer graphene is shown in Figure 9.1a, it shows that the carbon atoms are arranged in a hexagonal pattern and the graphene

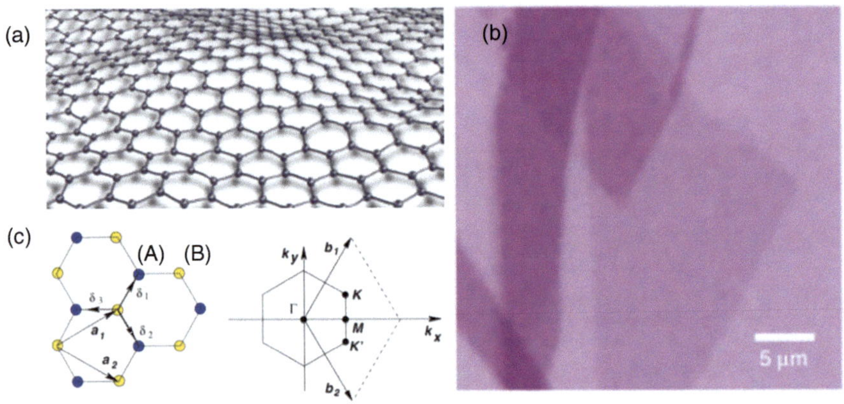

Figure 9.1 (a) Ball and stick model of the single-layer graphene layer. (b) Optical image of single-layer graphene sitting on SiO_2/Si substrate. Adapted with permission from Panchal et al. [1]. (c) Honeycomb lattice and the reciprocal first Brillouin zone. (A) Lattice structure of graphene, which is made out of two interpenetrating triangular lattices, $\boldsymbol{a}_1$ and $\boldsymbol{a}_2$ are the lattice unit vectors, and δ_i, $i = 1,2,3$ are the nearest-neighbor vectors. (B) The corresponding Brillouin zone. The Dirac cones are located at the K and K' points. *Source:* [2] / with permission of American Physical Society.

exhibits obvious ripple features, which is frequently observed in free-standing single-layer graphene [1, 2]. Monte Carlo simulations show that the ripples appear spontaneously due to the thermal fluctuations [3]; therefore, flat single-layer graphene without ripples could only be obtained by sitting on a substrate [1], such as on a SiO_2/Si substrate, as shown in Figure 9.1b, where the most transparent region represents the single-layer graphene, and other dark region means bilayer or triple-layer graphene. The triangular lattice unit cell of single-layer graphene is shown in Figure 9.1c, where A and B are two inequivalent carbon atoms [2]. The lattice vectors are then written as

$$\boldsymbol{a}_1 = \frac{a}{2}(3, \sqrt{3}) \tag{9.1}$$

$$\boldsymbol{a}_2 = \frac{a}{2}(3, -\sqrt{3}) \tag{9.2}$$

and the reciprocal lattice vectors are given by

$$\boldsymbol{b}_1 = \frac{2\pi}{3a}(1, \sqrt{3}) \tag{9.3}$$

$$\boldsymbol{b}_2 = \frac{2\pi}{3a}(1, -\sqrt{3}) \tag{9.4}$$

where a is the C—C bond length. Points K and K' in the Brillouin zone (BZ) are termed Dirac points, whose nomenclature origin will be discussed subsequently. The three nearest-neighbor vectors in real space δ_i ($i = 1$–3) are

$$\delta_1 = \frac{a}{2}(1, \sqrt{3}) \tag{9.5}$$

$$\delta_2 = \frac{a}{2}(1, -\sqrt{3}) \tag{9.6}$$

$$\delta_3 = -a(1, 0) \tag{9.7}$$

9.2.2 Extraordinary Properties of Graphene

9.2.2.1 Electronic Properties

A tight-binding model incorporating three nearest-neighbor and six second-nearest-neighbor interactions was employed to derive the fundamental electronic band structure of single-layer graphene, as illustrated in Figure 9.2. The plot reveals the π^* band (upper branch) and the π band (lower branch), which intersect at the Dirac points (designated as the K and K' points within the first Brillouin zone). Enlargement of the band structure near these Dirac points distinctly demonstrates the linear dependence of energy E_k on momentum components k_x and k_y, which could be approximated as

$$E_\pm(\boldsymbol{q}) \approx \pm v_\text{F}|\boldsymbol{q}| + O\left[\left(\frac{q}{k}\right)^2\right] \tag{9.8}$$

where $\boldsymbol{q}$ denotes the momentum measured relative to the Dirac points, and v_F represents the Fermi velocity ($\sim 1 \times 10^6$ m/s), approximately $c/300$ (c being the speed of light). The most significant distinction between this behavior and that of conventional materials lies in the energy–momentum relationship. Ordinarily, the velocity follows $v = p/m = \sqrt{2E/m}$, exhibiting a square root dependence on energy. In contrast, the Fermi velocity in single-layer graphene remains independent of energy or momentum, a characteristic shared by massless particles such as photons. Consequently, the system's behavior is described by the Dirac equation for massless particles, justifying the nomenclature "Dirac points" for the K and K' points.

Nevertheless, these massless electrons can acquire mass if atoms A and B become inequivalent due to an external potential. Researchers have demonstrated that by interfacing graphene with hexagonal boron nitride (h-BN), a 30-meV bandgap

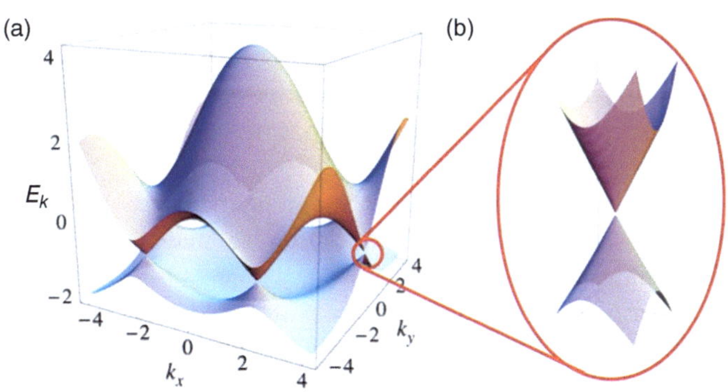

Figure 9.2 Electronic dispersion of graphene. (a) Energy spectrum (in units of t) for finite values of t and t', with $t = 2.7$ eV and $t' = -0.2\,t$. (b) Zoom in on the energy bands close to one of the Dirac points. *Source:* [2] / with permission of American Physical Society.

opens because atom A experiences a stronger external potential compared to atom B, imparting mass to the electrons [4]. Furthermore, the acquired electron mass in graphene can assume positive or negative values, contingent upon which sublattice atom (A or B) experiences the higher energy; a net positive mass arises when the energy shift on atom A exceeds that on atom B, and vice versa. Precise control of this mass characteristic could yield functional devices; for instance, enclosing a massless region within a massive region could create quantum dots (QDs) or quantum wires capable of carrying current without backscattering.

Single-layer graphene exhibits remarkably high electron mobility, reaching values of approximately 15,000 $cm^2/V/s$ at room temperature. Hole mobility achieves comparably high magnitudes, resulting in ambipolar charge transport. Remarkably, the mobility remains largely temperature-independent between 10 and 100 K, with only a marginal decrease observable at room temperature. This temperature independence signifies that carrier scattering is predominantly defect-mediated. Computational studies indicate that electrons in pristine, free-standing graphene are primarily scattered by acoustic phonons, potentially achieving mobilities near 200,000 $cm^2/V/s$ at carrier densities of $10^{12}/cm^2$. However, substrate interaction substantially reduces mobility; for example, placement on a SiO_2/Si substrate lowers electron mobility to approximately 40,000 $cm^2/V/s$ due to enhanced scattering from substrate optical phonons. Furthermore, adsorbed species such as oxygen or water molecules introduce nonreproducible and pronounced hysteresis in current–voltage (I–V) characteristics. To mitigate contamination effects, electrical measurements are typically conducted under high vacuum conditions. Due to this high mobility, graphene possesses an exceptionally low resistivity on the order of $10–^{6}$ Ω cm, surpassing even silver.

At this stage, it is pertinent to emphasize that infinitely large graphene constitutes a zero-gap semiconductor, as its conduction band and valence band converge at the Dirac points. However, when the in-plane dimensions of graphene are confined to the nanoscale (e.g., 5 nm), the crystal symmetry is disrupted. Consequently, the electronic band structure undergoes complex modifications. Research indicates that the electronic properties of graphene nanoribbons (GNRs)—single-layer graphene strips under 10 nm wide—are critically influenced by edge morphology. Experimental and computational studies demonstrate that nanoribbons with "zigzag" edges (Figure 9.3a) exhibit a zero bandgap. In contrast, "armchair"-edged configurations (Figure 9.3b) display a width-dependent nonzero bandgap, wherein reduced width correlates with increased bandgap magnitude.

Another notable electronic characteristic of single-layer graphene is the quantum Hall effect, which manifests Landau level formation under magnetic fields. The Hall conductivity σ_{xy} quantizes as $\sigma_{xy} = \pm 4N\, e^2/h$, where N is an integer, e denotes the elementary charge, and h is Planck's constant. Unlike conventional systems, graphene exhibits a half-integer–shifted quantum Hall effect expressed as $\sigma_{xy} = \pm 4(N+1/2)\, e^2/h$, where N represents the Landau level index and the prefactor 4 arises from valley and spin degeneracies. In Shubnikov–de Haas measurements, conventional metals exhibit peak longitudinal conductivity at integer Landau level filling factors. Graphene, conversely, displays minima—a

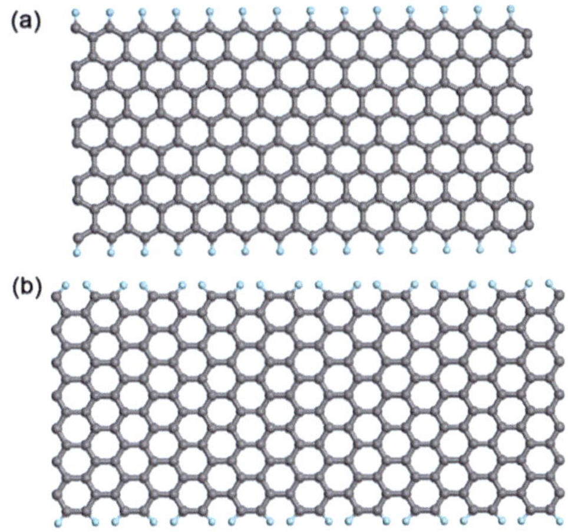

Figure 9.3 (a) Schematic of the "zigzag" and (b) "armchair" graphene nanoribbon. *Source:* [5] / with permission of IOP Publishing.

phenomenon attributed to Berry's phase. This phase results from the chirality (momentum-dependent pseudospin locking) of low-energy electrons near Dirac points. Furthermore, the fractional quantum Hall effect has been observed in single-layer graphene due to emergent fractional-charge quasiparticles in strongly correlated regimes. At low carrier densities under high magnetic fields, graphene transitions into a magnetic-field–tunable insulating state [6].

9.2.2.2 Optical Properties

The optical response of materials is generally dependent on the incident light intensity, particularly for two-dimensional materials characterized by atomic thickness. At low light intensities, single-layer graphene exhibits an ultrahigh absorption capacity of approximately $\pi\alpha \approx 2.3\%$ across the infrared to visible-light spectrum, with a thickness of only 3.35 Å, where α denotes the fine structure constant.

Graphene absorption undergoes saturation when the input optical intensity is sufficiently high. This saturation occurs readily over a broadband frequency range, attributable to graphene's universal optical absorption and zero bandgap. The saturation characteristic proves beneficial for applications such as mode-locking in fiber lasers, enabling full-band mode-locking through graphene-based saturable absorbers. Saturation effects have been observed in the frequency range extending from microwave to terahertz bands, thereby facilitating the utilization of graphene in photonic devices for microwave and terahertz applications.

Further increasing the light intensity to very high levels results in the dominance of the nonlinear Kerr effect. Experimental measurements demonstrate that graphene possesses a substantial nonlinear Kerr coefficient on the order of 10^{-7} cm^2/W, which is approximately nine orders of magnitude larger than that of conventional bulk dielectrics. This property establishes graphene as a promising ultrathin and efficient Kerr medium.

9.2.2.3 Mechanical Properties

The impressive mechanical properties of single-layer graphene were experimentally measured by Hone and coworkers, as illustrated in Figure 9.4 [7, 8]. A free-standing

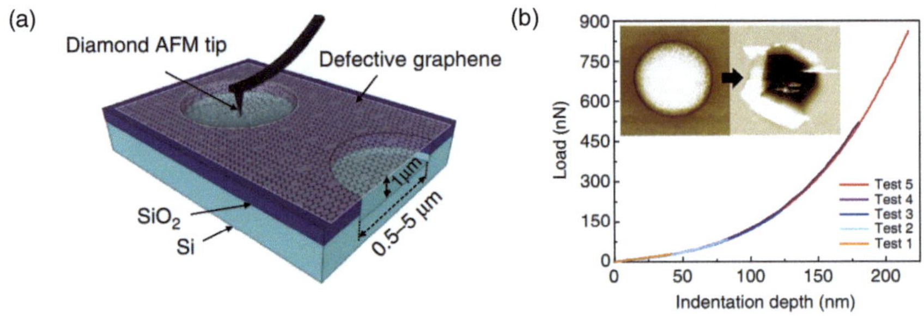

Figure 9.4 (a) Schematic representation of atomic force microscopy (AFM) nanoindentation test on suspended graphene sheets. (b) Typical force versus displacement curves of AFM nanoindentation test. Tests are repeated at increasing indentation depths until the sample breaks. Adapted with permission from Zandiatashbar et al. [8].

single-layer sample was placed over a hole etched into a SiO_2/Si substrate; an atomic force microscopy (AFM) tip was then used to measure the evolution of loading force with respect to indentation depth (Figure 9.4a), with the loading and unloading force curves plotted in Figure 9.4b. It was found that the relationship between force and indentation depth is independent of the AFM tip radius. However, the breaking force scales with tip size and is unaffected by sample size. Initially, the equation $\sigma_m^{2D} = \left(\frac{FE^{2D}}{4\pi R}\right)^{1/2}$ was employed to calculate the maximum stress, where E^{2D} is the second-order elastic stiffness, R is the tip radius, and F is the applied force. Subsequent numerical simulations incorporating the third-order elastic constants revealed that this formula is inaccurate. The corrected Young's modulus was then calculated as $E = 1.0 \pm 0.1$ TPa for a graphene thickness of 0.335 nm, corresponding to a strength of 130 GPa. This extraordinary stiffness and strength make graphene "the strongest material ever measured."

High stiffness and strength could only be achieved in pristine, defect-free single-crystal graphene. However, the toughness of polycrystalline graphene is 20–35% higher than that of its pristine counterpart. This enhanced toughness occurs because grain boundaries near a crack tip help dissipate stress, enabling deformation over a larger area. Additionally, grain boundaries can induce crack branching and more complex crack pathways, further aiding energy dissipation. It has been experimentally demonstrated that graphene-based springs can be elongated to 240% of their initial length without fracturing, indicating that super-tough macroscopic graphene objects are attainable through careful engineering of microscopic structures.

9.2.2.4 Thermal Properties

As illustrated in Figure 9.5a, suspended single-layer graphene was characterized using laser-induced thermal dissipation. A focused laser light served as the heat source, with measurements conducted on the graphite layers positioned at the periphery of the suspended graphene sample. Positional changes of the G peak, depicted in Figure 9.5b, provided the basis for calculating the thermal conductivity

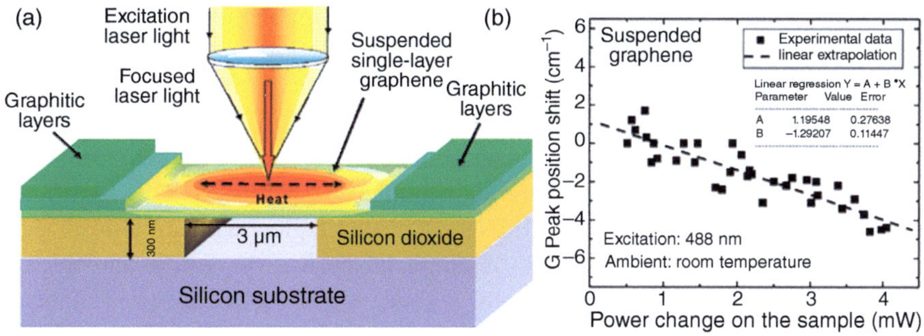

Figure 9.5 (a) Schematic diagram of the thermal conductivity measurement apparatus employing laser excitation. (b) Evolution of the G peak position as a function of dissipated power. *Source:* [9] / with permission of American Chemical Society.

of graphene. Initial experimental investigations revealed anomalously elevated thermal conductivity values approaching 5300 W/m/K [9]. This contrasts sharply with pyrolytic graphite, which exhibits a significantly lower thermal conductivity of approximately 2000 W/m/K. However, subsequent measurements of single-layer graphene specimens failed to consistently reproduce such high values, yielding conductivities within the range of 1500–2500 W/m/K. This disparity may be attributed to substantial measurement uncertainties as well as variations in material quality and processing conditions.

Substrate interaction exerts a profound influence on thermal conductivity. For single-layer graphene deposited on amorphous substrates, reported thermal conductivities decrease markedly to 500–600 W/m/K. Furthermore, in the case of CVD-grown graphene, the persistence of polymer residues left by the transfer process significantly degrades thermal conductivity. Additionally, the presence of carbon isotopes, specifically the C^{12}/C^{13} ratio, substantially impacts this property; even minimal incorporation of C^{13} into the graphene lattice acts to suppress thermal conductivity.

9.2.2.5 Chemical and Biological Properties

Compared to its bulk counterpart graphite, single-layer graphene features all carbon atoms exposed to the external reaction environment, yielding an extremely large specific surface area of 2630 m^2/g. This specific surface area substantially exceeds that of carbon black (approximately 900 m^2/g) or carbon nanotubes (ranging from 100 to 1000 m^2/g) and is comparable to that of activated carbon. It should be emphasized that carbon atoms at the edge sites exhibit higher reactivity than those within the basal plane. Owing to the complete exposure of surface atoms in single-layer graphene, its oxidation onset temperature is as low as 350 °C; moreover, it can be readily functionalized by oxygen- or nitrogen-containing chemicals. Consequently, infrared and X-ray photoelectron spectroscopy have been frequently employed to investigate the diverse functional groups in graphene layers. While it is anticipated that single-layer graphene possesses greater reactivity than multilayer graphene, a reactivity difference up to 100-fold was previously unanticipated.

In biological applications, the biocompatibility of graphene-based materials remains incompletely understood, despite extensive cellular studies, with contradictory results frequently reported. Research suggests that graphene has potential applications in neural cells, although its utility in cultured neural cell contexts is limited. Other investigations indicate that graphene may serve as a cancer biosensor or a DNA damage biomarker; however, substantial further research is required to explore potential applications in biological technologies. Critical unresolved questions concern the long-term fate of graphene derivatives within biological systems and their influence on intricate cellular signaling pathways. Notably, multiple studies demonstrate that lateral dimensions, layer number, and surface functionalization critically determine cytotoxic responses, ranging from negligible effects on some cell types to significant perturbations—including membrane damage, oxidative stress, and induction of inflammatory cascades—in others. Moreover, interactions with immune cells appear highly complex; evidence points to both pro-inflammatory activation and immunosuppressive effects, contingent on specific material characteristics. While the ultrahigh surface area facilitates exceptional biomolecule loading capacity—beneficial for sensing platforms—it concurrently amplifies unintended interactions with vital cellular components (e.g., proteins and lipids), potentially altering their native conformation and function. Consequently, reconciling these intrinsic material properties, which confer remarkable diagnostic and therapeutic potential, with profound biocompatibility challenges represents a central research imperative. A deeper mechanistic understanding is essential to advance safe and efficacious applications; this includes elucidating how pristine versus functionalized graphene interfaces with subcellular structures, triggers specific receptor-mediated pathways, and undergoes potential biotransformation or clearance.

9.3 Preparation Methods of Graphene

The synthesis of single-layer graphene can be classified into top-down and bottom-up approaches. The top-down method involves mechanically or chemically exfoliating single-layer graphene from bulk graphite, which yields material of the highest quality but is limited in sample size. Conversely, bottom-up approaches, such as CVD and epitaxial growth, enable the fabrication of large-area single-layer graphene. Current commonly employed synthesis methods are summarized in Figure 9.6, while the relative advantages and disadvantages of different techniques are compared in Table 9.1.

9.3.1 Mechanical Exfoliation

In fact, the seminal single-layer graphene sample, first isolated by Nobel laureates Andre Geim and Konstantin Novoselov, was achieved using Scotch tape to repeatedly cleave graphite. Single-layer graphene becomes obtainable fortuitously through this repetitive mechanical exfoliation. Graphene fabricated via this method

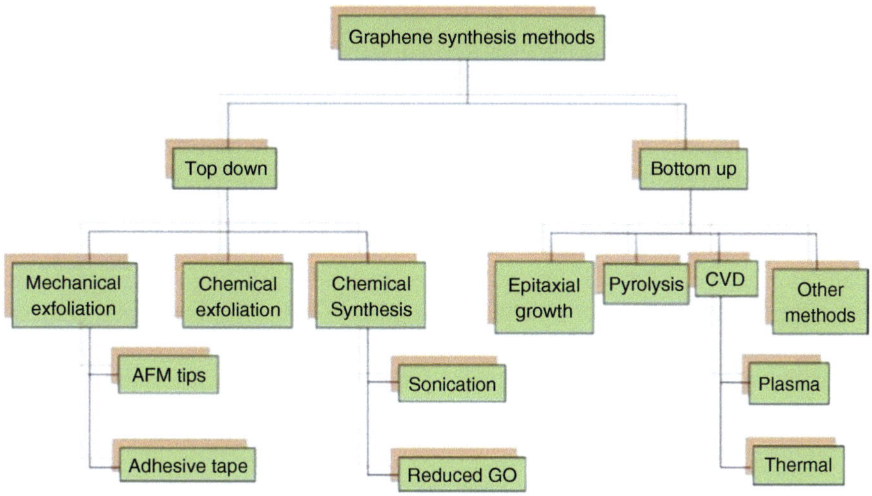

Figure 9.6 Currently developed synthesis methods for graphene. *Source:* [10] / with permission of Elsevier.

Table 9.1 Comparison between different methods to synthesis graphene.

Method	Thickness	Lateral size	Advantage	Disadvantage
Micro-mechanical exfoliation	Single to few layers	μm to cm	Pristine and high quality	Very small-scale production
Electro-chemical exfoliation	Single to few layers	500–700 nm	Functionalized with high electrical conductivity	High cost of ionic liquids
Direct sonication of graphene	Single to few layers	μm	Pristine and inexpensive	Low yield
Reduction of carbon monoxide (CO)	Multiple layers	Sub-μm	Un-oxidized	Contamination with α-Al_2S and α-Al_2O_3
Unzipping of carbon nanotubes	Single to few layers	Few μm long nanoribbons	Size dependent on the starting nanotubes	Expensive and oxidized graphene
Epitaxial growth on SiC	Single to few layers	Up to cm size	Large area	Very small scale
CVD	Single to few layers	Very large to meter size	Very large area and high quality	Lower crystalline quality

typically manifests as small flakes. Consequently, mechanically exfoliated graphene primarily serves to investigate intrinsic properties—such as electronic, mechanical, or thermal characteristics—and is not viable for large-scale applications. Furthermore, while AFM tips offer an alternative means for mechanical exfoliation of single-layer graphene, their efficiency remains considerably limited. To surmount the low yield inherent in the traditional tape approach, a more reliable method has emerged for obtaining relatively large-area single-layer graphene flakes, typically around 500 μm in diameter. This novel methodology comprises two distinct phases: initially, ambient adsorbates are effectively removed utilizing oxygen plasma treatment; subsequently, precise heating is applied to maximize conformal contact between substrates and target materials. Notably, this technique yields not only high-quality single-layer graphene but also single-layer bismuth strontium calcium copper oxide ($Bi_2Sr_2CaCu_2O_X$) [11].

9.3.2 Electrochemical Exfoliation

Although mechanical exfoliation excels at producing high-quality single-layer graphene, its yield is notably low. Consequently, chemical exfoliation has been proposed for large-scale graphene production. Figure 9.7 illustrates the fundamental steps involved in the chemical exfoliation of graphite. Bulk graphite forms, such as foil, rod, and plates, serve as either cathode or anode in an aqueous solution, for instance, ammonium sulfate ($(NH_4)_2SO_4$). Subsequently, a voltage of approximately 10 V is applied, facilitating simultaneous oxidation and intercalation.

As a result, molecular fragments, including H_2O, SO_4^{2-}, O_2, CO, and SO_2, penetrate the interlayer spaces, leading to graphite expansion and the generation of graphene layers. This technique demonstrates exceptional efficiency, with gram-scale production of high-quality few-layer graphene (≤3 layers) achievable within minutes, exhibiting a low oxidation degree, as indicated by a carbon–oxygen ratio of 17.3 [12].

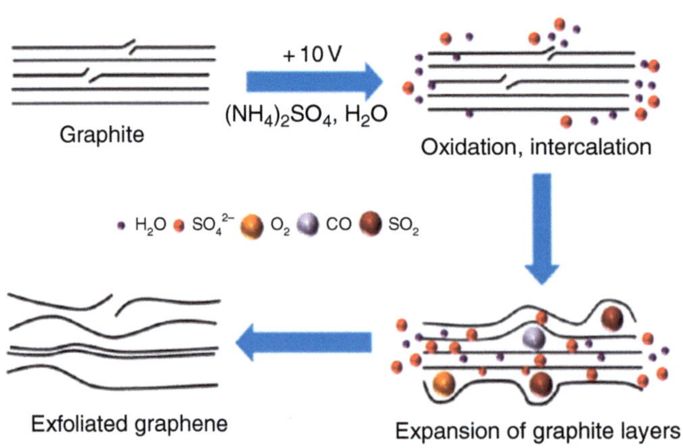

Figure 9.7 Exfoliated graphene by chemical intercalation. *Source:* [12] / with permission of American Chemical Society.

9.3.3 Direct Sonication Synthesis

Sonication is employed to assist in the exfoliation of graphene. Furthermore, it has been demonstrated that not only graphene but also other two-dimensional materials, such as h-BN, MoS_2, WS_2, and $MoSe_2$, can be directly obtained through high-power sonication in pure water without any chemical additives or surfactants. The key modification compared to conventional sonication is elevating the water temperature to 60 °C [13]. Surprisingly, the resulting dispersions of two-dimensional nanomaterials exhibit pronounced stability, exceeding one month under sustained thermal conditions. However, it should be noted that this purely sonication-based approach produces two-dimensional nanomaterials with significantly restricted lateral dimensions, typically less than 10 nm. Graphene oxide (GO) serves as an alternative precursor for graphene production via thermal reduction. Nevertheless, graphene synthesized through this method frequently contains numerous functional groups at both the edge sites and basal planes, leading to diminished mechanical properties and compromised electronic characteristics.

9.3.4 Epitaxial Growth

Silicon carbide (SiC) is a crystalline compound consisting of silicon and carbon atoms. Due to the negligible vapor pressure of carbon atoms under high-temperature annealing conditions, the preferential sublimation of silicon atoms leaves carbon atoms on the surface, ultimately transforming into graphene layers (Figure 9.8). Research demonstrates that the heating method—whether employing electron-beam heating or resistive heating—does not significantly impact the outcome and yields comparable results. Typically, the SiC sample is heated to temperatures exceeding 1500 °C under high vacuum to minimize potential contamination. Theoretical calculations indicate that three atomic layers of SiC are necessary to supply sufficient carbon atoms for forming monolayer graphene. It has been experimentally observed that silicon atoms sublime from the SiC surface at a high rate at elevated temperatures, leading to uncontrolled graphene growth. To mitigate this issue, argon gas is introduced during the heating process. The argon atoms collide with and partially reflect subliming silicon atoms, thereby reducing the sublimation rate of silicon and enabling controlled graphene growth.

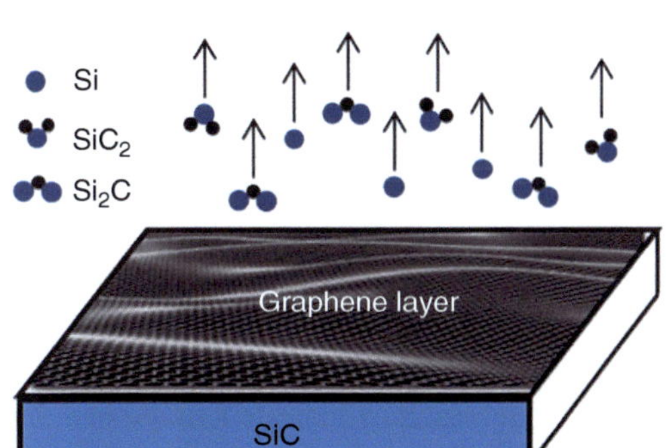

Figure 9.8 Schematic of the graphene epitaxial growth on SiC surface. *Source:* [14] / with permission of John Wiley & Sons.

Additionally, annealing the SiC sample in the presence of a silicon flux or disilane may facilitate graphene formation [15].

The crystallographic relationship between the SiC substrate and the epitaxial graphene layers is critically dependent on the surface termination—specifically, whether the surface is silicon-terminated or carbon-terminated. On the silicon-terminated SiC(0001) surface, single-layer graphene can be achieved at a reduced growth rate. It is important to note, however, that graphene does not nucleate directly on the bulk SiC surface. Instead, a $(6\sqrt{3} \times 6\sqrt{3})$ R30° carbon-rich buffer layer forms initially. This buffer layer is covalently bonded to the underlying substrate in a partially ordered manner. Subsequent graphene layers grow on top of this buffer layer without direct atomic contact with it, resulting in electronic properties comparable to those of free-standing graphene. The number of graphene layers can be controlled by adjusting the annealing temperature and duration. Despite the weak electronic coupling between the buffer layer and the graphene, a significant n-doping effect is consistently observed, presenting a major source of electronic scattering that limits device applications. In contrast, on the carbon-terminated SiC(000$\bar{1}$) surface, the graphene growth rate is significantly higher, typically yielding polycrystalline films with thicknesses ranging from 5 to 100 layers. Studies show that graphene layers grown on the SiC(000$\bar{1}$) surface also exhibit properties akin to isolated monolayer graphene. This behavior arises because individual graphene layers typically rotate relative to the substrate by angles between 0° and 30°, which breaks the unit cell symmetry and electronically decouples adjacent layers. For industrial implementation, the use of SiC wafers for graphene growth entails several limitations, including wafer size constraints, high wafer costs, and the demanding requirements of associated micromachining processes.

9.3.5 CVD Growth

CVD represents a highly promising method for graphene synthesis due to its reliability, scalability, high material quality, and cost-effectiveness. For instance, 30 inch graphene films have been successfully fabricated using a roll-to-roll (R2R) CVD process. Copper (Cu) has emerged as the standard substrate material for producing high-quality, predominantly single-layer graphene, primarily attributed to its extremely low carbon solubility. The crystallinity of the Cu substrate exerts a significant effect on the crystal orientation of the deposited graphene. Specifically, single-crystal Cu(111) surfaces facilitate the rapid growth of large-area, single-crystal, single-layer graphene. Graphene synthesized under these conditions exhibits exceptional properties, including a high mobility of ~23,000 $cm^2/V/s$ at 4 K and a room-temperature sheet resistance of ~230 Ω/□. Semiconductor substrates, such as germanium (Ge), also demonstrate characteristic metallic catalytic behavior and have been used to grow wafer-scale, single-crystal, single-layer graphene on 2 inch Ge wafers [17].

The CVD graphene growth mechanism on Cu involves several sequential elementary steps, as illustrated in Figure 9.9a. Initially, reactants like CH_4 and C_2H_4 are transported by carrier gases, typically Ar and H_2 (Step I). Simultaneously,

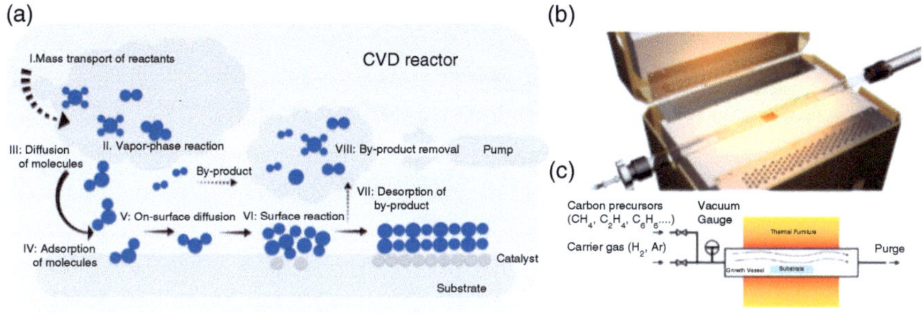

Figure 9.9 (a) Schematic of the elementary steps in the CVD process. (b) Typical furnace setup and (c) the carbon source and carrier gas used for graphene growth. *Source:* [16] / with permission of American Chemical Society.

gas-phase reactions may occur at the high process temperatures (typically ~1000 °C) (Step II). Reactant molecules subsequently diffuse through the boundary layer adjacent to the Cu substrate surface (Step III) and adsorb onto the active sites (Step IV). Following adsorption, surface diffusion of the adsorbed molecules occurs (Step V), leading to surface reactions (Step VI) that nucleate graphene domains. Concurrently, volatile by-products desorb (Step VII) and are removed from the chamber via pumping (Step VIII). Nucleated graphene domains then grow and merge, forming a continuous large-area single layer. The growth process is intrinsically self-limiting on Cu due to the catalytic passivation effect of the inert carbon monolayer. Figure 9.9b depicts a schematic of a typical CVD furnace for graphene synthesis, and Figure 9.9c details the typical gas flows and carbon sources employed.

Notably, graphene synthesized on Cu requires transfer to target substrates for most applications. This typically involves etching away the Cu substrate, often using a sacrificial polymer support layer. Poly(methyl methacrylate) (PMMA) is the most common transfer-assisting polymer, which is subsequently dissolved in acetone upon successful transfer of the graphene to the destination substrate. Despite the intrinsic high quality of CVD graphene grown on Cu, transfer-induced surface contamination (e.g., polymer residues) typically degrades the material's electronic properties, resulting in unintended performance deficits. While numerous alternative transfer methods and polymer supports beyond PMMA have been developed to mitigate contamination, this challenge remains incompletely resolved.

9.3.6 Others

Beyond the methods previously discussed, alternative approaches exist for obtaining graphene, including the unzipping of single-walled carbon nanotubes, as depicted in Figure 9.10. Conceptually, a single-walled carbon nanotube can be viewed as a single-layer graphene sheet rolled along a specific axis. Consequently, obtaining graphene by longitudinally cleaving a carbon nanotube is theoretically simple; however, breaking the robust carbon—carbon bonds necessitates the use of severe oxidative conditions. Specifically, reagents such as $KMnO_4$ are typically employed at

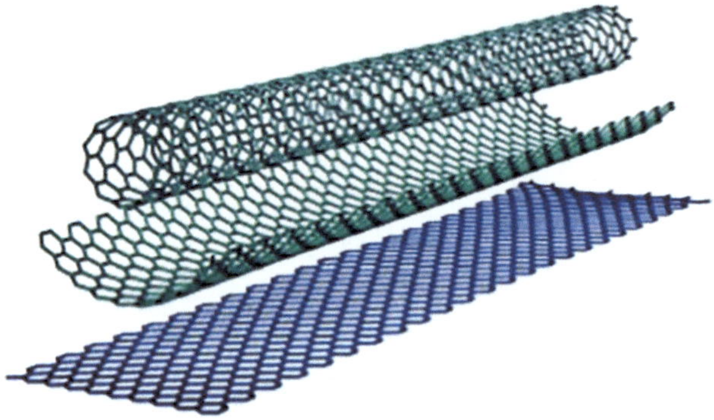

Figure 9.10 Graphene growth by unzipping carbon nanotube. *Source:* [18] / with permission of John Wiley & Sons.

elevated temperatures (55–70 °C) for approximately one hour [18]. GNR synthesized via this method exhibits high solubility in water, ethanol, and other polar solvents. This solubility behavior indicates the likely presence of numerous functional groups, such as hydroxyl or carboxyl groups, on the graphene surface.

9.4 Graphene-Based Devices

The remarkable properties of graphene unlocked in fundamental research have spurred intense efforts to translate them into functional devices across diverse technological domains. These devices leverage graphene's unique combination of ultrahigh charge-carrier mobility, exceptional thermal conductivity, outstanding mechanical strength and flexibility, broadband optical absorption, and quantum phenomena observable even at room temperature.

Devices fabricated from single-layer graphene can be classified into four fundamental categories based on their electrode configuration. The first type utilizes an FET structure, featuring three terminal electrodes: source, drain, and gate. In this configuration, the gate voltage modulates the current flow between the drain and source. A representative device of this category is the FET photodetector (Figure 9.11a).

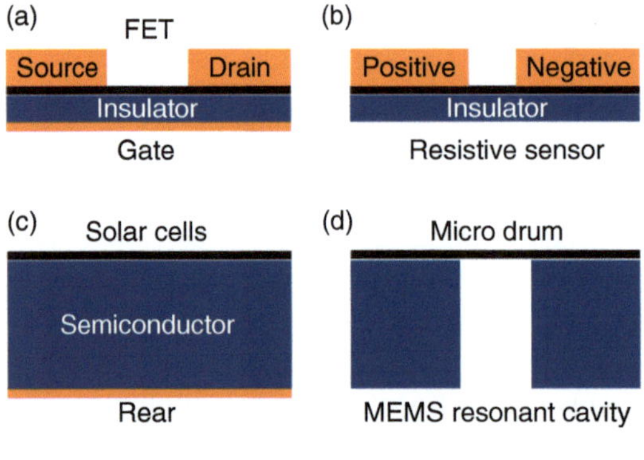

Figure 9.11 Single-layer graphene-based devices with (a) three electrodes, (b) two electrodes, (c) one electrode, and (d) without an electrode.

The second device type incorporates two terminal electrodes connected to the graphene, typically on the same side. Here, an electrical current passes through the graphene layer, and its magnitude can be modified by molecules adsorbed from the external environment onto the graphene surface. This category includes devices such as the resistive gas sensor (Figure 9.11b).

The third category comprises two-terminal devices where one terminal connects directly to the single-layer graphene front surface, while the counter electrode contacts the device from the opposing side. The graphene/silicon heterojunction solar cell exemplifies this electrode arrangement (Figure 9.11c).

The final type operates without any direct electrode connection to the single-layer graphene itself. In these devices, exemplified by certain nanoelectromechanical systems (NEMS), the intrinsic mechanical properties of the graphene sheet are instrumental (Figure 9.11d).

9.4.1 FET Devices

9.4.1.1 Electronic Transport Mechanism

Intrinsic graphene is known to be a zero-bandgap semiconductor, wherein the Fermi level resides at the Dirac point, the intersection between the valence and conduction bands. The application of a gate voltage (V_g) allows the continuous modulation of Dirac fermions from hole-dominated to electron-dominated conduction as V_g is swept from negative to positive values (Figure 9.12a). This modulation results in a resistance maximum at zero gate bias, followed by a decrease exhibiting $1/V_g$ dependence. The corresponding ON/OFF ratio exceeds 20 [21]. Figure 9.11b presents the Hall conductivity measured at various applied voltages, revealing a distinct linear relationship. Given the conductivity expression $\sigma = en\mu$ and the established fact that mobility μ remains constant at low temperatures (10 K) under low-injection rates, it can be deduced that the carrier density increases linearly with gate voltage.

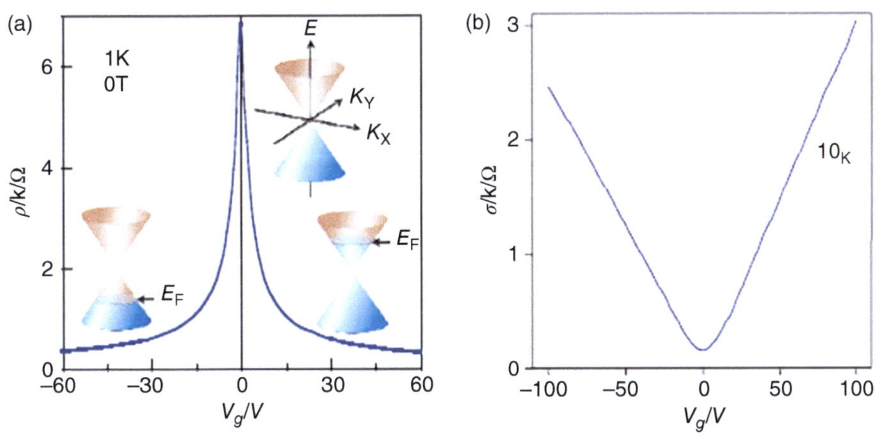

Figure 9.12 (a) Resistivity evolution concerning the voltage. Adapted with permission from Geim and Novoselov [19]. (b) Hall conductivity of graphene as a function of applied gate voltage. *Source:* [20] / with permission of Springer Nature.

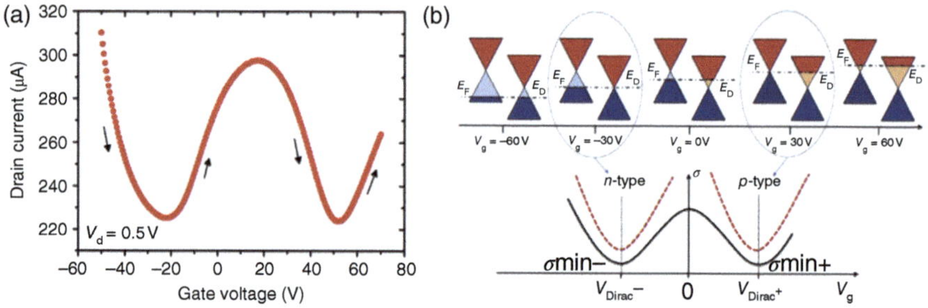

Figure 9.13 (a) "W"-shaped response of I_{ds}–V_g curve observed in graphene FET. (b) The possible formation mechanism for the unique behavior. *Source:* [22] / with permission of Elsevier.

This relationship is described by the equation $n_{e,h} = \alpha V_g$, where α represents the injection rate and $n_{e,h}$ denotes the electron or hole density, respectively. Figure 9.12 unambiguously demonstrates the ambipolar nature of graphene.

In contrast to the ambipolar conductivity of intrinsic graphene, unipolar transport, involving exclusively holes or electrons, holds greater interest for logic device applications. Furthermore, a thorough understanding of the transition mechanism between unipolar and ambipolar states is essential for developing graphene-based PN junctions, a critical step toward realizing electronic devices scaled to atomic thicknesses. To modulate the field effect in graphene and induce unipolar behavior, doping strategies employing metals (e.g., Al) or chemical dopants such as poly(ethyleneimine) (PEI), hydrazine, and decamethylcobaltocene (DMC) have been implemented. These dopants effectively shift the transport characteristics from ambipolar to unipolar. Figure 9.13a reveals a distinctive "W"-shaped current–gate voltage (I_{ds}–V_g) characteristic observed in bottom-gated devices [22]. This unique behavior is well accounted for by a PN junction model. It arises from the superimposition of two characteristic V-shaped σ–V_g curves, induced respectively by applied n-type (negative gate bias) and p-type (positive gate bias) doping conditions, as conceptualized in Figure 9.13b. Subsequent analysis confirmed that the graphene under metal contacts becomes n-doped, while the exposed channel region remains p-doped, primarily due to atmospheric oxygen adsorption. Various factors influence the manifestation of the "W"-shaped transport, including graphene quality, electrode metal selection, surface-adsorbed molecules, the underlying substrate properties, and conditions during preparation and measurement. Such "W"-shaped characteristics have also been documented in graphene FETs utilizing HfO_2 and SiO_2 gate dielectrics. In these cases, the combined effects of the dielectric constant, thickness, and graphene/dielectric interface properties contribute concurrently to the observed transport response.

9.4.1.2 Device Structure of Graphene FETs

Four fundamental structure designs exist for graphene-based FETs, as illustrated in Figure 9.14: bottom-gated, top-gated, dual-gated, and suspended configurations [23–26]. In the bottom-gated structure (Figure 9.14a), a heavily doped silicon substrate coated with a nanometer-thick aluminum oxide (Al_2O_3) insulating layer

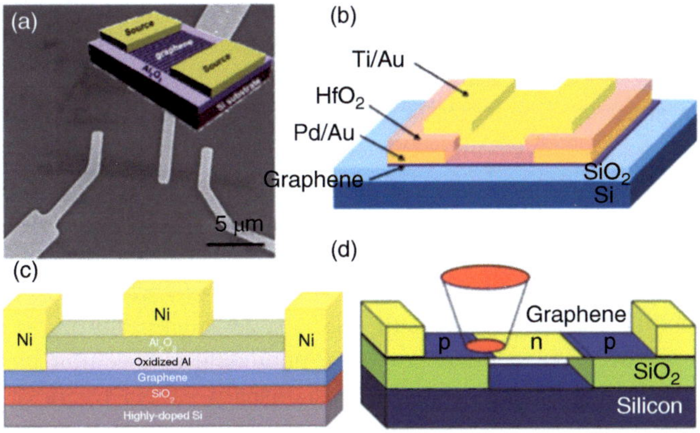

Figure 9.14 (a) Bottom-gated (*Source:* [23] / with permission of IOP Publis), (b) top-gated (*Source:* [24] / with permission of American Chemical Society), (c) dual-gated (*Source:* [25] / with permission of AIP Publishing), and (d) suspended graphene FETs (*Source:* [26] / with permission of American Chemical Society).

functions as the gate electrode [23]. Graphene, obtained via mechanical exfoliation or CVD, is subsequently transferred onto this substrate. Photolithography or electron-beam lithography then defines the drain and source (D/S) electrodes, typically composed of titanium/gold (Ti/Au) or chromium/gold (Cr/Au) metallization. The FETs are prepared for measurement following the standard liftoff process. Heavily doped silicon with a silicon dioxide (SiO_2) insulating layer is also frequently employed as the gate electrode. It has been demonstrated that the dielectric constant (k) exerts a marked effect on operating voltage and transconductance. Consequently, high-k dielectric materials are favorable; hafnium oxide (HfO_2) has been introduced to enhance the performance of bottom-gated graphene FETs. The voltage applied to the gate electrode modulates both carrier type and carrier density within the channel, thereby regulating the drain-to-source current.

In contrast to the bottom-gated configuration, the gate electrode in top-gated FETs (Figure 9.14b) is fabricated directly onto the upper surface of the graphene [24]. Although top-gated FETs offer advantages for circuit integration, carrier mobility is typically reduced due to scattering induced by the top-gate dielectric and electrode. Potential damage during the dielectric/electrode deposition process represents an additional source of mobility degradation. Careful selection of the top-gate dielectric material is therefore critical; PMMA, SiO_2, HfO_2, Al_2O_3, and silicon nitride (Si_3N_4) are commonly utilized by researchers.

To achieve enhanced control over carrier density in the graphene channel, dual-gated FET structures have been developed (Figure 9.14c) [25]. Leveraging dual gates enables modulation of the graphene layer simultaneously from both top and bottom surfaces. For example, an Al_2O_3/Ni top gate and a SiO_2/Si bottom gate can be employed. However, it should be noted that this design may introduce additional scattering pathways, necessitating comprehensive optimization of the fabrication process to mitigate such effects.

Physical contact between electrodes/gate dielectrics and graphene inevitably introduces charge-carrier scattering, complicating the study of graphene's intrinsic mobility. Suspended graphene FETs (Figure 9.14d) were devised to eliminate scattering originating from substrate impurities or substrate/gate dielectric phonons [26]. This configuration has achieved high carrier mobilities of approximately $2 \times 10^5\ cm^2/V/s$ at carrier concentrations near $2 \times 10^{11}/cm^2$. The fabrication process resembles that of bottom-gated FETs, with an additional step involving liquid etching to remove the underlying silicon/silicon dioxide (Si/SiO_2) support beneath the channel region, thereby suspending the graphene. Suspended FETs not only facilitate intrinsic mobility measurements but also enable high-performance graphene devices. For instance, suspended graphene photodetectors exhibit an approximately one-order-of-magnitude increase in responsivity—a key figure of merit—compared to their substrate-supported counterparts.

9.4.1.3 Graphene Nanoribbon FETs

Intrinsic single-layer graphene possesses a zero bandgap; consequently, the source–drain current cannot be completely suppressed upon application of a gate voltage, leading to a low ON/OFF ratio that is unfavorable for circuit applications. To induce a bandgap in intrinsic graphene, several strategies have been proposed, including substrate modulation, strain application, width reduction, and superlattice formation. Among these, width reduction—for instance, through the formation of GNRs with widths below 10 nm—is particularly prominent, as it preserves the high mobility of intrinsic graphene and enables bandgaps of 0.5–2.8 eV depending on the width. Furthermore, the electrical properties of GNRs depend on their edge structure: armchair GNRs exhibit semiconducting behavior, whereas zigzag GNRs demonstrate metallic characteristics. Figure 9.15a displays an atomic force microscopy (AFM) image of chemically derived GNRs with widths below 10 nm [27], revealing smooth edges that may correspond to well-defined armchair or zigzag configurations. Transfer measurements demonstrate that GNR FETs can be completely deactivated under positive gate bias, as illustrated in Figure 9.15b. The achieved ON/OFF ratio is shown in Figure 9.15c, thereby facilitating the

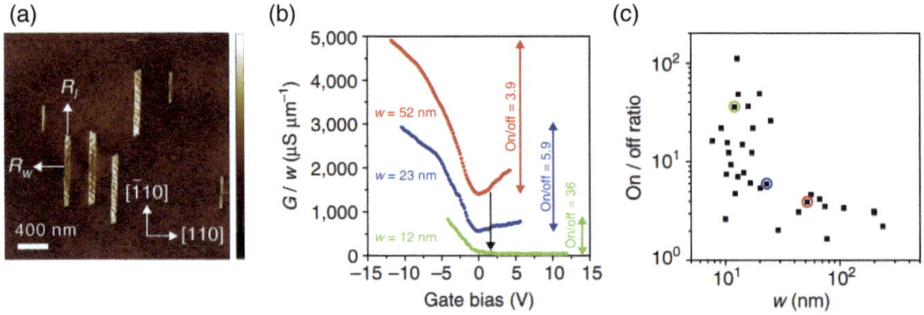

Figure 9.15 (a) AFM image of the graphene nanoribbon. (b) The transfer measurements of the GNR FETs with varied channel width. (c) ON/OFF ratios of GNRFETs with the difference in width. *Source:* [27] / Springer Nature / CC BY 4.0.

application of GNRs in logic circuits. However, fabricating GNRs on a wafer scale with well-defined edge configurations remains highly challenging.

9.4.1.4 Application of Graphene FETs

As stated earlier, FETs fabricated from intrinsic graphene cannot be completely turned off due to its gapless electronic band structure, rendering them unsuitable for large-scale logic circuits requiring low power consumption. However, graphene's implementation in radio-frequency devices, sensors, photodetectors, and plasmonic applications shows significant promise, owing to its inherent properties: high carrier mobility, large surface area, tunable optical absorption, and strong plasmonic resonance.

Radio-Frequency Devices High-speed radio-frequency devices exemplify this potential. IBM reported the first high-speed graphene transistors [28]. Figure 9.16a shows the optical image of the device layerout. Figure 9.16b depicts the schematic cross section. Measured output characteristics and transfer characteristics are shown in Figure 9.16c. These graphene FETs inherently lack current saturation at high drain voltages due to the absence of a bandgap. The cutoff frequency (f_T) of 4 GHz was characterized from the short-circuit current gain ($|h_{21}|$) in Figure 9.16d, defined as the frequency where $|h_{21}|$ equals unity. IBM's later work further demonstrates

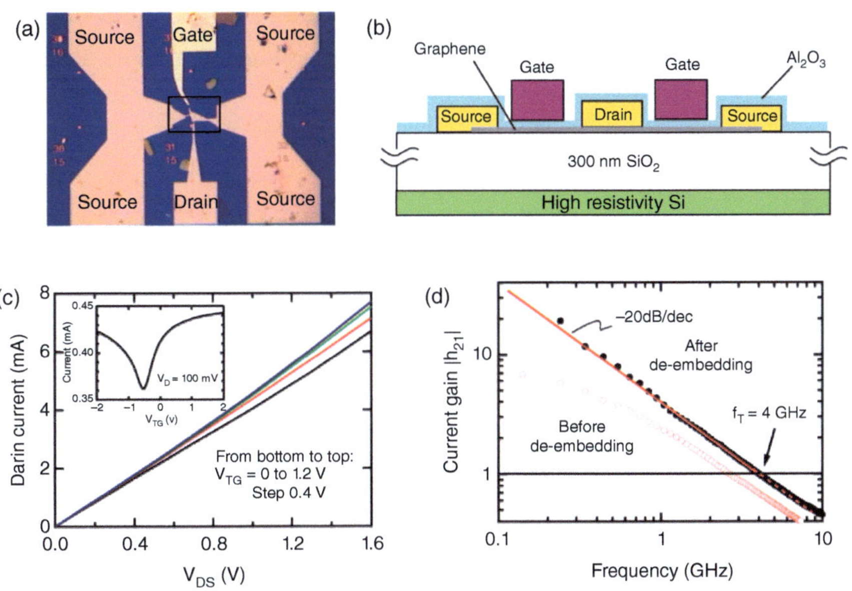

Figure 9.16 (a) Photograph of the top-gated graphene FETs on a 2 inch SiC wafer, inset is the cross-section schematic of the device. (b) Source drain current as a function of applied gate voltage, the drain voltage is fixed at 1 V, the device transconductance g_m is plotted on the right axis. (c) Drain current evolution concerning the drain voltage with different applied gate voltage. (d) Measured small-signal current gain $|h_{21}|$ with a gate length of 240 and 550 nm, whose cutoff frequency is 100 and 53 GHz, respectively. *Source:* [28] / with permission of American Chemical Society.

graphene FETs utilizing graphene synthesized on a silicon carbide (SiC) substrate as the channel material, by annealing the SiC wafer at approximately 1450 °C [29]. An electron density of $3 \times 10^{12}/cm^2$ was determined for the graphene channel. A hybrid dielectric gate layer, composed of a polymer and HfO_2, modulated the device conductance. To optimize the interface properties, a 1 nm thick polymer insulating layer was applied between the graphene and the gate electrode. Subsequently, a high-k HfO_2; dielectric was deposited as the gate oxide, yielding measured carrier mobilities ranging from 900 to 1520 cm^2/V s. The cutoff frequency (f_T) of 100 GHz was characterized for a channel length of 240 nm, which exceeds comparable silicon devices at identical gate lengths.

The cutoff frequency of graphene FETs may be enhanced to 100–300 GHz through the utilization of small-diameter nanowire-gate electrodes, facilitating significantly reduced channel lengths [30]. Initially, high-quality graphene obtained via mechanical exfoliation is transferred onto a SiO_2/Si substrate. A Co_2Si nanowire, uniformly coated with a 5 nm thick Al_2O_3 layer, subsequently serves as the gate electrode upon placement onto the graphene layer, as illustrated in Figure 9.17a. A 10 nm thick Pt layer is then deposited onto the device structure. The three-dimensional geometry of the nanowire causes the Pt layer to physically separate, thereby forming distinct drain and source electrodes (Figure 9.17b).

This nanowire gate configuration provides dual advantages. Firstly, because the nanowire gate is transferred after graphene deposition, integrity damage to the graphene layer is avoided. Secondly, the diameter of the nanowire can be minimized to dimensions below 200 nm. This extreme reduction enables a corresponding decrease in the channel length between the drain and source electrodes. Consequently, channel resistance is significantly lowered, leading to enhanced overall device performance.

FET Sensors Charge-carrier density in graphene can be dramatically altered by adsorbed molecules, enabling graphene-based FETs to function as ultrasensitive gas sensors. In traditional gas sensors, detection precision is significantly constrained by thermal noise, which typically exceeds the signal from a single gas molecule. In contrast, noise originating from defects or other sources in graphene is substantially lower than that in conventional sensors, facilitating detection at the single-molecule level. Figure 9.18a illustrates an FET gas sensor capable of detecting individual

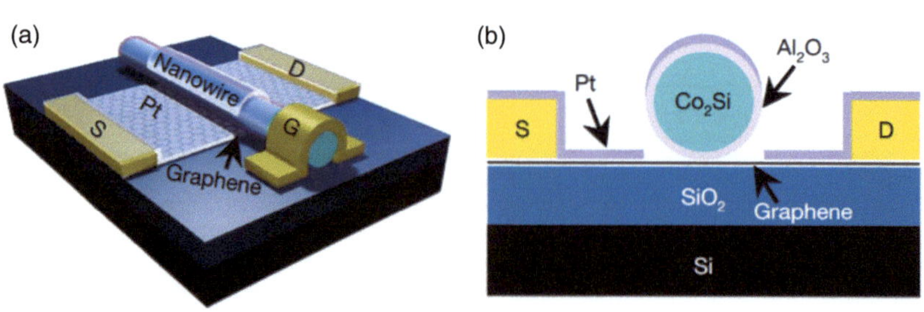

Figure 9.17 (a) Three-dimensional and (b) cross section of the nanowire-gated graphene FETs. *Source:* [30] / with permission of Springer Nature.

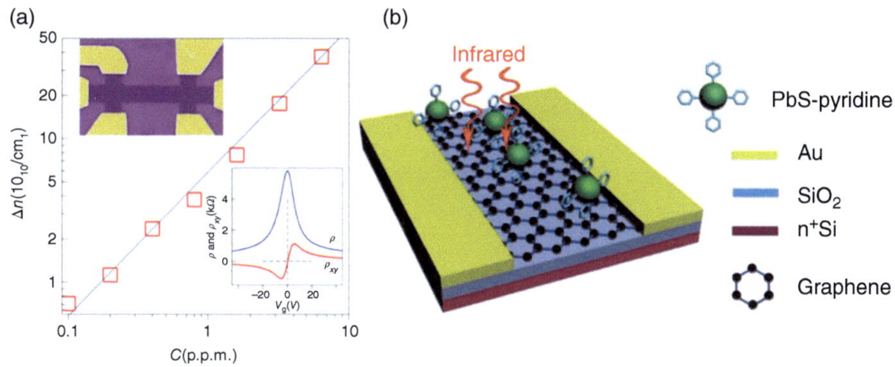

Figure 9.18 (a) NO_2 sensor based on graphene FETs. *Source:* [31] / with permission of Springer Nature. (b) Hybrid graphene/PbS infrared photodetector. *Source:* [32] / with permission of John Wiley & Sons.

NO_2 molecules. The carrier density in graphene exhibits a linear increase with NO_2 concentration; the inset displays the evolution of resistivity (ρ) and Hall resistivity (ρ_{XY}) versus applied gate voltage [31]. FET sensors are also applicable to nonpolar gas molecules, such as hydrogen, permitting easy identification at concentrations as low as 0.05% in air. Beyond gas sensing, graphene FETs are employed in tracking biological molecular signals. For instance, single-stranded probe DNA can be identified without labeling, and DNA at concentrations of 1 fM can be readily detected. Owing to its broadband electromagnetic absorption, graphene is suitable for high-sensitivity photodetectors. Figure 9.18b depicts a graphene/PbS QD hybrid FET configured for infrared light detection [32]. Charge transfer between PbS QDs and graphene induces the photoresponse under illumination, achieving a responsivity as high as 10^7 A/W.

FET Memory Despite graphene's intrinsic zero bandgap rendering it unsuitable for logical devices, it has demonstrated high promise in nonvolatile memory devices owing to its excellent electronic properties. Graphene can function as either a channel material or a charge storage layer depending on the working mechanism. Owing to its high electron mobility, graphene is employed as channel material to enable rapid read/write speeds. Concurrently, GO is utilized in memory devices due to its charge storage capability. Hybrid graphene/ferroelectric configurations have been extensively investigated for memory applications [33]. As illustrated in Figure 9.19a, graphene serves as the channel material, while the ferroelectric layer P(VDF-TrFE) (Poly(vinylidene fluoride-co-trifluoroethylene)) functions as the gate insulator. The polarized state is readily modulated by the applied voltage, inducing charge density variations in graphene that are detected via current flow from drain to source. Binary information "1" and "0" are encoded by the voltage and decoded by the current; the polarized state persists after voltage removal, achieving nonvolatility. Graphene is also applicable in flash memory devices (Figure 9.19b), where a monolayer MoS_2 acts as channel material, graphene serves as the source and drain electrodes, and few-layer graphene forms the floating gate [34]. A 6 nm thick HfO_2 insulating layer

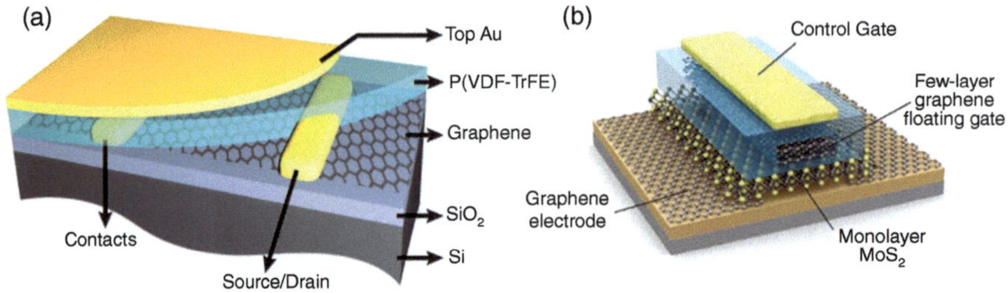

Figure 9.19 (a) Hybrid graphene/P(VDF-TrFE) nonvolatile memory device. *Source:* [33] / with permission of AIP Publishing. (b) MoS_2/graphene floating gate memory device. *Source:* [34] / with permission of American Chemical Society.

is interposed between the MoS_2 and the floating gate. Under applied voltage, charges are injected or extracted through the tunneling layer for write or erase operations, inducing a threshold voltage shift; information is read from the current, analogous to nonvolatile memory devices.

FET Antennas FET antennas leverage the tunable properties of graphene. Single-layer graphene exhibits readily adjustable doping via positive or negative electrostatic gating, displacing the chemical potential (μ_c) across the Dirac point and thereby modifying its conductivity. Due to Pauli blocking suppressing interband transitions, graphene manifests metallic behavior, supporting strong, broadband plasmonic effects spanning visible to infrared and terahertz wavelengths. Precise optical response tuning of graphene-based antennas is achievable through gradual shifts in μ_c induced by applied voltage adjustments.

The intrinsic optical absorption of pristine single-layer graphene is limited to 2.3%, an insufficient level for effective electromagnetic wave detection, where high absorption is a prerequisite for significant responsivity. Consequently, strategies including graphene plasmon excitation, photonic mode engineering, and critical coupling principles have been employed to enhance absorption in graphene antennas. For instance, log-periodic toothed graphene antennas achieve tunable multi-resonant behavior and spectral positioning through chemical potential modulation. Additionally, closely packed graphene nanodisk arrays achieve up to 30% absorption by optimizing geometric parameters, interparticle spacing, and electrostatic gating voltage. Incorporating metallic structures provides further tuning capability; integrating dual split-cross antennas enables approximately 30% absorption over a broadband spectrum from 780 to 1760 nm.

The voltage-dependent dielectric constant of graphene serves as a key mechanism for antenna performance modulation. A graphene-loaded silver ribbon antenna demonstrates substantial amplitude modulation, achieving a transmission modulation depth up to 90% by shifting between resonant and nonresonant states. Dynamically varying graphene's surface conductivity yields intensity modulation ratios of 100 and phase shifts exceeding 240°. Adjustment of the gate voltage in a graphene-integrated mid-infrared antenna facilitates a resonance wavelength shift exceeding 600 nm alongside an intensity modulation depth surpassing 30%.

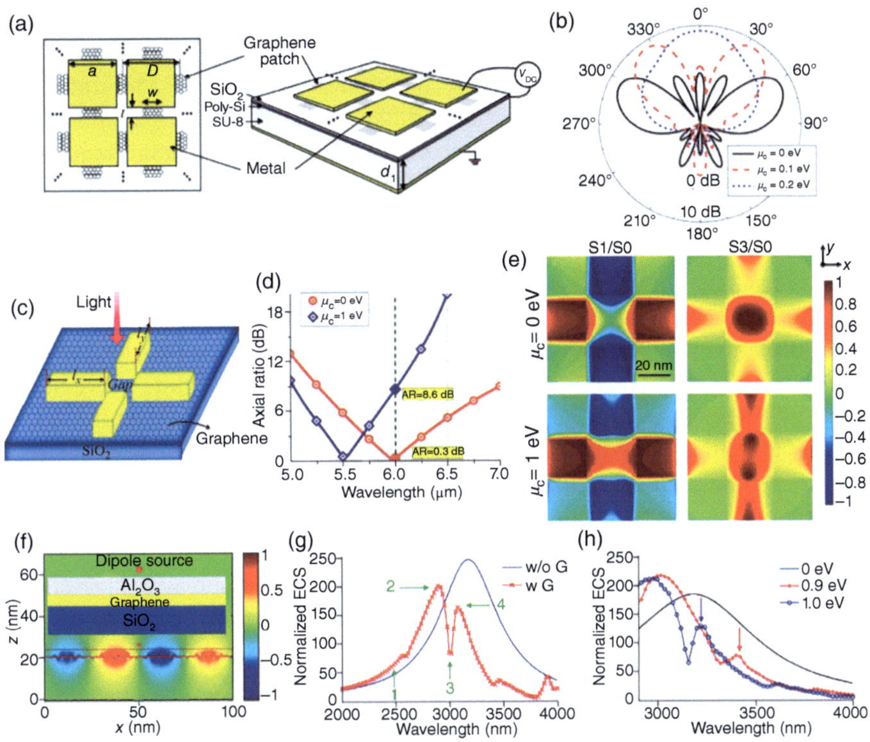

Figure 9.20 (a) Hybrid metal graphene 2D leaky-wave antenna, where the graphene is used as a high-impedance surface. (b) Intensity and phase modulation of the 2D leaky-wave antenna with shifted μ_c. *Source:* [35] / with permission of IEEE. (c) Integrated graphene metal polarizer. (d) Tunable axial ratio and (e) circular–linear polarized state transformation with shifted μ_c. *Source:* [36] / with permission of IOP Publishing. (f) Coupled plasmon between graphene and metal. (g) Splitting of the peak with and without graphene. (h) ON and OFF state of the coupling enabled by shifted μ_c. *Source:* [37] / with permission of Optical Society of America.

Significantly, both the operating frequency and beam-steering direction of terahertz waves are modulated in a two-dimensional leaky-wave antenna through voltage control (see Figure 9.20a). Beam steering over 60° is realized with a μ_c shift of 0.2 eV (Figure 9.20b) [35].

Polarization control, vital for numerous applications, is feasible with graphene antennas. Figure 9.20c demonstrates this capability: single-layer graphene beneath orthogonally arranged dipole antennas allows dynamic polarization state adjustment via gate-voltage-modulated μ_c. At an operational wavelength of 6 μm, the asymmetric cross antenna generates a highly circularly polarized reflected beam, characterized by an axial ratio of 0.3 dB. Biasing the μ_c to 1 eV modifies the axial ratio by 9 dB (Figure 9.20d). Stokes parameter analysis reveals dominance of the normalized S3/S0 (indicating the degree of circular polarization) under μ_c variation, while S1/S0 reflects linear polarization components (Figure 9.20e) [36]. These results confirm highly circular polarization at zero μ_c, transitioning to linear polarization along the x-axis with applied voltage.

Graphene antenna modulation extends to diverse configurations, including Fabry–Perot cavity leaky-wave antennas, terahertz Yagi–Uda antennas, split-ring resonator-based Fano antennas, and gold nanorod pair antennas. The underlying mechanism remains consistent: electrostatic gating-induced μ_c modulation. Furthermore, graphene serves as a photoconductive material for radio-frequency antennas and as an impedance-matching load in terahertz patch antennas, exhibiting unique time-domain modulation characteristics under dynamic biasing.

Gate–voltage tunability of graphene surface plasmons provides spectral resonance control. Studies demonstrate graphene's plasmonic performance rivals noble metals like gold and silver. Strong coupling between graphene plasmons and metal plasmons is attainable (Figure 9.20f). To enable this coupling while preventing quantum tunneling, a thin Al_2O_3 spacer isolates the graphene and metallic dipole antenna. Near-field profiles confirm graphene plasmon existence; resonance splitting emerges under μ_c modulation via voltage (Figure 9.20g). Crucially, voltage-controlled shifting of μ_c enables ON/OFF switching of the plasmonic coupling effect (Figure 9.20h) [37]. This system potentially facilitates plasmon-generated hot electron tunneling into the graphene monolayer, suggesting utility for highly sensitive photodetectors. The performance of such plasmonic graphene photodetectors can thus be optimized by incident laser power and bias voltage across the electrode terminals.

The typical size of graphene, ranging from several to hundreds of micrometers, is well-suited for integration with terahertz antennas, such as dipole antennas. By modulating graphene's conductivity, the radiation patterns and beam direction of these antennas can be actively controlled. For instance, voltage application to the graphene layer enables radiation pattern reconfigurability in leaky-wave antennas. Furthermore, beam direction can be modulated using circular antennas, where a graphene–metal loop functions as a reflector to dynamically manage resonant locations.

Graphene antennas exhibit the capability to generate both linearly polarized and vortex waves. Zhu et al. demonstrated that a cross-shaped structure beneath graphene can generate plasmon vortices (Figure 9.21a) [38]. Under excitation by linearly polarized light at 45° and −45°, the electric field component E_z amplitude displays substantial variation along the rotational direction (Figure 9.21b), while Figure 9.21c exhibits characteristic vortex features. Notably, the topological charge sign of the graphene plasmon vortex can be controlled via the incident light's polarization direction.

Ma et al. proposed a broadband vortex beam generator based on the graphene metasurface reflectarray in terahertz region [39]. The schematic diagram of the graphene metasurface reflectarray structure is shown in Figure 9.21d. The consisting graphene metasurface reflective unit is shown in Figure 9.21e, each unit is composed of a circular graphene patch, a dielectric layer and a perfect electric conductor. The designed graphene metasurface reflectarray structure realizes the generation of orbital angular momentum vortex wave with topological charges of $l = \pm 1$, $l = \pm 2$ and $l = \pm 3$ ranging from 2 to 2.4 THz by modulating the chemical

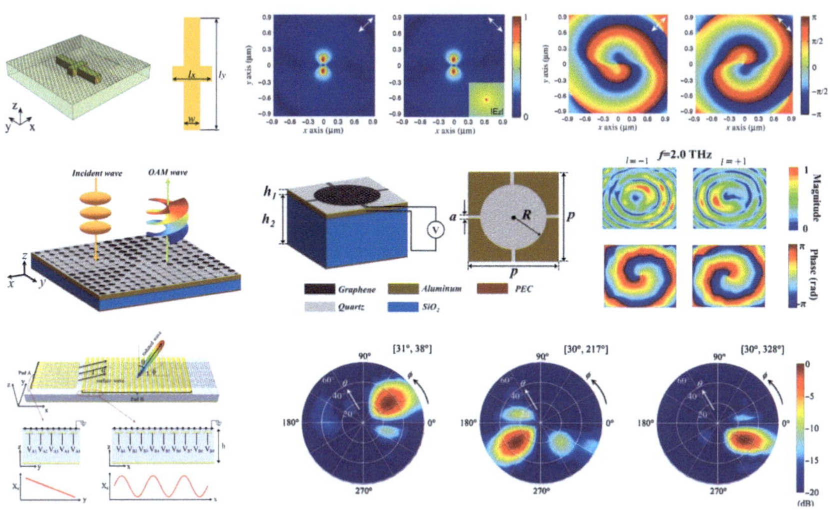

Figure 9.21 (a) Schematic and geometrical parameters of the cross-shaped antenna. (b) Simulated normalized amplitude and (c) phase profiles of electric field component Ez excited by the cross-shaped antenna under 45° and 45°linearly polarized light. Adapted with permission from Zhu et al. [38]. (d) The schematic diagram of the graphene metasurface reflectarray structure and (e) the consisting graphene periodic unit structure located on the silicon dioxide. (f) The phase and the magnitude of the reflected waves with topologic charge $l = \pm 1$ under 2 THz. Adapted with permission from Ma et al. [39]. (g) Schematic representation of a graphene leaky-wave antenna for dynamic beam scanning toward any spatial 2-D direction. The bottom panels show the cross-section configurations and the monolithic impedance variation along y direction for both pad group A and B. (h) Radiation pattern from the graphene leaky-wave antenna with different designed beam radiation. Adapted with permission from Cheng et al. [40].

potential and the radius of the graphene unit. The results for $l = \pm 1$ under 2 THz are shown in Figure 9.21f with a high purity.

Graphene monolayers also show potential for microwave antenna applications. Utilizing graphene's metal–insulator transition—characterized by low surface resistance under electrostatic bias ("on" state) and high surface resistance in its absence ("off" state)—allows efficient control of microwave antenna radiation patterns.

Spatial modulation of graphene reveals compelling antenna applications. Cheng et al. demonstrated a leaky-wave antenna that features two-dimensional beam scanning in both elevation and azimuth planes via electrical control in real time, and at a single frequency [40]. Figure 9.21g shows the schematic representation. The structure consists of a graphene sheet on a metal-backed substrate. The two-dimensional beam-scanning performance is achieved through the proper biasing configuration of graphene by two pad groups, which are orthogonal and decoupled, with one group offering monotonic impedance variation along one direction, and the other offering sinusoidal impedance modulation along the other direction. Real-time control of the beam direction can be realized, as demonstrated by the radiation pattern from the graphene leaky-wave antenna with different designed two-dimensional beam radiation in Figure 9.21h.

FET Terahertz Modulators Within optical imaging and communication systems, optical modulators manipulate the intensity, phase, and polarization of light, typically operating at wavelengths around 1.55 μm. Conversely, modulators operating within the terahertz regime target longer wavelengths. Graphene has garnered significant attention for terahertz modulation owing to its substantial conductivity variation. Since 2010, numerous studies have reported graphene-based terahertz modulators achieving large modulation depths via active control of carrier concentration using an applied voltage. For instance, Sensale-Rodriguez et al. in 2012 demonstrated that significant terahertz frequency modulation depth could be achieved in a large-area graphene FET by manipulating the Fermi level of graphene, resulting in a terahertz electroabsorption modulator. It should be noted that despite the large achievable modulation depth enabled by the atomic thickness of the graphene layer, the overall device performance remains significantly below that required for practical applications.

To enhance performance, a reflection-mode structure is employed, involving deposition of a metallic reflecting layer on the substrate backside. Absorption within the graphene layer can be maximized when the substrate thickness corresponds to an odd multiple of a quarter wavelength. Experimental results indicate that amplitude modulation greater than 64% with 2 dB insertion loss is achievable under proper voltage application. However, this modulator exhibits narrowband response wavelengths due to its reflection-based architecture.

Graphene plasmonic structures also achieve terahertz modulation. The plasmonic resonance of graphene stripes can be tuned by altering the graphene carrier density; incorporating metamaterials further enhances performance via augmented light-matter interactions. In 2012, Lee et al. reported a metamaterial modulator achieving a high modulation depth (86%) but at the cost of substantial insertion loss (exceeding 10 dB) [42]. Figure 9.22 depicts their design, where an active terahertz graphene metamaterial is sandwiched between two extraordinary transmission (EOT) electrodes (Figure 9.22a). Periodically arranged meta-atoms, consisting of a hexagonal metallic frame or asymmetric double-split rings (aDSRs), exhibit Fano-like resonance (Figure 9.22b). The fabricated graphene metamaterial is flexible and can be integrated into a printed circuit board (PCB) over large areas (15 × 15 mm) (Figure 9.22c) and can be wrapped onto a glass rod with a 2 mm diameter (Figure 9.22d). Experimental findings consistently demonstrate a trade-off between modulation depth, insertion loss, and modulation bandwidth.

Figure 9.23 depicts a graphene modulator exhibiting substantial modulation depth, wherein single-layer graphene is transferred onto a gold slit device with a width of 2 μm and a period of 20 μm [43]. Field enhancement arises from coherent radiation of the enhanced in-plane local field within the slits. Consequently, broadband terahertz modulation by bare graphene is preserved in this hybrid structure, schematically illustrated in Figure 9.23a. Measurements indicate this device achieves large modulation depth across a broad frequency range. Specifically, a 90% modulation depth in terahertz wave transmission occurs upon application of a negative bias voltage such as −1.73 V (Figure 9.23b). Figure 9.23c presents

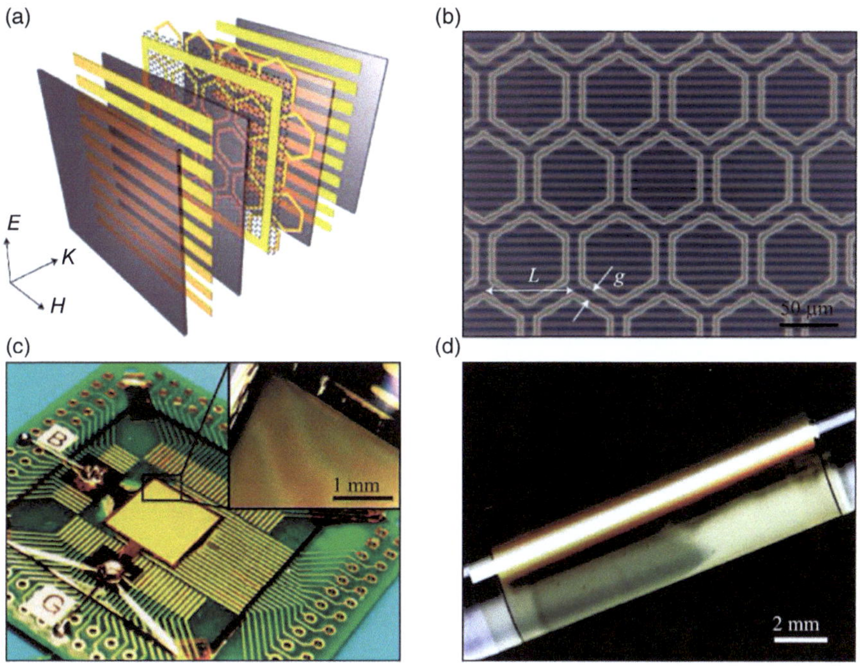

Figure 9.22 (a) Schematic illustration of the active-controllable graphene-based terahertz modulator, in which the single-layer graphene is deposited onto a layer of hexagonal metallic meta-atoms (size of $60 \times 5\,\mu m$) and sandwiched between two EOT electrodes. The polarization of the incident terahertz wave is perpendicular to the line electrode, as indicated by the arrows. Colors, the gray represents dielectric materials, the yellow represents the metallic patterns. (b) Optical image of the hexagonal-shaped metallic frame. (c) PCB board for THz–TDS measurement with fully integrated graphene metamaterial modulator, where the B is connected to the bottom EOT electrode, and the G is connected to the graphene layer. Inset, the close gave the gate-controlled graphene metamaterial. (d) Optical image of the flexible graphene metamaterial wrapped around a glass rod with a diameter of 2 mm. *Source:* [42] / with permission of Springer Nature.

resistance measurements versus applied voltage, with the inset displaying a side view of the ion gel–gated device. Figure 9.23d shows the normalized terahertz field transmission at various gate voltages, confirming the attainment of significant modulation depth.

Although InAs nanowire FETs demonstrate terahertz wave detection capabilities within the 0.3–3 THz frequency range, offering rapid response and high responsivity, graphene plasmonics provides an alternative avenue for robust room-temperature terahertz photodetection of low-energy photons. This approach circumvents limitations inherent to traditional semiconductor terahertz devices, including slow response, cryogenic cooling requirements, large footprint, and low operational speed. Several plasmonic devices employing monolayer or bilayer graphene configured in lateral, vertical, or sandwiched geometries have been developed for terahertz wave detection. Notably, graphene/insulator/graphene heterostructure tunnel FET structures exhibit strong resonant terahertz detection

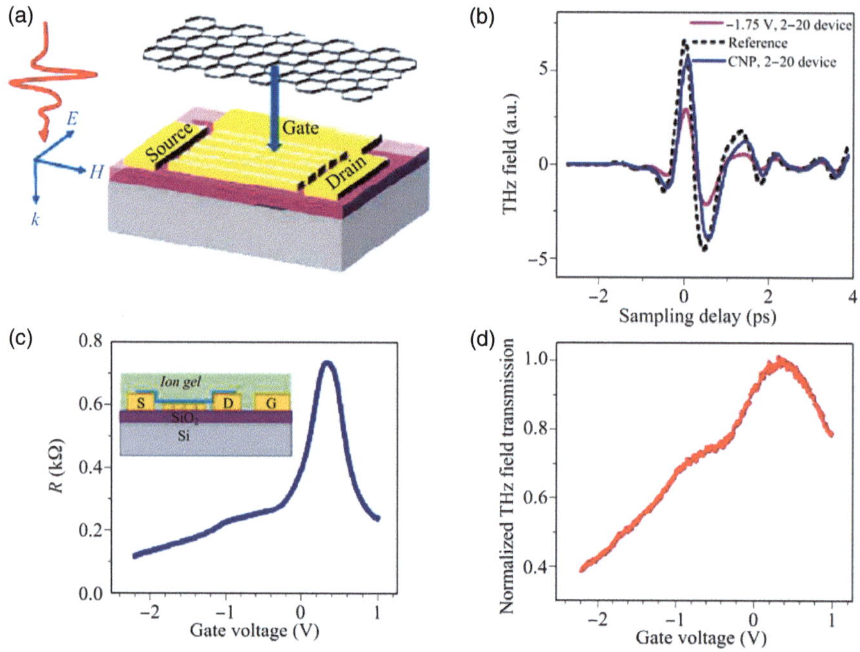

Figure 9.23 (a) Schematic of the graphene/metamaterial hybrid device, the graphene is transferred onto the gold slit device with a width of 2 μm and a period of 20 μm, and the incoming terahertz wave is polarized perpendicular to the slit orientation. (b) Transmitted terahertz wave for the reference (dashed black), the hybrid device at charge neutral point (CNP, solid blue curve), and negative bias voltage −1.75 V (solid magenta curve). (c) Resistance measurement shows the CNP at the positive-gated voltage at 0.33 V, inset is the side view of the ion gel–gated hybrid device. (d) Normalized terahertz wave transmission as a function of gated voltage concerning the graphene at CNP. *Source:* [43] / with permission of American Chemical Society.

characteristics. Additionally, top-gated antenna structures, such as split bow-tie antennas, have been integrated into graphene-based terahertz detectors to enhance coupling efficiency and responsivity at frequencies including 0.34 and 0.6 THz, enabling broadband response.

9.4.2 Type I Two-Terminal Devices

Type I two-terminal devices constitute a specific class of electronic components characterized predominantly by linear or near-linear current–voltage characteristics under standard operating conditions. These devices, exemplified by resistive elements, typically exhibit minimal hysteresis and maintain stability across a range of temperatures and frequencies, distinguishing them from nonlinear counterparts such as Type II devices. The high-sensitivity gas sensor represents a quintessential functional Type I two-terminal device. These sensors can be fabricated with a two-terminal structure featuring interdigital electrodes deposited on a SiO_2/Si substrate, followed by the transfer of a graphene layer onto the electrode array;

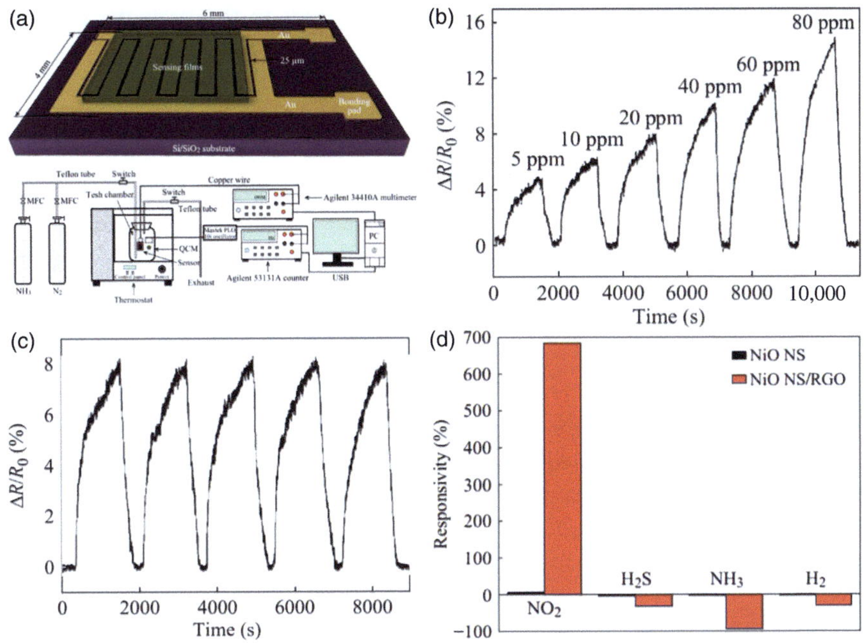

Figure 9.24 (a) Schematic of the resistance gas sensors (top panel), and the measurement set up (lower panel). *Source:* [44] / with permission of Elsevier. (b) Resistance response of the rGO sensor under DMMP vapor concentration 5–80 ppm. (c) Response of the sensor in the time domain. (d) The responsivity of the NiO/rGO hybrid sensor under different gas types. *Source:* [45] / Springer Nature / CC BY 4.0.

the electrode gap is typically less than 50 μm. Figure 9.24a depicts the structure of a typical two-terminal gas sensor (upper panel) and schematizes the testing setup (lower panel) [44].

Device performance is evaluated via the resistance change (ΔR), with the figure of merit defined as $\Delta R/R_0$, where R_0 denotes the baseline device resistance prior to exposure to the target gas molecules. Introduction of a dimethyl methylphosphonate (DMMP) gas flow decreases the device resistance due to a doping effect, as illustrated in Figure 9.24b, achieving a detection limit as low as 5 parts per million (ppm). Detection repeatability constitutes a key factor for practical application; Figure 9.24c demonstrates excellent repeatability, where the resistance variation (ΔR) remains consistent across five sequential measurements. However, the response time progressively increases due to incomplete DMMP molecule desorption between cycles [45].

The physicochemical properties of target molecules significantly influence sensor responsivity. As shown in Figure 9.24d, NO_2 induces a substantial positive response exceeding 700% in the NiO/rGO hybrid sensor. Conversely, exposure to H_2S, NH_3, and H_2 results in resistance increases, yielding negative responses with markedly lower responsivity magnitudes. Compared to FET-based gas sensor architectures, the resistive gas sensor structure is fundamentally simpler. Nevertheless, it lacks the degree of freedom afforded by a back gate voltage to modulate graphene resistance, a limitation constraining the achievable detection limit to the parts-per-million level [45].

9.4.3 Type II Two-Terminal Devices

Type II two-terminal devices, often classified based on their nonlinear electrical characteristics under high-field conditions, are semiconductor or superconductor components with distinct operational dynamics compared to their Type I counterparts.

Photodetectors Photodetectors can be evaluated using several key parameters. The photocurrent (ΔI_λ) is defined as (ΔI_λ) = $I_\mathrm{p} - I_\mathrm{d}$, where the I_p is current with light illumination, while the I_d is the current in the dark. It should be noted that a reverse bias voltage is commonly applied to accelerate the carrier dissociation, and the performance of photo detectors is dependent on the wavelength λ of incident light. Spectra responsivity (R_λ) and external quantum efficiency (EQE) are two important parameters, which are defined as

$$R_\lambda = \frac{\Delta I_\lambda}{P_\lambda} \tag{9.9}$$

$$\mathrm{EQE} = \frac{hcR_\lambda}{q\lambda} \tag{9.10}$$

where the P_λ is incident irradiation power, and the h, c, and q represent the Plank constant, the velocity of light, and the charge of electrons, respectively. A specific detectivity is another important parameter to characterize the detector performance, which is defined by

$$D^* = \frac{(A\Delta f)^{0.5} R_\lambda}{i_\mathrm{noise}} \tag{9.11}$$

where A is detector area, Δf is bandwidth, and i_noise is the noise current. If the noise in the detectors is mainly the shot noise $i_\mathrm{n,s}$, then the i_noise can be approximated as

$$i_\mathrm{noise} \approx i_\mathrm{n,s} = (2qI_\mathrm{d}\Delta f)^{0.5} \tag{9.12}$$

Then the specific detectivity is reformulated as

$$D^* = \frac{R_\lambda}{\left(\frac{2qI_\mathrm{d}}{A}\right)^{0.5}} \tag{9.13}$$

Photons absorbed within a photodetector induce electrical responses through distinct mechanisms: (i) The photovoltaic effect generates a built-in electric field upon photon absorption within a PN junction. (ii) The photothermoelectric effect creates a voltage in thermoelectric materials via light-induced temperature gradients. (iii) The bolometric effect produces conductance changes due to light absorption and heating. (iv) The photo-gating effect modulates carrier density in materials like graphene via light-induced interactions with traps or adsorbed molecules. (v) A

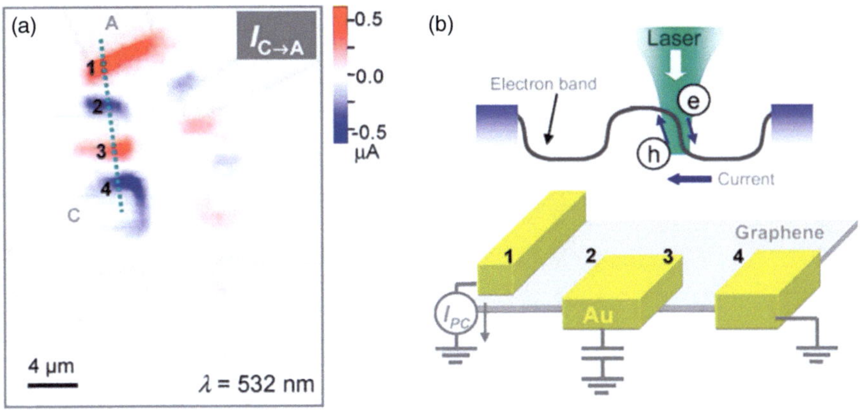

Figure 9.25 (a) Superposition image of light reflection and photocurrent measured with a 532 nm laser (230 μW). (b) Illustrated mechanism of photocurrent generation in graphene. *Source:* [46] / with permission fo American Chemical Society.

plasma wave–assisted mechanism occurs when the two-dimensional electron gas in structures such as graphene FETs forms a resonant cavity for plasma waves.

Theoretical models predict Schottky barrier formation at graphene–metal interfaces. Scanning photocurrent microscopy (SPCM) measurements map photocurrent generation at these interfaces, revealing strong local electric fields near the junction, Figure 9.25a. However, measurable net current output is absent due to the alternating polarity of the induced photocurrent. Figure 9.25b illustrates the photocurrent generation mechanism: photoexcited electron–hole pairs undergo separation and transport in opposite directions driven by band bending, resulting in directional photocurrent. Consequently, asymmetric electrode configuration is essential to generate a net photocurrent output [46].

Compared to metal/graphene junctions, the semiconductor/graphene junction exhibits rectifying characteristics analogous to those of a traditional Schottky junction formed between an n-type semiconductor and monolayer graphene. For this device, the active area is defined through conventional photolithography, involving the initial etching of the SiO_2 insulating layer and subsequent deposition of the front Au/Ti electrode. The graphene monolayer is then transferred to establish the Si/graphene junction, followed by deposition of the Au/Ti rear electrode (Figure 9.26a). Incident light passes through the transparent monolayer graphene and is absorbed by the underlying n-type silicon, generating electron–hole pairs in the bulk silicon. These charge carriers are separated by the built-in electric field at the Si/graphene interface, producing a measurable photocurrent. The corresponding energy band alignment is depicted in Figure 9.26b. Figure 9.26c presents the photovoltage responsivity for a typical Si/graphene photodetector with an active area of 5000 μm^2. A high responsivity exceeding 10^7 V/W is achieved under low incident optical power (10 nW); its performance decreases linearly with increasing incident power [47]. To enhance the built-in electric field, silicon QDs (Si QDs) are coated onto the graphene top surface. This modification significantly increases the

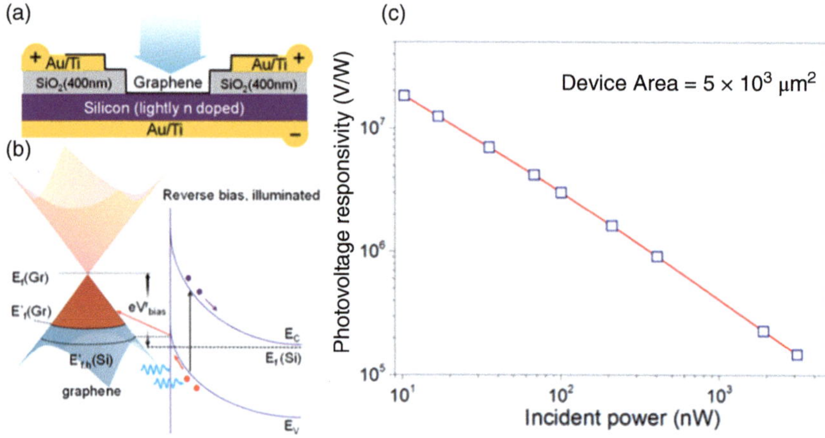

Figure 9.26 (a) Illustration of the Si/graphene photodetectors. (b) Energy diagram and Fermi level of the device under reverse bias. (c) Photovoltage responsivity variation concerning the incident power. *Source:* [47] / with permission of American Chemical Society.

built-in electric field and concurrently reduces surface reflection, resulting in an improved responsivity of ≈0.495 A/W and a short response time of ≤25 ns.

Intrinsic graphene exhibits a low optical response. Consequently, hybrid structures combining graphene with other two-dimensional monolayer materials, such as transition metal dichalcogenide (TMD) molybdenum disulfide (MoS_2), have been proposed to enhance responsivity. In such hybrid devices, a built-in electric field forms between the graphene and MoS_2 layers. Upon illumination, photogenerated carriers transfer to graphene facilitated by the electrostatic band bending. Figure 9.27a illustrates the schematic of the graphene/MoS_2 hybrid device; the dashed outline indicates the graphene layer. Application of a negative gate bias ($V_g \ll V_T$) injects generated electrons into the *p*-doped graphene, increasing its resistance and consequently reducing the drain–source current I_{ds} (Figure 9.27b, left panel). Conversely, under a positive gate bias ($V_g \gg V_T$), the near-metallic density of states in MoS_2 flattens its bands, thereby suppressing the unidirectional electron

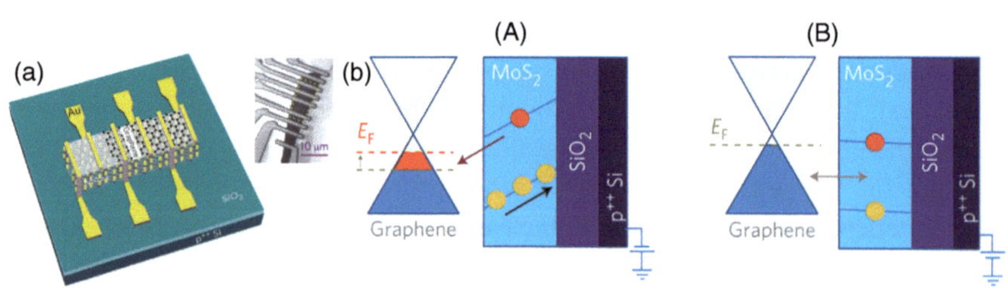

Figure 9.27 (a) Schematic of the device structure; inset displays the SEM image with dashed outline indicating graphene. (b) Schematic of the charge exchange process for (A) $V_g \ll V_T$ and (B) $V_g \gg V_T$; yellow and red circles represent holes and electrons in the valence and conduction bands of MoS_2, respectively. *Source:* [48] / with permission of Springer Nature.

transfer to graphene. Owing to the exceptional performance of the graphene/MoS_2 heterojunction, the responsivity of this hybrid device reaches 5×10^8 A/W at room temperature [48]. An alternative approach for constructing hybrid graphene photodetectors involves employing photosensitizers, such as colloidal PbS QDs. Their strong light absorption and tunable bandgap, achievable by modifying QD size or shape, enable photogeneration within the QDs followed by charge transfer to the graphene layer, yielding a high responsivity of $\sim 10^7$ A/W.

Schottky Solar Cells When the active area of a photodetector is sufficiently large, its output current can be utilized to power external electronic devices. Under these conditions, the device is termed a solar cell. Intrinsically, graphene exhibits semimetallic behavior. Consequently, a Schottky junction forms upon contact with a semiconductor, typically n-type silicon similar to that used in photodetectors [49-50]. As depicted in Figure 9.28a, the work function of graphene ($\emptyset_G$) is lower than the Fermi level of n-type silicon. To achieve Fermi level equilibrium across the system, the band edges bend, establishing a built-in electric field (V_{bias}). Upon illumination, electron–hole pairs generated within the silicon are dissociated by this built-in electric field and are ultimately transported in opposite directions. Specifically, electrons move through the silicon while holes transport across the graphene.

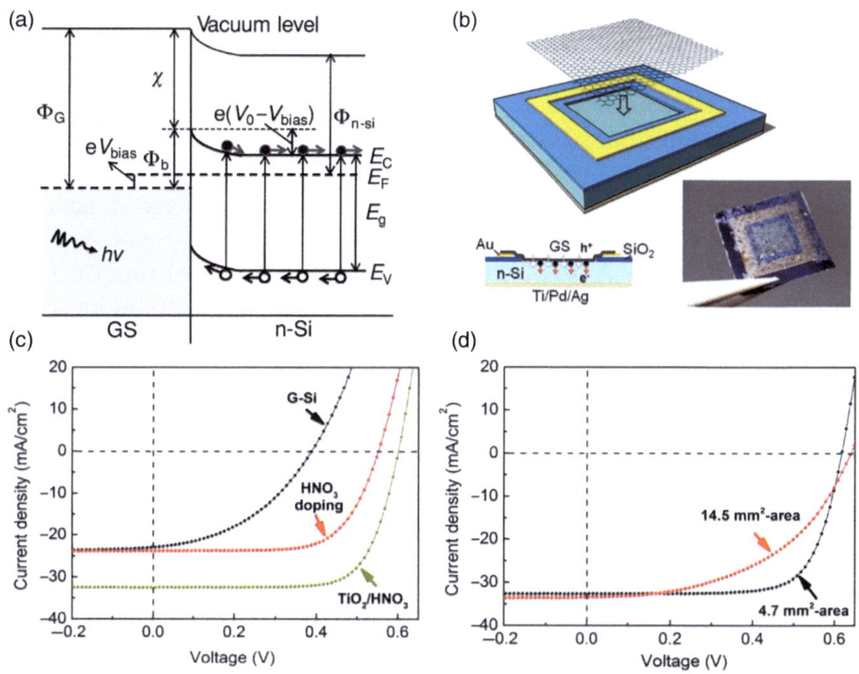

Figure 9.28 (a) Energy level diagram of the graphene/n-Si Schottky junction. (b) Schematic of the device structure and photograph of fabricated solar cells. *Source:* [49] / with permission of John Wiley & Sons (c) Influence of antireflection coating and doping on cell performance. (d) J–V characteristics of graphene/silicon solar cells with small and large active areas. *Source:* [50] / with permission of American Chemical Society.

The typical device structure is illustrated in Figure 9.28b. The active area of the solar cell is defined by lithography, and the SiO_2 insulating layer is subsequently etched using HF (hydrofluoric) acid. A front Au electrode is then deposited. A graphene film is transferred onto the freshly etched silicon surface. Device fabrication concludes with the evaporation of a Ti/Pd/Ag rear electrode. Voltage sweep measurements indicate that pristine graphene/silicon solar cells exhibit an open-circuit voltage (V_{oc}) below 0.4 V. As shown in Figure 9.28c, this V_{oc} can be increased to over 0.5 V through HNO_3 doping. Doping is achieved by exposing the device to concentrated HNO_3 vapor. NO_2 species adsorb onto the graphene, shifting its work function downward and thereby enhancing the built-in electric field. Application of an antireflection layer further improves device performance, yielding a higher short-circuit current density (J_{sc}).

It is noteworthy that graphene conductivity critically influences the performance of silicon/graphene solar cells. Devices with small active areas (e.g., 4.7 mm^2) exhibit low series resistance. However, as shown in Figure 9.28d, resistance increases significantly when the area is scaled up to 14.5 mm^2. Schottky junctions can also form between graphene and other semiconductors, such as PbS. Nevertheless, the resultant V_{oc} and J_{sc} values are considerably lower, leading to poor overall device performance in these systems.

Organic Solar Cells Organic solar cells are typically fabricated using a bulk-heterojunction (BHJ) structure comprising donor and acceptor materials, such as P3HT and PCBM. In this configuration, electron–hole excitons diffuse to the donor/acceptor interface, dissociate, and subsequently generate photocurrent. However, the high cost of PCBM necessitates cheaper alternatives. Solution-processable functionalized graphene (SPF Graphene), containing abundant oxygen functional groups (Figure 9.29a, offers a promising alternative for organic solar cells. A device with the layer structure ITO/PEDOT:PSS/P3HT:Graphene/LiF/Al (from bottom to top; Figure 9.29b) utilizes graphene as the electron acceptor within the active layer, achieving a power conversion efficiency (PCE) of 1.1%. Additionally, functionalized multi-walled carbon nanotubes (f-MWNTs) have been incorporated into organic solar cells to enhance hole transport, resulting in increased short-circuit current [51].

Dye-Sensitized Solar Cells Dye-sensitized solar cells (DSSCs) extensively utilize titanium dioxide (TiO_2) as a photoanode material. The transport of photogenerated carriers and the rates of recombination within the TiO_2 electrode significantly influence the overall solar cell performance. Incorporation of graphene into TiO_2 nanoparticles to form a hybrid TiO_2/graphene composite significantly improves charge transfer kinetics at the interface. In this composite, graphene functions as a highly conductive bridge, reducing the interfacial resistance associated with TiO_2/TiO_2 contact impedance. The addition of graphene also enhances the concentration of charge carriers within the anode, contributing to improved light absorption and collection efficiency.

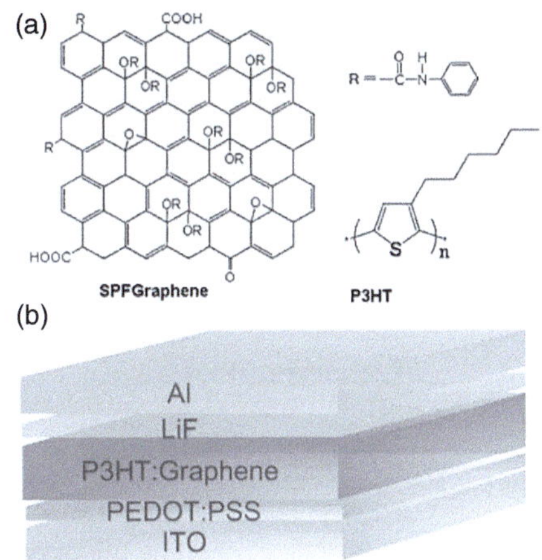

Figure 9.29 (a) Chemical structures of SPF Graphene and P3HT. (b) Device structure of the P3HT/graphene organic solar cell. *Source:* [51] / with permission of John Wiley & Sons.

Furthermore, graphene finds application in DSSCs as a material for counter electrodes. These counter electrodes are fabricated by annealing films composed of graphene nanosheets at different temperatures. The resultant graphene counter electrode possesses a three-dimensional (3D) network structure, which facilitates efficient electron transport. An efficiency of 6.81% was recorded for a solar cell employing a counter electrode annealed at 400 °C. Annealing at elevated temperatures compared yields higher conductivity within the graphene network, which is advantageous for counter electrode performance.

Perovskite Solar Cells Perovskite solar cells typically employ organic–inorganic hybrid perovskite materials with the general chemical formula ABX_3, where A and B represent cations and X is an anion. Methylammonium lead iodide ($FAPbI_3$) represents a widely studied composition within this category. Graphene has emerged as a functional material applicable to both electron transport layers (ETLs) and hole transport layers (HTLs) in these devices. Titanium dioxide (TiO_2) is frequently employed as the ETL; however, an inherent energy barrier exists between TiO_2 and the indium tin oxide (ITO) electrode. Incorporating graphene into the TiO_2 matrix effectively reduces this energy barrier, thereby enhancing electron collection efficiency. The superior conductivity of graphene simultaneously diminishes the device's series resistance. Consequently, the performance of TiO_2/graphene nanocomposites can approach that of high-temperature sintered TiO_2 films. Additionally, graphene and its derivatives, such as GO, demonstrate efficacy as HTLs within inverted perovskite solar cell architectures (see Figure 9.30) [52].

9.4.4 Graphene Nanoelectromechanical Systems

NEMS represent the further miniaturization of traditional microelectromechanical systems (MEMS). Owing to their integrated electrical and mechanical properties, NEMS are considered essential components for future computing and sensing applications. As the dimensions of NEMS devices are reduced to the nanometer scale (e.g., below 100 nm), their operational frequency can exceed 1 GHz, yielding

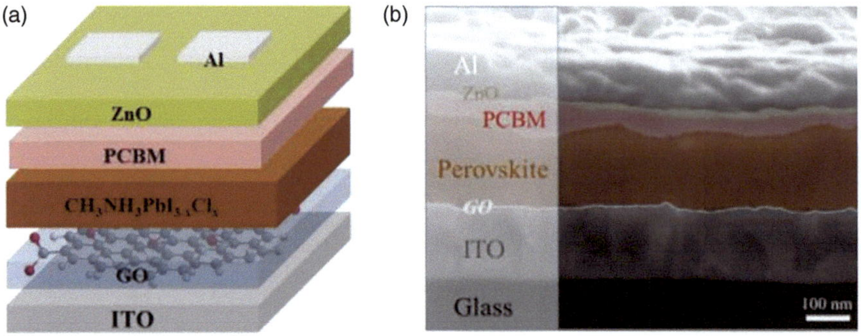

Figure 9.30 Schematic illustration of a perovskite solar cell employing a graphene oxide (GO) hole transport layer (a) and the corresponding cross-sectional scanning electron microscope (SEM) image revealing the device structure, highlighting the extremely thin GO layer (b). *Source:* [52] / with permission of Royal Society of Chemistry.

exceptionally high sensitivity. Traditional semiconductors such as silicon (Si) and gallium arsenide (GaAs) have been widely utilized for NEMS fabrication due to their high Young's modulus (>100 GPa) and scalability enabled by integrated circuit processes. Nevertheless, Si and GaAs cannot fully satisfy the requirements of NEMS. The resonant frequency (f) of NEMS is governed by the material's density (ρ) and Young's modulus (E), expressed as:

$$f = \left(\frac{E}{\rho}\right)^{\frac{1}{2}} \tag{9.14}$$

where a high operational frequency necessitates materials with low density and high Young's modulus. Carbon-based materials, including diamond, carbon nanotubes, and graphene, fulfill these criteria simultaneously. Graphene, in particular, has garnered significant attention due to its exceptional hardness, high Young's modulus, large surface area, superior thermal conductivity, and intrinsically low mass.

Figure 9.31 depicts a nanometer-scale graphene drum resonator composed of reduced graphene oxide (rGO) suspended over an Si substrate. Scanning electron microscopy (SEM) images illustrate graphene layers transferred onto the substrate with thicknesses of 20 nm (Figure 9.31a) and 4 nm (Figure 9.31b). The solution-processed graphene layer exhibits uniformity over large areas, completely covering the underlying micropillars. A representative drum NEMS structure, fabricated and characterized, is shown via optical (Figure 9.31c), AFM (Figure 9.31d), and SEM (Figure 9.31e) images. The device fabrication involved etching a hole pattern and transferring graphene onto the substrate; the devices were ready for measurement following desiccation. However, during the drying process, surface tension forces drew portions of the graphene layer into the holes, causing cracking in some regions. Figure 9.31f presents the measured intrinsic resonant frequency of the drum resonator, compared against theoretical calculations assuming Young's moduli of 1 TPa (red) and 0.5 TPa for graphene. The

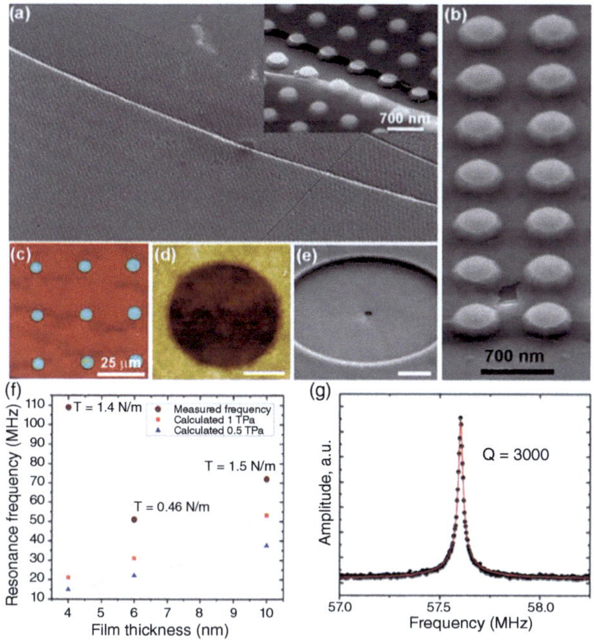

Figure 9.31 The resonator is based on reduced graphene oxide (rGO). SEM image of the rGO transferred to the substrate with a thickness of 20 nm (a) and 4 nm (b). Optical (c), AFM (d), and SEM (e) images of the microscope drum. (f) The resonant frequency evolution concerning the drum thickness. (g) The Q factor of the drum with a thickness of 6 nm. *Source:* [53] / with permission of American Chemical Society.

measured resonant frequencies substantially surpass the theoretical predictions, indicating that the graphene layer experiences tension rather than behaving as a free-standing membrane. Additionally, the presence of oxygen functional groups from the solution process enhances adhesion between the graphene and the underlying substrate. This increased adhesion substantially elevates the quality factor (Q) of the resonator, achieving a Q value exceeding 3000 at room temperature [53].

Hybrid optomechanical devices integrating light emitters and mechanical systems exhibit promising applications in single-photon control. Reserbat-Plantey et al. proposed a novel optomechanical device (Figure 9.32a), wherein nitrogen-vacancy (NV) centers, recognized as stable single-photon sources operable at room temperature, are positioned at the base of a cavity formed on a SiO_2/Si substrate. The emission intensity of these NV centers is modulated by the distance between a metallic graphene layer and the NV center surface. Electrostatic force is applied by adjusting DC and AC bias voltages (V_g^{dc} and δV_g). Nanoscale motion is optically measured using both single-photon avalanche diodes and interferometry. A 532 nm laser pumps the NV centers, and both reflected and emitted lights are simultaneously detected by a fast photodiode. Figure 9.32b illustrates graphene suspended 10–50 nm above the NV centers, while Figure 9.32c depicts one or more fluorescent NV centers deposited within the cavity. When graphene is positioned within the optical near-field of the NV centers at nanoscale separation, the relaxation rate

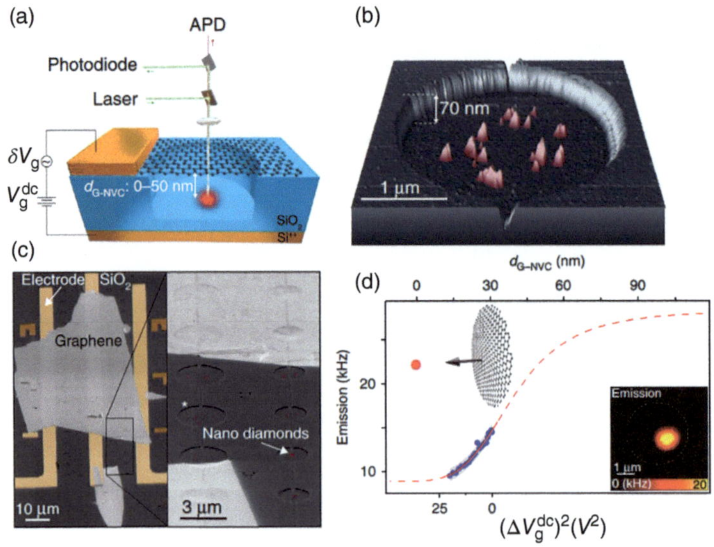

Figure 9.32 (a) Schematic of the optomechanical device for emission control. (b) AFM image of a nanodiamond (highlighted in red) deposited in the Si substrate cavity. (c) SEM image of the nanodiamond-based hybrid device. (d) In situ electromechanical control of dipolar coupling; the inset shows the localized NV center emission, with the dashed circle indicating the suspended graphene region. Adapted with permission from Reserbat-Plantey et al. [54] / Springer Nature / CC BY 4.0.

and light emission become strongly modulated through transduction between nanomechanical motion and the optical field. Consequently, the emission intensity of the NV centers is controllably suppressed. Figure 9.32d demonstrates the emission frequency variation as a function of the applied DC voltage ΔV_g^{dc}. Increasing the voltage reduces the graphene-to-NV center distance, resulting in progressive quenching of NV center emission [54].

9.5 Future of Graphene-Based Devices

Monolayer graphene offers an emerging platform integrating materials science, condensed matter physics, and chemical interactions. It exhibits exceptionally high carrier mobility, superior thermal conductivity, outstanding mechanical strength and toughness, as well as significant broadband light absorption and nonlinear Kerr effects, establishing its potential across electronic, optical, plasmonic, and mechanical devices. However, the intrinsic absence of a bandgap fundamentally limits its application in large-scale logic circuits requiring low power consumption. Nevertheless, its high carrier mobility enables promising applications within high-frequency radio-frequency devices. Furthermore, graphene's remarkable broadband absorption capability facilitates ultrasensitive photodetectors operating across a wide spectrum from visible to infrared, terahertz, and even microwave frequencies, particularly within hybrid device architectures.

Intrinsically, graphene supports highly tunable plasmons. Their performance is substantially enhanced through integration with noble metal nanostructures, enabling novel ultracompact optical devices featuring high operational speed, low driving voltage, and minimal power consumption. The exceptional mechanical properties of monolayer graphene render graphene-based MEMS exceptionally promising. Integration with optical and electronic components could transform these into fully coupled optoelectromechanical microsystems.

Significant challenges must be addressed before these exciting potential applications transition into commercially viable technologies. Foremost is the scalable synthesis of high-quality, large-area monolayer graphene, as achieving this would profoundly impact the entire field. Enhancing carrier mobility, for instance, directly prolongs plasmon lifetimes and is fundamentally correlated with critical device metrics including light–matter interaction strength, spectral resolution, detectivity, modulation efficiency, sensitivity, and absorbance. The second challenge involves developing integration methodologies that merge high-quality monolayer graphene with heterogeneous conventional electronic and optical components without compromising its atomic thickness during transfer or fabrication processes. A third critical requirement is achieving a comprehensive understanding of the fundamental interface physics governing the electronic and optical characteristics of diverse hybrid graphene devices.

While the ultimate realization of graphene-based technologies remains ongoing, graphene embodies the strongest candidate for future integrated, multifunctional, and ultracompact devices and systems, provided the core challenges in synthesis precision, reliable integration, and underpinning fundamental physics are definitively resolved.

References

1 Panchal, V., Yang, Y., Cheng, G. et al. (2018). *Commun. Phys.* 1: 83.

2 Castro Neto, A.H., Guinea, F., Peres, N.M.R. et al. (2009). *Rev. Mod. Phys.* 81: 109–162.

3 Fasolino, A., Los, J.H., Katsnelson, M.I. et al. (2007). *Nat. Mater.* 6: 858–861.

4 Fuhrer, M.S. (2013). *Science* 340: 1413–1414.

5 Terrones, H., Lv, R., Terrones, M. et al. (2012). *Rep. Prog. Phys.* 75: 062501.

6 Bolotin, K.I., Ghahari, F., Shulman, M.D. et al. (2009). *Nature* 462: 196–199.

7 Lee, C., Wei, X., Kysar, J.W. et al. (2008). *Science* 321: 385–388.

8 Zandiatashbar, A., Lee, GH., An, S. et al. (2014). *Nat Commun* 5: 3186.

9 Balandin, A.A., Ghosh, S., Bao, W. et al. (2008). *Nano Lett.* 8: 902–907.

10 Ghany, N.A.A., Elsherif, S.A., Handal, H.T. et al. (2017). *Surf. Interfaces* 9: 93–106.

11 Huang, Y., Sutter, E., Shi, N.N. et al. (2015). *ACS Nano* 9: 10612–10620.

12 Parvez, K., Wu, Z.S., Li, R. et al. (2014). *J. Am. Chem. Soc.* 136: 6083–6091.

13 Kim, J., Kwon, S., Cho, D.-H. et al. (2015). *Nat. Commun.* 6: 8294.

14 Mishra, N., Boeckl, J., Motta, N. et al. (2016). *Phys. Status Solidi A* 213: 2277–2289.

15 Tromp, R.M. and Hannon, J.B. (2009). *Phys. Rev. Lett.* 102: 106104.

16 Lin, L., Deng, B., Sun, J. et al. (2018). *Chem. Rev.* 118: 9281–9343.

17 Lee, J.-H., Lee, E.K., Joo, W.-J. et al. (2014). *Science* 344: 286–289.

18 Hirsch, A. (2009). *Angew. Chem. Int. Ed.* 48: 6594–6596.

19 Geim, A.K. and Novoselov, K.S. (2007). *Nat. Mater.* 6: 183–191.

20 Novoselov, K.S., Geim, A.K., Morozov, S.V. et al. (2005). *Nature* 438: 197–200.

21 Novoselov, K.S., Geim, A.K., Morozov, S.V. et al. (2004). *Science* 306: 666–669.

22 Feng, T., Xie, D., Xu, J. et al. (2014). *Carbon* 79: 363–368.

23 Liao, L., Bai, J., Qu, Y. et al. (2009). *Nanotechnology* 21: 015705.

24 Bai, J., Liao, L., Zhou, H. et al. (2011). *Nano Lett.* 11: 2555–2559.

25 Kim, S., Nah, J., Jo, I. et al. (2009). *Appl. Phys. Lett.* 94: 062107.

26 Freitag, M., Low, T., and Avouris, P. (2013). *Nano Lett.* 13: 1644–1648.

27 Jacobberger, R., Kiraly, B., Fortin-Deschenes, M. et al. (2015). *Nat. Commun.* 6: 8006.

28 Lin, Y., Jenkins, K. A., Valdes-Garcia, A. et al. (2009). *Nano Lett.* 9: 422.

29 Lin, Y.-M., Dimitrakopoulos, C., Jenkins, K.A. et al. (2010). *Science* 327: 662.

30 Liao, L., Lin, Y.C., Bao, M. et al. (2010). *Nature* 467: 305–308.

31 Schedin, F., Geim, A.K., Morozov, S.V. et al. (2007). *Nat. Mater.* 6: 652–655.

32 Sun, Z., Liu, Z., Li, J. et al. (2012). *Adv. Mater.* 24: 5878–5883.

33 Zheng, Y., Ni, G.-X., Toh, C.-T. et al. (2009). *Appl. Phys. Lett.* 94: 163505.

34 Bertolazzi, S., Krasnozhon, D., and Kis, A. (2013). *ACS Nano* 7: 3246–3252.

35 Wang, X.C., Zhao, W.S., Hu, J. et al. (2015). *IEEE Trans. Nanotechnol.* 14: 62–69.

36 Qin, Y., Xiong, X.Y.Z., Sha, W.E.I. et al. (2018). *J. Phys. Condens. Matter* 30: 144007.

37 Ren, X., Sha, W.E.I., and Choy, W.C.H. (2013). *Opt. Express* 21: 31824–31829.

38 Zhu, B., Ren, G., Gao, Y. et al. (2017). *Plasmonics* 12: 863–868.

39 Ma, H., Pan, G., Yuan, G. et al. (2022). *Phys. Scr.* 97: 125506.

40 Cheng, J., Jafar-Zanjani, S., Mosallaei, H. et al. (2016). *Appl. Optics* 55: 6137.

41 Cheng, Y., Wu, L., Tang, M. et al. (2017). *IEEE Antennas Wirel. Propag. Lett.* 16: 3000–3004.

42 Lee, S.H., Choi, M., Kim, T.T. et al. (2012). *Nat. Mater.* 11: 936–941.

43 Shi, S.F., Zeng, B., Han, H.L. et al. (2015). *Nano Lett.* 15: 372–377.

44 Wu, Z., Chen, X., Zhu, S. et al. (2013). *Sens. Actuators B Chem.* 178: 485–493.

45 Wang, T., Huang, D., Yang, Z. et al. (2016). *Nano Micro Lett.* 8: 95–119.

46 Park, J., Ahn, Y.H., and Ruiz-Vargas, C. (2009). *Nano Lett.* 9: 1742–1746. https://pubs.acs.org/doi/10.1021/nl8029493.

47 An, X., Liu, F., Jung, Y.J. et al. (2013). *Nano Lett.* 13: 909–916.

48 Roy, K., Padmanabhan, M., Goswami, S. et al. (2013). *Nat. Nanotechnol.* 8: 826–830.

49 Li, X., Zhu, H., Wang, K. et al. (2010). *Adv. Mater.* 22: 2743–2748. https://onlinelibrary.wiley.com/doi/abs/10.1002/adma.200904383.

50 Shi, E., Li, H., Yang, L. et al. (2013). *Nano Lett.* 13: 1776–1781.

51 Liu, Q., Liu, Z., Zhang, X. et al. (2009). *Adv. Funct. Mater.* 19: 894–904.

52 Wu, Z., Bai, S., Xiang, J. et al. (2014). *Nanoscale* 6: 10505–10510.

53 Robinson, J.T., Zalalutdinov, M., Baldwin, J.W. et al. (2008). *Nano Lett.* 8: 3441–3445.

54 Reserbat-Plantey, A., Schadler, K.G., Gaudreau, L. et al. (2016). *Nat. Commun.* 7: 10218.

Index